Airport Operations

Other McGraw-Hill Books of Interest

In order to receive additional information on these or any other McGraw-Hill titles, in the United States please call 1-800-822-8158. In other countries, contact your local McGraw-Hill representative.

Airport Operations

Second Edition

Norman Ashford

H. P. Martin Stanton

Clifton A. Moore

McGraw-Hill, Inc.

New York San Francisco Washington, D.C. Auckland Bogotá
Caracas Lisbon London Madrid Mexico City Milan
Montreal New Delhi San Juan Singapore
Sydney Tokyo Toronto

Library of Contress Cataloging-in-Publication Data

Ashford, Norman.
    Airport operations / Norman Ashford, H.P. Martin Stanton, Clifton
A. Moore. — 2nd ed.
      p.   cm.
    Includes bibliographical references.
    ISBN 0-07-003077-4 (alk. paper)
    1. Airports—Management.   I. Stanton, H.P. Martin.   II. Moore.
Clifton A.   III. Title.
IN PROCESS
387.7'36'068—dc21   96-48398
                                                  CIP

McGraw-Hill

*A Division of The **McGraw·Hill** Companies*

1 2 3 4 5 6 7 8 9 0 DOC/DOC 9 0 1 0 9 8 7 6

ISBN 0-07-003077-4

*The sponsoring editor for this book was Shelley IC. Chevalier, the
technical editor was Charles Spence, and the production supervisor
was Don Schmidt. This book was set in ITC Century Light. It was
composed by Wanda Ditch through the services of Barry E. Brown
(Broker—Editing, Design, and Production).*

Printed and bound by R.R. Donnelley & Sons Company.

This book is printed on acid-free paper.

Contents

Preface to first edition

This book, which is a sequel to *Airport Engineering* (first published in 1979), is the result of both a series of postexperience short courses and seminars conducted in Britain by Loughborough University and the authors' joint teaching experience of airport planning and design courses in the United States, notably at the University of California—Berkeley and the Georgia Institute of Technology. For the purpose of discussion, our definition of airport operations is broad. Modern airports can be extremely complex organizations, often involving the activities of many governmental organizations, private companies, airlines, and aircraft operators, and, of course, the airport operators' own organization. There is a great need, therefore, for inter-relation of the organizational skills found within the confines of a large airport.

Although many of the world's airfields are grass strips with few facilities, there are large and increasing numbers of airports in the truer sense of the word, where airport operations have a wider scope than aircraft operations. The authors have drawn on their own broad experience and the broader experience of many other individuals and organizations in a wide range of countries to put together an introductory text in this area. It is likely to be useful to all levels of airport management, to the planners and designers of these facilities, and to those involved with airports from local and central government organizations as well as to individuals with a general interest in how airports run. For academic purposes, the material is suitable as a basic text for a senior undergraduate or an introductory postgraduate-level course or as a supplementary text for an airport planning and design course.

In working closely with industry in the expanding area of postexperience training, we found that there is a worldwide recognition of training and education needs for middle and upper management in the area of airport operations. Postexperience education has been found necessary for airport authorities, civil aviation organizations, and, not least, airlines. The material contained has been assembled with the help of many individuals and organizations involved with the teaching of a postgraduate program designed in conjunction with the personnel commission of the International Civil

Airports Association in Paris, with substantial material assistance from the International Civil Aviation Organization in Montreal. It is hoped that this text, although necessarily limited in content, will provide basic instructional material for those involved in a burgeoning industry.

NORMAN ASHFORD
Loughborough, Leicestershire

H. P. MARTIN STANTON
Crowthorne, Berks

CLIFTON A. MOORE
Llano, California
November 1983

Preface to second edition

Since this book was first published 12 years ago, much in the civil air transport world has remained much the same but in other areas very significant changes have taken place. The full impact of deregulation in the United States has been felt and with it liberalization of civil air transport in Europe and many other parts of the world. With deregulation and liberalization, many long-familiar names in the airline industry have disappeared (Pan American, and Eastern, for example), and new names have appeared, many for only a short time. The airlines left free to provide service with minimal or no regulation have greatly varied and altered their services and schedules, with significant impact where, for example, airline *hubs* have been introduced in the United States and elsewhere.

The new wave of privatization and commercialization of airports has changed the appearance and the mode of operation of many airports almost beyond recognition, and many more airport facilities have come under pressure and expanded to accommodate the doubling of passenger and freight traffic, which has taken place worldwide since the early 1980s.

This edition has been rewritten to take these changes into account while still recognizing that most of the world's airports are still simple facilities that have little in common with the sophisticated complexity of such giants as Denver, Chicago, London Heathrow, Changi, and Seoul.

NORMAN ASHFORD
Loughborough, Leicestershire

H. P. MARTIN STANTON
Crowthorne, Berks

CLIFTON A. MOORE
Llano, California
July 1995

Acknowledgments

In the preparation of this book we are greatly indebted to the following organizations who have given freely of their time in interviews and discussions to assist in the gathering of data. Without their help it would not have been possible to complete this book.

Aeroportos e Navegaïïo Area (Portugal)
Airports Council International
Atlanta Hartsfield International Airport
Birmingham International Airport
BAA
British Airways
Civil Aviation Authority (United Kingdom)
Matthew Coogan, Consultant, White River JCT Junction, Vermont
Dallas-Fort Worth International Airport
East Midlands International Airport
Federal Aviation Administration
Frankfurt Airport Authority
Georgia Institute of Technology
International Air Transport Association
International Civil Aviation Organization
Lufthansa German Airlines
Manchester International Airport
Munich Airport
Siemens Plessey Airports
Rio de Janeiro International Airport
Schiphol Airport Authority
Singapore International Airport
Spanish Airports Authority
Tampa International Airport
United Airlines
Vienna Airport Authority
Westinghouse.

We are also greatly indebted to many other organizations and individuals too numerous to name.

Norman Ashford
H. P. Martin Stanton
Clifton A. Moore

The Airport as an Operational System

1.1 The Airport as a System

The airport forms an essential part of the air transport system, because it is the physical site at which a modal transfer is made from the air mode to land modes. Therefore, it is the point of interaction of the three major components of the air transport system:

- The airport, including for these discussion purposes the airways control system
- The airline
- The user

The planning and operation of airports must, if they are to be successful, take into account the interaction of these three major components, or system actors. For the system to operate well, each of the actors must reach some form of equilibrium with the other two. Failure to do so will result in suboptimal conditions, exemplified by a number of undesirable phenomena that are indicators of inadequate operation. Each can, in a state of unrestrained competition, lead to an eventual decline in scale of operation at the airport facility, as traffic is attracted elsewhere, and in the absence of a competitive option, to depressed demand levels. Such conditions become manifest in a variety of ways:

- Deficit operations by the airport
- Deficit operations by the airline at the airport

1

- Unsatisfactory working conditions for airline and airport employees
- Inadequate passenger accommodation
- Insufficient flight supply
- Unsafe operations
- High operational costs to users
- Inadequate support facilities for airlines
- High delay levels for airline and passenger
- Inadequate access facilities
- Sluggish passenger demand levels

Figure 1.1 displays a simplified hierarchical system diagram of the primary interactions between the airport, the airline, and the user (Ashford

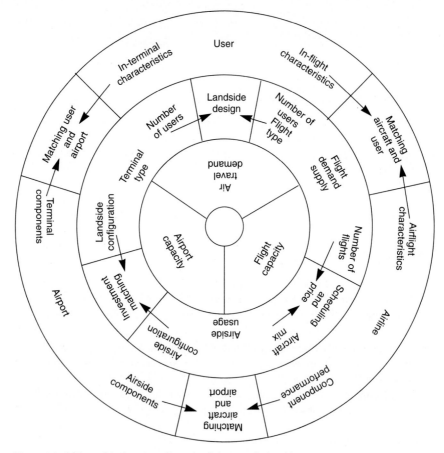

Figure 1.1 A hierarchical system diagram of airport relationships.

**TABLE 1.1   Organizations Affected by the
Operation of a Large Airport**

Principal actor	Associated organizations
Airport operator	Local authorities and municipalities
	Central government
	Concessionaires
	Suppliers
	Utilities
	Police
	Fire service
	Ambulance and medical services
	Air traffic control
	Meteorology
Airline	Fuel supplies
	Engineering
	Catering/duty free
	Sanitary services
	Other airlines and operators
Users	Visitors
	Meeters and senders
Non-users	Airport neighbor organizations
	Local community groups
	Local chambers of commerce
	Anti-noise groups
	Environmental activists
	Neighborhood residents

and Boothby 1979). The diagram attempts to show how these interactions produce the prime parameters of operational scale, passenger demand, airport capacity, and flight capacity. Although a simplified diagram helps the conceptualization of the main factors of airport operation, large airports are in fact very complex organizational structures. This is not surprising since a large airport might well be one of the largest generators of employment in a metropolitan region. The very largest hub airports, such as Chicago O'Hare, Los Angeles, London Heathrow, and Atlanta all have total site employment levels of more than 50,000. To place this in some context of scale, the number of persons could well be the total central business district employment of a city of a quarter- to half-million population. Large systems such as this are necessarily more complex than the simple trichotomy expressed in Figure 1.1. A more complete listing of involvement for a large airport is contained in Table 1.1. This now includes an important fourth actor, the nonuser, who can have an important impact on airport operation and who is also greatly affected by large-scale operations.

1.2 National Airport System

Modern airports, with their long runways and taxiways, extensive apron and passenger terminal areas, and expensive ground handling and flight navigation equipment, constitute substantial infrastructure investments. All over the world, they are now seen as facilities requiring public investment. As such, they are frequently part of a national airport system designed and financed to produce maximum benefit from public investment. Each country with its own particular geography, economic structure, and political philosophy will develop a national airport system peculiar to its own needs. This system is important to the operation of the individual airport in that national constraints will determine the nature of current and future traffic handled at the facility in terms of such parameters as volume, aircraft type, international/domestic split, number of airlines served, and growth rates. Two different national systems will be discussed briefly, those of the United States and the United Kingdom. They serve to illustrate how systems vary to suit different national needs.

The United States is a very large industrialized nation with more than 18,000 airports (including heliports, STOLports, seaplane bases, and joint-use civil-military airports), of which approximately 12,800 are closed to the public or of limited public use (Figure 1.2a). Of the remainder, approximately 670 provide commercial service through passenger transport aircraft. Some of the more than 5000 airports in the U.S. not served by air carriers have more flight operations than do a number of the airports served by major or commuter airlines. General aviation airports like Van Nuys in California and Fort Lauderdale in Florida have in excess of 200,000 itinerant operations annually. Van Nuys is the ninth busiest airport in the U.S. Publicly owned airports in the National Plan of Integrated Airport Systems (NPIAS) are eligible for federal aid in the construction of most facilities required at the airport, other than those related to commercial activities. Ownership of the large and medium sized airport is almost entirely in the hands of local communities. The two Washington, DC, medium sized airports were formerly operated by the federal government, but these are now under the operation of the Metropolitan Washington Airports Authority. Privately owned commercial transport airports now are few in number and are not significant to the national system, although by the mid-1990s there is significant pressure for some of the largest facilities to be privatized. The scale of monies derived by each publicly owned airport from the Airport and Airway Trust fund is largely related to the functional role of the facility. The system also consists of geographically widely separated hub, regional, and municipal airports that have the opportunity to expand as traffic increases. As a consequence, the functional classification system of the U.S. National Airport System plan (Figure 1.2b) is a relatively free structure related to passenger throughput and aviation activity. Airports are free to

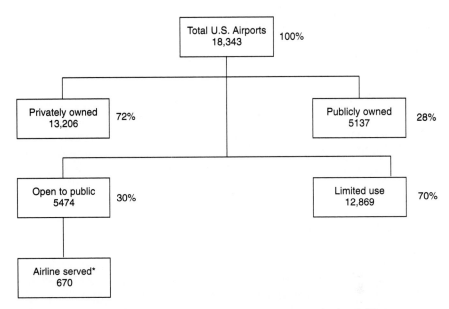

*Regional airline association reports 758 airports served by commuter/regional airlines

(a)

Figure 1.2a A classification of the U.S. airport system, number of airports by ownership and public or limited use, including heliports, STOLports, seaplane bases, and joint-use civil-military airports as of December 31, 1994. *(FAA)*

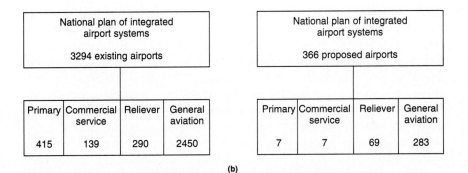

(b)

Figure 1.2b The U.S. National Plan of Integrated Airports System, NPIAS report, 1993 to 1997. *(FAA)*

move their classification, which is a de facto demand classification (NPIAS 1987).

An entirely different national airport system exists in the United Kingdom, which has, in addition to a few major airports, a large number of small regional facilities spread over a small area. Economic appraisal of the performance of the British airport system indicated that most of this system was comprised of loss-making facilities that were unlikely to be able to recoup past investments made in the hope of attracting traffic (Doganis and Thompson 1973; Doganis and Pearson 1977). To encourage the development of a national airport system in which a few major hubs would cater to international traffic and the rest of the airports would assume a support role, the British government in 1978 developed an airports policy to guide central government investment into the airport system. The policy recognized four distinct categories of airports and indicated that governmental approval in the form of finance and permission to develop facilities would be based on this categorization. These were: Gateway International Airports, which supplied a wide range of international and intercontinental services; Regional Airports, which provided short-haul international and domestic services; Local Airports providing third level services, (scheduled passenger services with aircraft having fewer than 25 seats); and General Aviation Airports.

For a number of reasons both political and economic, in 1987 all publicly owned airports with an annual revenue in excess of $1 million were privatized. This meant that the British Airports Authority, which owned seven airports including the two largest, London Heathrow and London Gatwick, was floated on the London share market in a public flotation in that year (Ashford and Moore 1993). All other airports at that time remained as private companies but with all shares entirely owned by local governmental authorities. Since the general privatization occurred in 1987, several other airports have been brought into private hands: e.g., Prestwick, Liverpool, East Midlands, Belfast, and Southampton. The major elements of the British airport system are shown in Table 1.2, although de facto deregulation since privatization has considerably blurred the distinctions between the categories. Five major airports can still be considered to have "gateway" status: the three London airports (Heathrow, Gatwick, and Stansted), Glasgow, and Manchester. Some long-range intercontinental services are available from airports such as Birmingham, which might otherwise be classified as Regional airports. By the mid-1990s, the neat classifications of 1978 had been allowed to blur into a laissez faire system where airport capacity and ability to generate traffic were the prime determinants of function. For example, a number of airports that would otherwise fall into the Regional airport category provided scheduled intercontinental service particularly on transatlantic routes. The development of traffic and the change of status of the airports to a privatized status have reduced the number of airports that are overall loss makers.

TABLE 1.2 The British National Airport System (as of 1978)

Gateway International Airports

Airports supplying a wide range and frequency of international services, including intercontinental services and a full range of domestic services.

Regional Airports

Airports catering to the main air traffic demand of individual regions. They are concerned with the provision of a network of short-haul international services (mainly to Scandinavia and other parts of Europe), a range of charter services and domestic services, including links with gateway airports.

Local Airports

Airports providing third-level services, (scheduled passenger services operated by aircraft with fewer than 25 seats), catering privately for local needs, concentrating on general aviation with some feeder services and some charter flights.

General Aviation Airports

Airports concerned primarily with the provision of general aviation facilities.

1.3   The Function of the Airport

An airport is either an intermediate or terminal point of an aircraft on the air portion of a trip. In simple functional terms, the facility must be designed to enable an aircraft to land and take off. In between these two operations, it might, if required, unload and load payload and crew and be serviced. It is customary to divide the airport's operation between the *airside* and *landside* functions shown in the simplified system diagram in Figure 1.3. More detailed system diagrams are provided for passenger and cargo processing in Figures 8.1 and 10.7. The overall system diagram indicates that after approach an aircraft uses the runway, taxiway, and apron prior to docking at a gate position, where its payload is processed through the terminal to the access/egress system. Departing passengers make their way through the landside operation to the departing gates.

The airport passenger and freight terminal is itself a facility that has three distinct functions (Ashford and Wright 1992).

1. *Change of mode.* To provide a physical linkage between the air vehicle and the surface vehicle designed to accommodate the operating characteristics of the vehicles on airside and landside, respectively.

2. *Processing.* To provide the necessary facilities for ticketing, documentation, and control of passengers and freight.

3. *Change of movement type.* To convert continuous arrivals of freight by trucks and of departing passengers by car, bus, and train into aircraft-sized batches that generally depart according to a preplanned schedule: to reverse this process for arriving aircraft.

Many small airports that provide little more than a simple passenger terminal for low-volume passenger operations, provide very little more than a

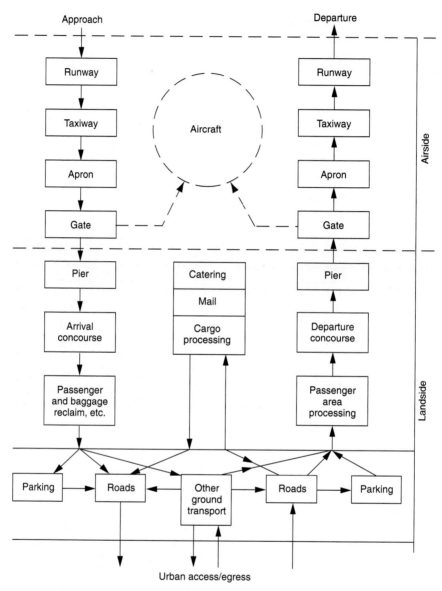

Figure 1.3 The airport system.

passenger terminal facility. The operation of the airport is not significantly more complex than that of a railroad station or an interurban bus station. The medium- or large-scale airport is, however, very much more complex and requires an organization that can cope with such complexity. Airports of a significant size must have an organization that can either supply or administer the following facilities:

- Handling of passengers
- Servicing, maintenance, and engineering of aircraft
- Airline operations including aircrew, cabin attendants, ground crew, terminal and office staffs
- Businesses necessary for the economic stability of the airport (concessions, leasing companies, etc.)
- Aviation support facilities (air traffic control, meteorology, etc.)
- Government functions—agricultural inspection, customs, immigration, health

Large international hubs are complex entities that have all the problems of any large organization with many employees. In some cases, the airport itself is a large employer (e.g., Frankfurt, with a throughput of 30 million and an airport authority staff of 12,000). In other cases the authority acts more as a broker of services, resulting in a low level of direct authority employment (e.g., Los Angeles, with 51 million passengers and 1335 staff in 1994). Regardless of the chosen modus operandi, overall staffing levels at major airports will be high, and there will be complex interactions between the various employing organizations. Inefficient operations and operational failures in such large systems, whether through incompetence, disorganization, or industrial action, involve huge expenses in terms of additional wages, lost passenger time, and delayed freight costs.

1.4 Centralized and Decentralized Terminal Systems

The way in which an airport is operated and the administrative structure of the operating authority are affected by the physical design of the airport itself. It is convenient to classify airports into two broad and very different operational classes: *centralized* and *decentralized.* Most older airport terminals were designed using the centralized concept, where processing was carried out in the main terminal building and access to the aircraft gates was attained by piers and satellites or by apron transporters. Many airports still operate quite satisfactorily using centralized facilities (e.g., Tampa and Brussels). Other airports started as centralized facilities but became decentralized when additional terminals were added to cope with increased traffic (e.g., London Heathrow, Paris Orly, Madrid, and Johannesburg). Other airports were designed ab initio as decentralized facilities operating with a number of unit terminals each with a complete set of facilities (e.g., Dallas-Fort Worth, Paris Charles de Gaulle, Kansas City, Toronto, and New

York JFK). Another form of decentralization occurs with extensive remote pier developments (e.g., Atlanta) and remote satellites (e.g., Pittsburgh). Figures 1.4a and 1.4b show examples of a centralized and decentralized layout, respectively.

Up to the early 1960s, passenger traffic through even the world's largest airports was so small that centralized operation was commonplace. There are economies of scale on the use of fixed equipment such as baggage systems and check-in desks and with movable apron equipment. Similar economies are found with airport authority, airline, and airport tenants' staff requirements. Subsequently it has been found that fewer security personnel are required in centralized designs.

As traffic grew at the major hubs, the physical size of the centralized facilities grew with additions until they became extraordinarily large in scale. For example, in the 1980s, before the redevelopment of Chicago O'Hare within a single terminal building, the distance between extreme gates was just under 1 mile (1.5 km). Parking facilities also grew in scale. Therefore, travelers could face very long walks, either interlining or leaving the terminal at the end of a trip.

In order to overcome the problem of unsatisfactory walking distances, decentralized designs were developed to keep walking distances down to the region of 984 feet (300 m) as recommended by International Air

(a)

Figure 1.4a Schiphol Airport, Amsterdam—a centralized terminal layout. *(Schiphol Amsterdam Airport)*

(b)

Figure 1.4b JFK Airport, New York

Transport Association (IATA), with a maximum distance between curbside and the furthest check-in counter set at 328 feet (100 m). Decentralization was carried as far as the gate arrival concept of Dallas-Fort Worth and Kansas City, where total walking distances from car to aircraft are approximately 328 feet (100 m). The advantages of decentralization are significant. Terminals are kept on a human scale, passenger volumes never become uncomfortably high, and walking distances are kept low (Ashford and Moore 1993). Parking lots are small, keeping walking distances reasonable. Therefore, the lots are easier to supervise and thus safer from a crime viewpoint, and curbside drop-off areas are simple to design. Operationally, however, decentralization can lead to higher airport staff requirements, since the same functions, such as administration and security, must be carried out separately in each terminal. Because the scale of such facilities is very large, each unit requires a full range of passenger and staff facilities. It is possible, therefore, to have poor economy of scale in terms of the fixed facilities, such as baggage rooms, baggage claims, and check-in areas, and the mobile facilities, such as apron handling equipment.

For a large airport, the scale of separation between units can be very large. For example, if the 14-unit terminals at Dallas-Fort Worth (DFW) were to be constructed according to the original master plan the distance between the two most distant units would be 3 miles (4.5 km). Completely decentralized designs mean that interlining passengers must have some form of transit system to permit interterminal movement. At DFW, this is achieved by an automatic transit vehicle; at older terminals, such as New York JFK and London Heathrow, and newer airports such as Charles de Gaulle, this is achieved by a simple bus service. Neither method is particularly convenient.

An often ignored operational difficulty arising from decentralization is the loss of daily capacity that occurs when a given terminal area is broken down into a number of subareas. Capacity is determined by peak hour operation, and demand peaks are more easily smoothed for one large terminal than three smaller terminals Figures 1.5a through 1.5e show total passenger flows at Terminals 1, 2, 3, and 4 at London Heathrow. These would be smoothed to the much less peaky pattern shown in Figure 1.5f if all operations were in one terminal.

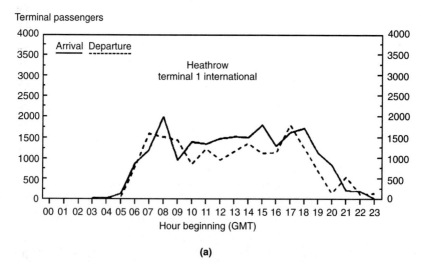

Terminal passengers

(a)

Figures 1.5a to 1.5e Passenger flows through terminals 1, 2, 3, and 4 at London Heathrow. *(a)* arrival/departure Terminal 1 International; *(b)* arrival/departure Terminal 1 domestic; *(c)*arrival/departure Terminal 2; *(d)* arrival/departure Terminal 3; *(e)* arrival/departure Terminal 4. **Figure1.5f** Effect of multiple terminal operations: Heathrow peak month departure and arrival demands for individual terminals and totals.

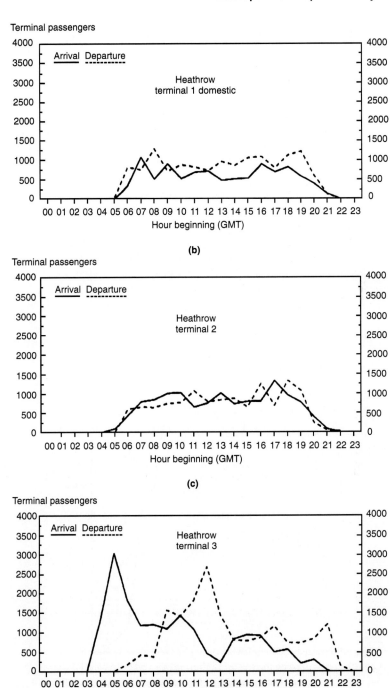

Figures 1.5 Continued.

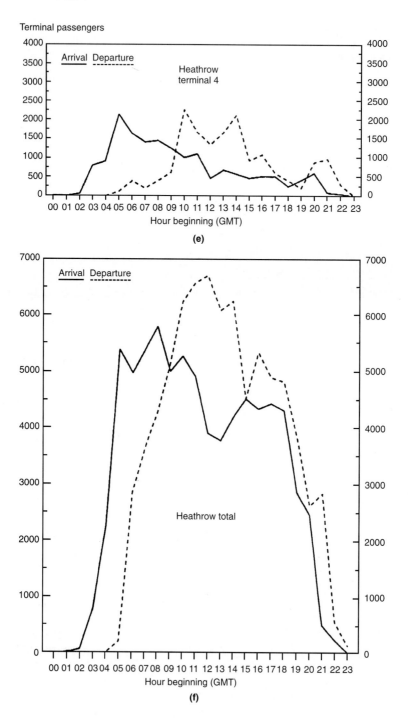

Figures 1.5 Continued.

1.5 The Complexity of Airport Operation

Airports are complex businesses with functions that extend substantially beyond the airfield or "traffic" side of operations. As airports increase in size in terms of passenger throughput, the nonaviation revenues become increasingly important (Ashford and Moore 1993). It is also clear that in most countries, airports maintain economic viability by developing a broadly based revenue capability. In general, the organizational structure of the operating authority changes to reflect the increasing importance of commercial revenues with increasing passenger throughput. As the relative and absolute size of the non-traffic element of an airport's revenue increase, much more attention must be paid to developing a commercial expertise. Many airport authorities around the world have developed a considerable expertise in maximizing this kind of revenue.

1.6 Management and Operational Structures

There is no unique form of administrative structure that is ideal for every airport. Airports differ in their type and scale of throughput, vary in their interrelationship with other governmental and quasi-governmental bodies, and also fit within differing matrices of allied and associated organizations at central and local governmental levels. Organizational structures must also be recognized as being evolutionary in nature, depending on the previously existing structure and on the pressures for change, not the least of which are those that relate to the personalities and abilities of individuals with directional responsibilities within the organization.

Organizational structure will undergo radical reform should an airport become privatized. Structure also depends on the role the airport authority assumes, varying from a brokering function with minimal operational involvement in many on-airport activities (the U.S. model) to direct operational involvement in many of the airport's functions (the European model). It must also be borne in mind that in common with other commercial and governmental organizations that operate facilities, the management structure of airports may be divided into staff and line functions. The way these functions are accommodated also varies between airports. Staff departments are those that provide direct managerial support to the administrator. Often relatively small in size in terms of personnel, they are engaged in decision making that has a major impact on the whole organization. Line departments, on the other hand, are those portions of the organization engaged in the day-to-day operation of the facility. In comparison with the staff departments, they usually require heavy staffing. The manner in which staff and line departments report to the airport director differs greatly between airports. Figure 1.6 shows three different formalized structures that cover the range of what might occur at any particular airport.

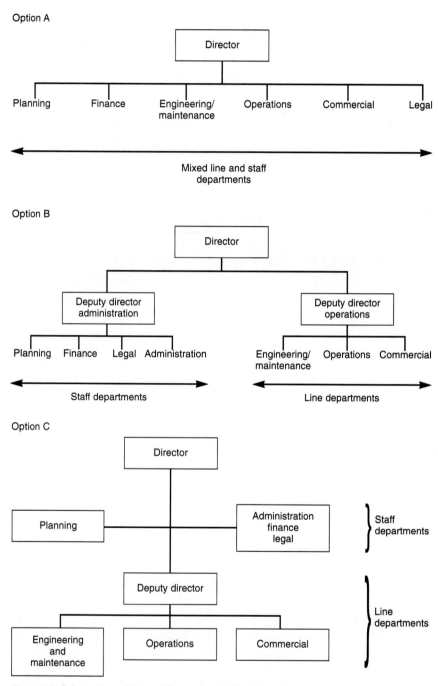

Figure 1.6 Schematic positions of line and staff departments in the structure of airport administrations.

Option A is the structure where staff and line departments all report directly to the airport director. This is the normal situation at a small airport where the staff functions are not excessive and the airport director, as a matter of course, is closely involved with day-to-day operations. Option B is likely to occur at larger airports. The increasing busy line departments report through a deputy director while the staff departments are in a close supportive role to the director. At large airports, Option C is likely to occur, with line and staff departments reporting through two separate deputy directors. Examples with minor variants are shown later in this chapter.

Figures 1.7 and 1.8 show the organizational structures of two autonomous West European airports. Both structures reflect the fact that the organization is involved with the operation of a single airport on the European model and that much of the ground handling of passengers and of aircraft is carried out by airport authority employees. The Munich structure is somewhat conventional in its categorization of activities into operations, technical, and

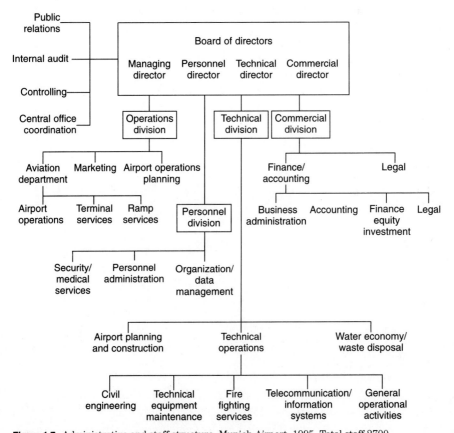

Figure 1.7 Administrative and staff structure, Munich Airport, 1995. Total staff 2700.

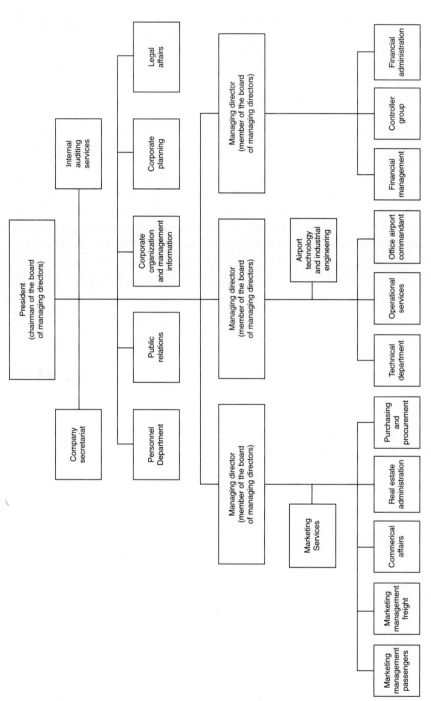

Figure 1.8 Administrative and staff structure, Schiphol Amsterdam Airport, 1995.

commercial divisions. Less usual is the arrangement whereby each division reports to an individual director who serves as an individual member on the board of directors. With an annual passenger throughput of more than 14 million, the number of annual passengers per airport authority employee is approximately 3500. This reflects the high level of airport authority activity on the operational side, with approximately 50 percent of the airport staff being employed in the operations division.

The total number of persons employed on the Schiphol Airport site in 1995 was 41,900, making it the second largest employer in the Amsterdam region, the municipality being the largest. Of these, close to 1550 persons were airport authority employees, giving a relationship of approximately 15,160 annual passengers per authority employee, indicating a much lower level of operational involvement in the airport. It is interesting to note that the Schiphol structure reflects the airport's reputation for aggressively pursuing the commercial and marketing aspects of passenger traffic. Of the eight divisions, two are concerned with external relations and commercial exploitation. Partly as a result of the commercial policy of the airport, 80 percent of all passengers passing through Schiphol take advantage of the shopping facilities.[1]

A very different form of functional arrangement exists at Los Angeles International Airport (Figure 1.9), where the throughput of annual passengers was 51 million in 1994, and the airport authority employs directly a staff of only 1335 giving a ratio of 38,239 annual passengers per airport authority employee. This high ratio does not necessarily imply a greater efficiency or productivity at the U.S. airports; it reflects instead the philosophy that in the United States the airlines are required to provide and staff more of their own facilities and equipment at airports and that more areas are designated to the uses of particular airlines, with fewer common areas and common facilities staffed by airport authority personnel. The implications of airport authority staffing are shown in Figure 1.9, where the line relationships indicate how much operational activity is delegated by the U.S. airport authorities.

Sacramento is a smaller metropolitan airport in northern California. With an airport authority staff of only slightly more than 250 in the mid 1990s, a very simple organization structure was feasible (Figure 1.10). This structure is frequently used at smaller airports around the world. Where the airport director is routinely involved in the day-to-day decisions of running the airport and the technical complexities of the various departments are not of a nature that require extensive range of skilled staff, a structure with essentially two operational divisions (planning/development and operations/

---

[1]The commercial possibilities at large airports are substantial. It is of interest to note that the commercial areas at Heathrow generate a greater turnover per square meter than the world's most successful department stores.

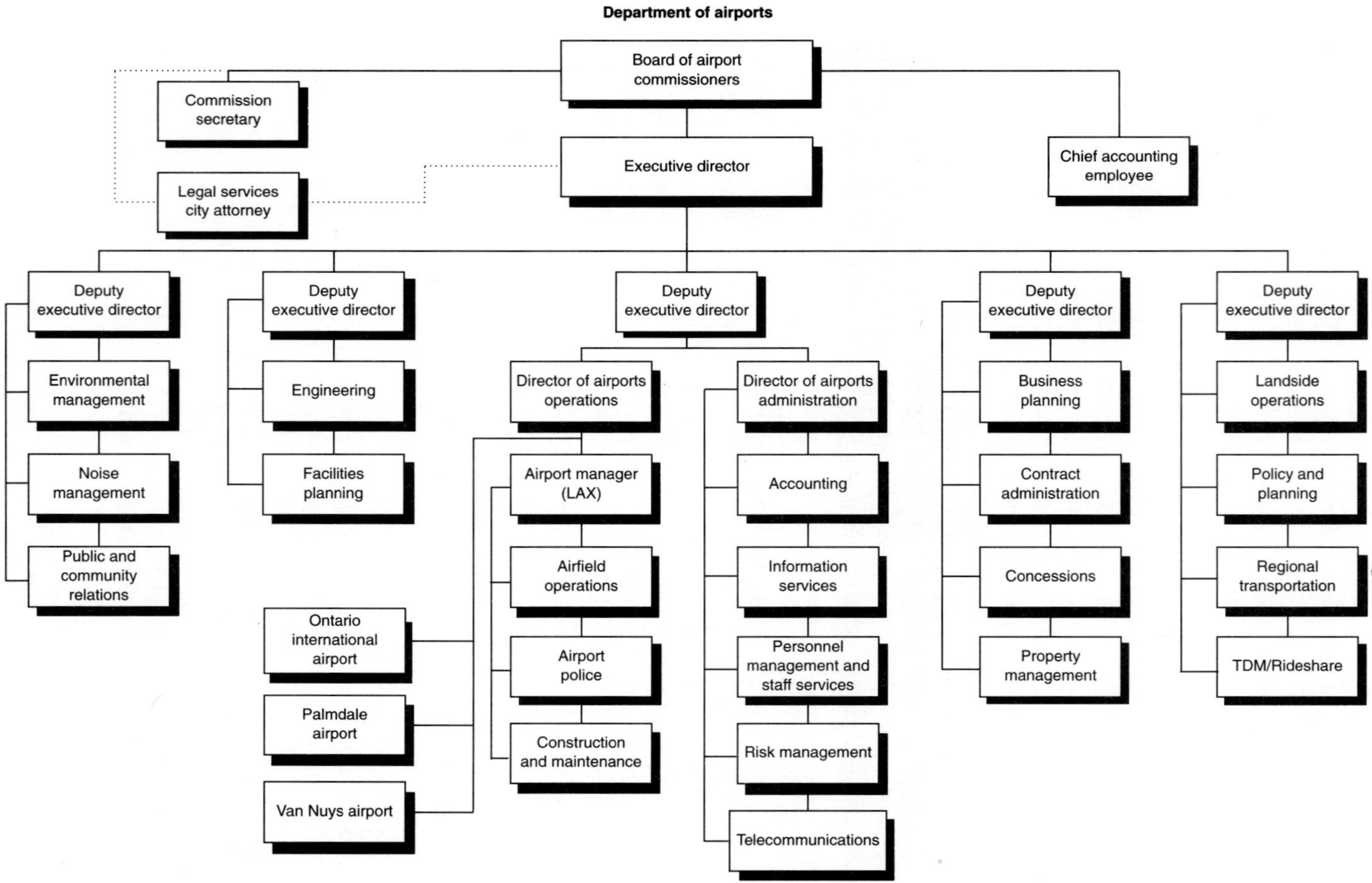

Figure 1.9 Organizational structure of Los Angeles International Airport. Note: *For simplicity, the administrative structure of Van Nuys and Ontario International Airport are omitted from this chart.*

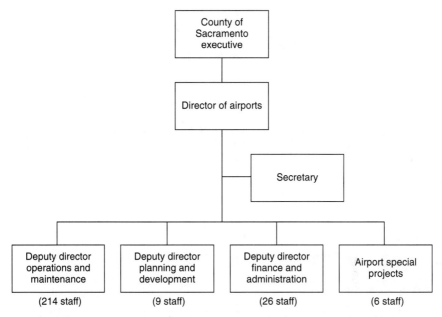

Figure 1.10 Organizational structure of Sacramento Airport.

maintenance) and one staff division (finance and administration) works well. As the complexity of the airport operation increases with size, a structure of this nature often breaks down because the individual divisions usually become too large, and the division directors are overloaded with day-to-day details. This is not always the case, as can be seen from the administrative structure of San Francisco International Airport, where divisional responsibility accrues to only three divisional directors who in conjunction with five heads of bureaus report to the Director (Figure 1.11).

In many countries, governmental or quasi-governmental authorities are charged with the operation of a number of airports (e.g., the Port Authority of New York and New Jersey [PANYNJ], Aeroport de Paris, and Aeroportos e Navegacao Aerea [ANA] in Portugal). The organizational structure of such multiairport authorities is usually designed for achieving systemwide objectives. Therefore, policies are directed by an overall executive to whom the usual system staff functions give support. Individual airports become operational elements in the overall structure. Wiley (Wiley 1981) has developed a typical pro forma organization for a three-airport authority that is shown in Figure 1.12. Practical examples of structure are shown in Figures 1.13a and 1.13b, which show the actual organizational structures of The Port Authority of New York (1995), and the Civil Aviation Authority of Portugal (1995). In each case, the organization permits the development of systemwide policies affecting a number of airports; this is a requirement that is

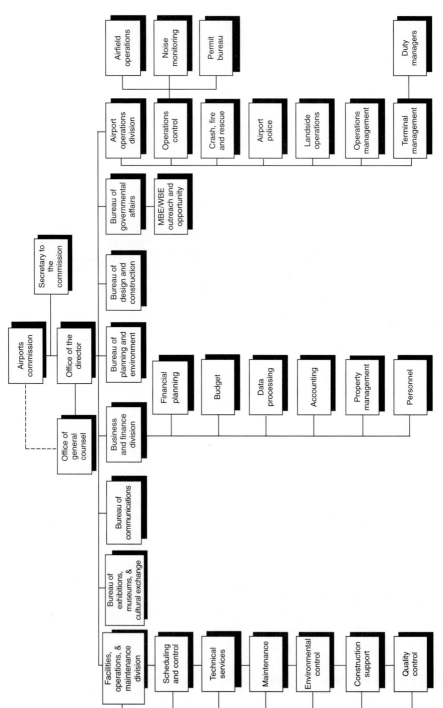

Figure 1.11 Organizational structure of San Francisco Airport.

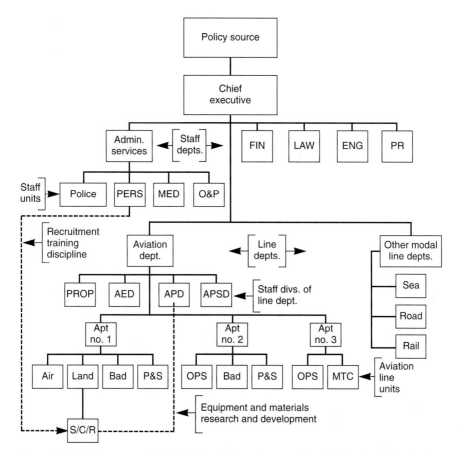

Figure 1.12 Proforma organization chart for a three-airport multimodal planning and operating authority. (*Ashford and Wright 1992*)

Air	Airside operations
Apt No 1	Airport no. 1, etc.
AED	Aviation economics division
APD	Aviation planning division
APSD	Aviation public services division
Bad	Business administration
ENG	Engineering department
FIN	Finance department
Land	Landside operations
LAW	Law department
MTC	Maintenance
MED	Medical department
O&P	Organization & procedures departments
OPS	Airport operations (both air and landside)
PERS	Personnel department
P&S	Plant and structures
PROP	Properties
PR	Public relations department
S/C/R	Security/crash/rescue unit

clearly not necessary in the case of the autonomously operated single airport. The PANYNJ structure is especially interesting because of the multimodal interests of the authority. Aviation constitutes only one department within the complex structure even though this department operates the three major airports in the New York metropolitan area.

The variety of structures shown in this chapter when taken in conjunction with the great differences of roles undertaken by the various authorities means that it is not possible to determine or even impute any strong relationship between passenger throughput and the size of the airport staff. Where the airport authority "brokers out" most activities, staff requirements will be low. As more activities are undertaken by the airport itself, the staff requirements naturally increase. Figure 1.14 shows the annual throughput for a number of airports in the mid 1990s, plotted against the

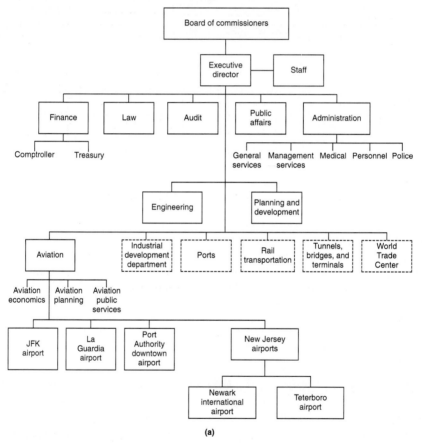

(a)

Figure 1.13a The organizational structure of the Port Authority of New York and New Jersey.

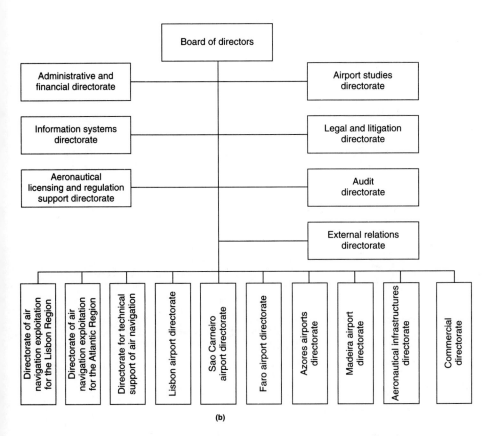

Figure 1.13b The organizational structure of the Portuguese Civil Aviation System.

airport staff at that time. As expected, there is a very large variation of the points about any single "fitted" line, indicating that there is little correlation between the two variables. However, considerably more distinct relationships do appear if the data are divided into three categories:

- North American
- British
- Other

Reasonable strong correlation can be determined for each category and the log-log regression lines are plotted on the diagrams. It is emphasized that this cross-sectional data should not be used in terms of comparing or determining the efficiency of an airport authority. Each point on the graph represents a different situation with different responsibilities. However, the

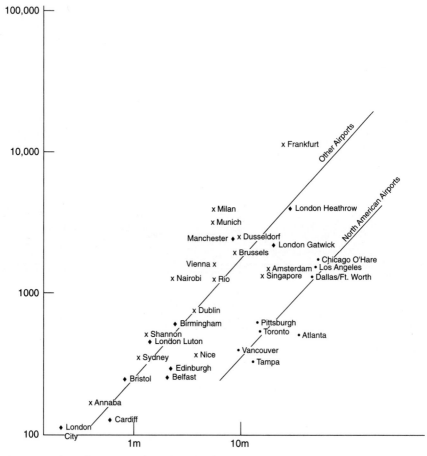

Figure 1.14 Annual passenger throughput in relation to airport authority staff.

graph does dramatically indicate the increased labor requirements of airport authorities that retain a proportion of handling activities in lieu of delegating these to the airlines.

References

Ashford, N.J., and J. Boothby. 1979. *Difficulties in modeling airport systems.* Transportation Research Record. Washington, DC: Transportation Research Board.

Ashford N., and C.A. Moore. 1993. *Airport Finance.* New York: van Nostrand Reinhold.

Ashford, N.J., and P.H. Wright. 1992. *Airport Engineering, 3rd edition.* New York: Wiley-Interscience.

Doganis, R., and R. Pearson. 1977. *The Financial and Economic Characteristics of the U.K. Airport Industry.* London: Polytechnic of Central London.

Doganis, R., and G. Thompson. 1973. *Economics of British Airports.* London: Polytechnic of Central London.

National Plan of Integrated Airport Systems (NPIAS) 1986–1995. 1987 Washington DC: Federal Aviation Administration.

Wiley, John R. 1981. *Airport Administration.* Westport, Conn.: Eno Foundation.

2

Airport Peaks and Airline Rescheduling

2.1 The Problem

Airport operators usually speak enthusiastically of the business of the airport in terms of the throughput of passengers and cargo as represented by the annual number of passengers processed or the annual turnover of tons of air freight. This is entirely understandable, since annual income is to a large degree determined by these parameters; moreover, the numbers are impressively large. However, it is important when considering annual figures to bear in mind that whereas annual flows are the prime determinant of revenue, it is the peak flows that to a large degree determine the physical and operational costs involved in running a facility. Staffing and physical facilities are naturally keyed much more closely to hourly and daily requirements than to annual throughput.

In common with other transportation facilities, airports display very large variations in demand levels with time. These variations can be described in terms of:

- Annual variation over time
- Monthly peaks within a particular year
- Daily peaks within a particular month or week
- Hourly peaks within a particular day

The first of these is extremely important from the viewpoint of the planning and provision of facilities. Air transportation is still the fastest-growing mode, and there is little indication that this situation is likely to change. Consequently, operators of airport facilities often are faced with growing volumes that approach or exceed capacity. During the period 1970 to 1995, the average worldwide rate of air passenger growth was close to 7 percent (ICAO 1995). Even during the difficult period of oil price increases in the 1970s and early 1980s the average world growth rate was also approximately 7 percent. Air transport passenger travel is expected to continue to grow at over 5 percent up to the year 2005 and air cargo to grow at 6.5 percent during the same period (World Traffic 1994). Although the operator must be closely involved in the long-term planning of the airport, it is not the function of this text to deal with planning aspects that are covered elsewhere (Ashford and Wright 1992). Emphasis here will be on the short-term considerations of day-to-day operation. Therefore, the discussion will concentrate on monthly, daily, and hourly variations of flows. In the operational context, this is natural because many of the marginal costs associated with the day-to-day provision of staffing and rapidly amortized equipment are not really related to long-term variations of traffic, but rather to variations within a 12-month span.

At most, if not all, air transport airports, the major consideration must be passenger flow. At many of the larger airports, cargo operations are becoming increasingly important, partially because cargo transport continues to outstrip passenger traffic in terms of growth rate. In the planning and operation of air cargo facilities, however, it must be noted that the peaks for air cargo operations do not coincide with those for air passenger transport. The two submodes can usually be physically separated to the necessary degree, even though proximity of the freight and passenger aprons is desirable because most freight is carried in the bellies of passenger transport aircraft. The particular problems of cargo operations will be discussed in Chapter 10.

When considering the characteristics of the peaking of passenger flows, it is always important to bear in mind that the "passenger" is not a homogeneous entity. Passenger traffic is built up from the individual journey demands of many passengers. These passengers are traveling under different conditions, they have different needs and consequently place different overall demands on the system. Not surprisingly, this is reflected in different peaking characteristics, depending, for example, on whether the passenger is domestic or international, scheduled or charter, leisure or business, full fare or special fare.

Complicating the whole matter of peaking is the fact that, unlike the situation in most other modes of transportation where the passenger is dealing with only one operator, in air transport there is the complex interrelationship of the passenger, the airport, and often several airlines. In the matter of peaking, the aims of the airline and the airport operator do not necessarily coincide. The airport operator would like to spread demand more evenly over the operating day in order to decrease the need for the supply of facilities governed by the peak. The airline, on the other hand, is looking to maximize fleet utilization and to improve load factors by offering services in the most attractive time slots. There is, therefore, a potential conflict between the airline satisfying its customer, the passenger, and the airport attempting to influence the demands of its customer, the airline.

2.2 Methods of Describing Peaking

Even the busiest airport operates over a wide range of traffic flows. Many of the world's largest air transport terminals are virtually deserted for many hours of the year; these same facilities only a few hours later might be operating at flows that strain or surpass capacities. Few facilities are designed to cope with the very highest flow volume that occurs in the design year of operation. Most are designed such that for a few hours of the year there will be an acceptable level of capacity overload. Different airport and aviation authorities approach this problem in different ways. Figure 2.1 shows one of the characteristics of traffic peaking for a typical airport, that is, the curve of passenger traffic volumes in ranked order of magnitude. It can be seen that for a few hours per year there are very high peaked volumes of traffic. Operational practice tends to accept that for a few hours of each year facilities must be operated at some level of overload (i.e., volumes that exceed physical and operational capacity) with resultant delays and inconvenience. To do otherwise and to attempt to provide capacity for all volumes would result in uneconomical and wasteful operation.

The Standard Busy Rate (SBR)

The *Standard Busy Rate* measure or a variation of it is a design standard used in the United Kingdom and elsewhere in Europe, most notably by the

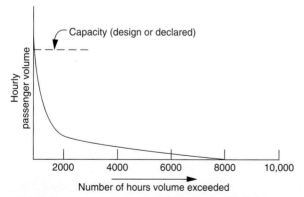

Figure 2.1 Typical distribution of hourly passenger traffic volumes at an air transport airport throughout the year.

former British Airports Authority. It is frequently defined as the 30th highest hour of passenger flow, or that rate of flow that is surpassed by only 29 hours of operation at higher flows. The concept of the 30th highest hour is one that is well rooted in civil engineering practice in that this form of design criteria has been used for many years to determine design volumes of highways. Design for the SBR ensures that facilities will not operate at or beyond capacity for more than 30 hours per year in the design year, which was felt to be a reasonable number of hours of overload. The method does not, however, take explicit note of the relationship of the SBR to the actual observed annual peak volume. Although in practice this relationship is likely to be on the order of

$$\text{absolute peak hour volume} = 1.2 \times \text{standard busy rate} \qquad (2.1)$$

there is no guarantee that this will be so.

Table 2.1 shows that in terms of aircraft movements the ratio of the SBR to the absolute peak increases with increasing annual volume. This reflects the fact that as the traffic of an airport develops, extreme peakness of flows tends to disappear.

The table indicates that use of the SBR method in low-volume airports could result in high (peak/SBR) ratios that in turn could lead to severe overcrowding for a few hours per year. The location of the standard busy hour is shown in Figure 2.2.

Busy Hour Rate (BHR)

A modification of the SBR, which has been used for some time is the *Busy Hour Rate* or the 5 percent busy hour. This is the hourly rate above which 5 percent of the traffic at the airport is handled. This measure was introduced to overcome some of the problems involved with using the SBR, where the implied level of congestion was not common between airports. The BHR is easily computed by ranking the operational volumes in order of

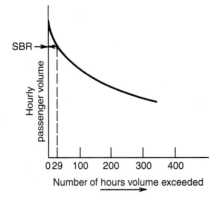

Figure 2.2 Location of the standard busy rate.

TABLE 2.1 Relationship Between Annual, Peak Hourly, SBR and Peak Day Aircraft Movement Rates

Annual movements	Peak day to average day ratio	number	Peak hour to peak day ratio	number	SBR to peak hour ratio	number
10,000	2.666	73	0.1125	8	0.688	6
20,000	2.255	124	0.1051	13	0.732	10
30,000	2.045	168	0.1011	17	0.759	13
40,000	1.907	209	0.0983	21	0.779	16
50,000	1.807	248	0.0961	24	0.794	19
60,000	1.729	284	0.0944	27	0.807	22
70,000	1.666	320	0.0930	30	0.819	24
80,000	1.613	354	0.0918	32	0.828	27
90,000	1.568	387	0.0908	35	0.837	29
100,000	1.529	419	0.0898	38	0.845	32
110,000	1.494	450	0.0890	40	0.852	34
120,000	1.463	481	0.0883	42	0.859	36
130,000	1.435	511	0.0876	45	0.865	39
140,000	1.409	541	0.0869	47	0.871	41
150,000	1.386	570	0.0863	49	0.876	43
160,000	1.365	598	0.0858	51	0.881	45
170,000	1.345	626	0.0853	53	0.886	47
180,000	1.326	654	0.0848	55	0.891	49
190,000	1.309	681	0.0844	57	0.895	51
200,000	1.293	708	0.0840	59	0.899	53
210,000	1.278	735	0.0836	61	0.903	55
220,000	1.264	762	0.0832	63	0.907	57
230,000	1.250	788	0.0828	65	0.910	59
240,000	1.237	814	0.0825	67	0.914	61
250,000	1.225	839	0.0821	69	0.917	63
260,000	1.214	864	0.0818	71	0.920	65
270,000	1.203	890	0.0815	73	0.924	67
280,000	1.192	914	0.0812	74	0.927	69
290,000	1.182	939	0.0810	76	0.929	71
300,000	1.172	964	0.0807	78	0.932	72
310,000	1.163	988	0.0804	79	0.935	74
320,000	1.154	1012	0.0802	81	0.938	76
330,000	1.146	1036	0.0799	83	0.940	78
340,000	1.137	1060	0.0797	84	0.943	80
350,000	1.130	1083	0.0795	86	0.945	81
360,000	1.122	1106	0.0793	88	0.948	83
370,000	1.114	1130	0.0791	89	0.950	85
380,000	1.107	1153	0.0788	91	0.952	87
390,000	1.100	1176	0.0786	92	0.954	88
400,000	1.094	1199	0.0785	94	0.957	90
410,000	1.087	1221	0.0783	96	0.959	92
420,000	1.081	1244	0.0781	97	0.961	93
430,000	1.075	1266	0.0779	99	0.963	95
440,000	1.069	1288	0.0777	100	0.965	97
450,000	1.063	1311	0.0776	102	0.967	98
460,000	1.057	1333	0.0774	103	0.969	100
470,000	1.052	1354	0.0772	105	0.971	102
480,000	1.047	1376	0.0771	106	0.972	103
490,000	1.041	1398	0.0769	108	0.974	105
500,000	1.036	1420	0.0768	109	0.976	106

SOURCE: UK Civil Aviation Authority.

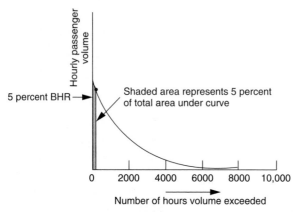

Figure 2.3 Five percent busy hour rate.

magnitude and computing the cumulative sum of volumes that amount to 5 percent of the annual volume. The next ranked volume is the busy hour rate. This is shown graphically in Figure 2.3. One great disadvantage of the BHR method is that so many data must be collected and analyzed that this method might be beyond the resources of a small airport.

Typical Peak Hour Passengers (TPHP)

The Federal Aviation Administration (FAA) uses a peak measure called *Typical Peak Hour Passengers* that is defined as the peak hour of the average peak day of the peak month. In absolute terms this approximates very closely with the SBR. To compute the TPHP from annual flows, the FAA recommends the relationships shown in Table 2.2. Stated in this form it is apparent that the peak is more pronounced with respect to annual flows at small airports. As airports grow larger the peaks flatten and the troughs between peaks become less pronounced.

TABLE 2.2 FAA Recommended Relationships for TPHP Computations from Annual Figures

Total annual passengers	TPHP as a percentage of annual flows
30 million and over	0.035
20,000,000 to 29,999,999	0.040
10,000,000 to 19,999,999	0.045
1,000,000 to 9,999,999	0.050
500,000 to 999,999	0.080
100,000 to 499,999	0.130
Under 100,000	0.200

Busiest Timetable Hour (BTH)

This simple method is applicable to small airports with limited data bases. Using average load factors and existing or projected timetables, the *Busiest Timetable Hour* (BTH) can be calculated. The method is subject to errors in forecasting, the rescheduling and reequipping vagaries of the airlines and variations in average load factors.

Peak Profile Hour

Sometimes called the average daily peak, the *Peak Profile Hour* method is fairly straightforward to understand. First the peak month is selected. Then for each hour, the average hourly volume is computed across the month, using the month (i.e., 28, 30, or 31 days as applicable). This gives an average hourly volume for an "average peak day". The peak profile hour is the largest hourly value in the average peak day. Experience has shown that for many airports the peak profile hour is close to the standard busy rate.

Other methods

Although many outside the United States use some form of the SBR method to define the peak, there is little uniformity in method. In West Germany, for example, most airport authorities use the 30th highest hour. Prior to the introduction of the BHR, the British Airports Authority used the 20th highest hour, while most other British airports used the 30th highest hour. In France, the Aeroport de Paris bases its design on a 3 percent overload standard. (In Paris, studies have shown that the 30th busy hour tends to occur on the 15th busiest day.) Dutch airports use the 6th busiest hour, which is approximated by the average of the 20 highest hours.

Airport differences

The shape of the volume curve shown in Figure 2.1 differs between airports. The nature of these differences can be seen by examining the form of the curves for three airports with widely differing functions, as shown in Figure 2.4.

Airport A	A high-volume airport with a large amount of short-haul domestic traffic (typical U.S. or European hub).
Airport B	A medium-volume airport with balanced international/domestic traffic and balanced short-haul/long-haul operations (typical North European metropolitan).
Airport C	A medium-volume airport with high proportion of international traffic concentrated in a vacation season (typical Mediterranean airport serving resort areas).

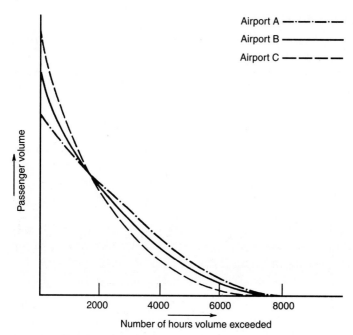

Figure 2.4 Variation of passenger volume distribution curves for airports of different traffic characters.

Airport C will carry a higher proportion of its traffic during peak periods, and there is therefore a leftward skew to the graph in comparison with Airport B. A typical U.S. or European hub, on the other hand, with larger amounts of domestic short-haul traffic carries more even volumes of passengers across the period 0700 to 1900 hours, decreasing the leftward skew of the graph.

Nature of peaks

Airport traffic displays peaking characteristics by the month of the year, by the day of the week, and by the hour of the day. The form and time of the peaks is very much dependent on the nature of the airport traffic and the nature of the hinterland served.

The following factors are among the most important affecting peaking characteristics:

1. *Domestic/international ratio.* Domestic flights will tend to operate in a manner that reflects the working day pattern because of the large proportion of business travelers using domestic flights.

2. *Charter/scheduled ratio.* Charter flights are timetabled for maximum aircraft usage and are not necessarily operated at the peak periods found most commercially competitive by scheduled airlines.

3. *Long-haul/short-haul.* Short-haul flights are frequently scheduled to maximize the usefulness of the day either after or prior to the flight. Therefore, they peak in early morning (0700 to 0900) and late afternoon (1630 to 1830). Long-haul flights are scheduled mainly for a convenient arrival time, allowing for reasonable rest periods for the travelers and crew and to avoid night curfews.

4. *Geographical location.* Schedules are set to allow passengers to arrive at a time when transportation and hotels are operating and can be conveniently used. For example, the six to eight hour eastward transatlantic crossing is most conveniently scheduled for early morning arrivals at the Western European airports avoiding curfews. Allowing for the time differences between North America and Europe, this means an evening departure from the Eastern Seaboard.

5. *Nature of catchment area.* The nature of the region served has a strong influence on the nature of traffic peaking throughout the year. Areas serving heterogeneous industrial-commercial metropolitan areas such as Chicago, Los Angeles, London, and Paris show steady flows throughout the year with surges at the Christmas, Easter, and summer holiday periods reflecting increased leisure travel. Airports in the vicinity of highly seasonal vacation areas, such as the Mediterranean, display very significant peaks in the vacation months.

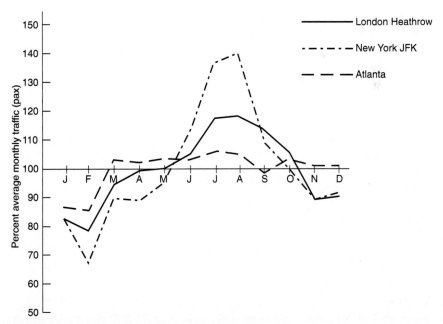

Figure 2.5 Monthly variations of passenger traffic for various airports. *(BAA plc and Los Angeles Department of Airports.)*

Figure 2.5 shows the monthly variations in traffic at several airports serving widely different geographical areas. Daily variations in the peak week are shown in Figure 2.6. The analysis is carried further by Figures 2.7 and 2.8, which show hourly passenger movements and arrival/departures, respectively. It is clear from the last two figures that hourly peaks are significant and differ substantially between airport terminals depending on the airport's function.

In spite of the difference between peaks caused by the many factors that affect peaking, there is in some aspects, in fact, great overall similarity between airports. It is therefore possible to deduce general relationships between peak and annual flows at airports, largely because no airport is entirely unifunctional, just as no town is entirely industrial, governmental, educational, or leisure-structured in its makeup.

Figure 2.9 shows the relationship between peak flows as represented by SBR and annual flows for a number of rather diversely selected airports. Also shown on this graph are the FAA peak/annual recommended ratios, as embodied in the typical peak-hour passengers (TPHP) concept. The great similarity of the two approaches becomes apparent when they are presented graphically.

2.3 Implications of Variations in Volumes

It can easily be demonstrated that the demand for peak-hour schedules affects the amount of infrastructure that must be supplied by the airport.

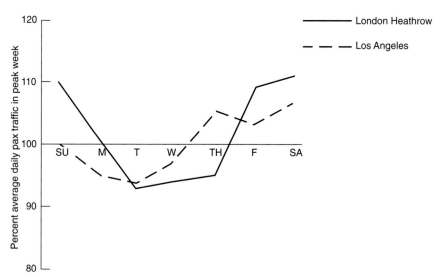

Figure 2.6 Daily variations of passenger traffic in peak week. *(BAA plc and Los Angeles Department of Airports.)*

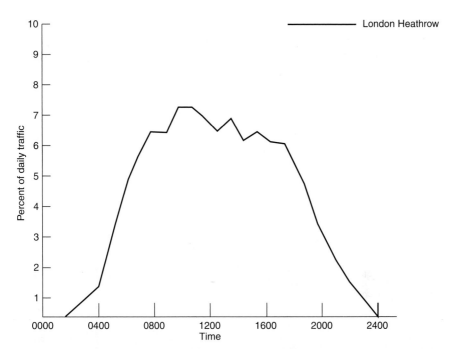

Figure 2.7 Hourly variations of passenger traffic in a typical day in peak season. *(BAA plc.)*

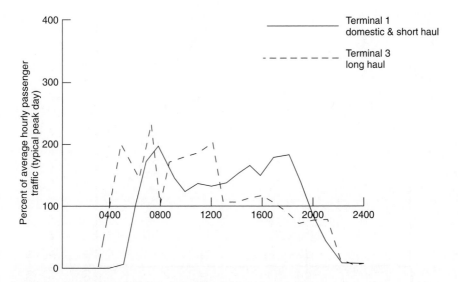

Figure 2.8 Hourly variations of arriving and departing passengers at London Heathrow for long- and short-haul terminals. *(BAA plc.)*

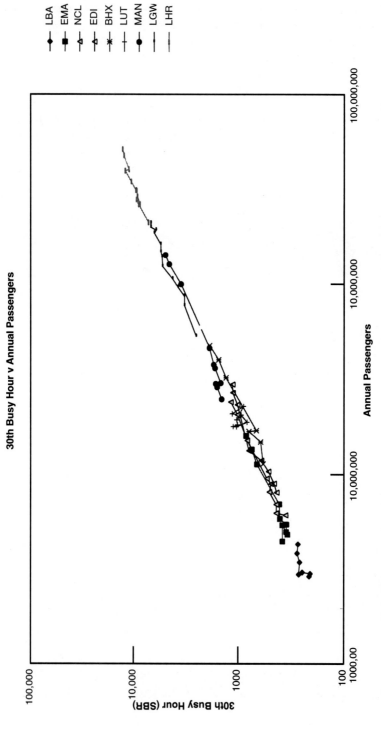

Figure 2.9 Relationship between standard busy rate, *typical peak hour passenger* volume, and annual passenger volume for selected airports. LHR = Heathrow; LGW = Gatwick; MAN = Manchester; LUT = Luton; BHX = Birmingham; NCL = Newcastle; EMA = East Midlands; LBA = Leeds Bradford. (*Civil Aviation Authority.*)

Whereas the need to implement service in an off-peak period will not necessarily involve the airport in significant marginal costs, at a crowded airport the decision to take another service in the peak hour might well add significant marginal costs. There are, however, economies of scale that result from peak hour operations.

Figure 2.10a shows the relationship between passenger flows and air carrier aircraft operations. It can be seen that whereas passenger volumes vary significantly between peak and nonpeak hours, the same scale of variation is not observed in aircraft movement volumes. This reflects the fact that during off-peak periods aircraft operate at lower load factors than during the attractive peak-hour slots. The implications in terms of costs and rev-

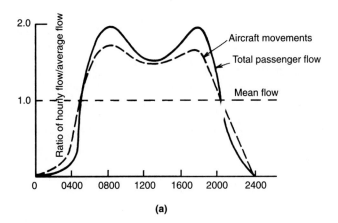

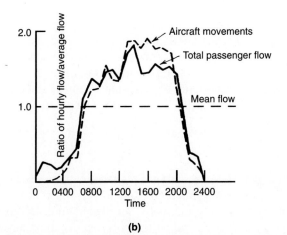

Figure 2.10 *(a)* Idealized relationship of air carrier movements to passenger flow under varying load factors throughout a typical day. *(b)* Observed relationship between aircraft movements and passenger flow at Chicago O'Hare. *(Federal Aviation Administration.)*

enues that are operations-oriented rather than unit passenger-oriented are obvious. Services such as ramp handling, emergency services, air traffic control, runway and taxiway handling, and even some terminal services (e.g., announcements and baggage check) are based on the aircraft unit rather than the passenger. In off-peak hours, these services are provided at a less economic rate per passenger than during peak periods due to low load factors during off-peak periods. Therefore, the airport is faced with a dilemma. Although peak operations would appear to involve high marginal costs in terms of infrastructure, operation at close to peak volumes is highly economic once this infrastructure is provided. This effect is not immediately discernible from an examination of graphs showing total aircraft movements and total passenger flows at a particular airport. Figure 2.10b, which shows total data for Chicago O'Hare on a peak summer day, does not reflect the theoretical graph of Figure 2.10a. It must be remembered that total figures include general aviation aircraft, which might seriously distort the pattern obtained from air carrier aircraft.

2.4   Factors and Constraints on Airline Scheduling Policies

The development of a schedule, especially at a major hub with capacity problems, is a complex problem for the airline. The process involves considerable skill and a clear understanding of company policies and operating procedures. Among the factors to be considered, the following are most important.

Utilization and load factors

Aircraft are expensive items of equipment that can earn revenues only when being flown. Clearly, all other factors being equal, high utilization factors are desirable. However, utilization alone cannot be used as the criterion for schedule development; it must be accompanied by high load factors. Without the second element, aircraft would be scheduled to fly at less than break-even level, which typically is close to 70 percent on long-haul operation of a modern wide-bodied aircraft.

Reliability

No airline would attempt to schedule using the sole criterion of maximizing utilization of aircraft. Utilization can be maximized, however, subject to the double constraints of load factors and punctuality. As attempted utilization increases, the reliability of the service will suffer in terms of punctuality. Schedule adherence is a function of two random variables: equipment serviceability and late arrivals or departures of aircraft due to en route factors. Computer models are used to predict the effect of schedules on punctuality, and the result is compared to target levels of punctuality set in advance for each season.

Long-haul scheduling windows

A schedule must take account of the departure and arrival times at the various airports at origin, en route, and at destination. For example, a departure from London destined eventually for Sydney, Australia, via Bombay and Perth could leave in the late afternoon to ensure arrival just after the Sydney curfew lifts at 0600. A morning departure from Heathrow at 0930 would give an arrival time the following evening at 2140, which is only 80 minutes before the 2300 curfew that has no exemptions except emergencies. This is clearly too small a margin for error. Departure times are also set recognizing that passengers must travel from city centers to the airport and must arrive at the airport some reasonable time before the scheduled time of departure.

Figure 2.11 gives examples of scheduling windows for flights to and from London. On the London to Tokyo route, two night jet bans could be involved: Hong Kong (1600 to 2230 GMT) and Tokyo (1400 to 2100 GMT). The schedule is constructed within narrow constraints to ensure arrival at Dubai before the immediate postmidnight hours and avoiding the night bans at both Hong Kong and Tokyo. Long distance schedules of this nature must be constructed to absorb reasonable delays before departure and en route. The Seychelles to London example shows a schedule that might be built around the Zurich night jet ban and a reasonably convenient arrival time at Bahrain. The Zurich night jet ban would determine arrival time in London and departure time from Bahrain. Departure from the Seychelles would be determined by an arrival time just after midnight local time at Bahrain. Even so, the aircraft would be involved in an unnecessarily long turnaround in Bahrain due to a schedule on which there is very little possible variation if commercially attractive times are to be used.

Airport (runway) slots. Runway takeoff and landing slots must also be considered. In many airports, especially in Europe, North America, and S.E. Asia, existing runways are running near to capacity during peak periods of the day. This capacity is due to safety margins required in the separation of arriving and departing aircraft. Many airports near their slot capacity are coordinated. This means that a regulatory authority, such as the FAA or the CAA, has to check and to allocate a number of slots as being available to arriving or departing flights. Actual coordination is carried out semi-annually at IATA slot conferences. A carrier will often have the right only to its historical slots, provided that these are being used. Consequently, at a coordinated airport any carrier will be uncertain whether it will be possible to move from its historic slots or gain more slots. This situation poses prob-

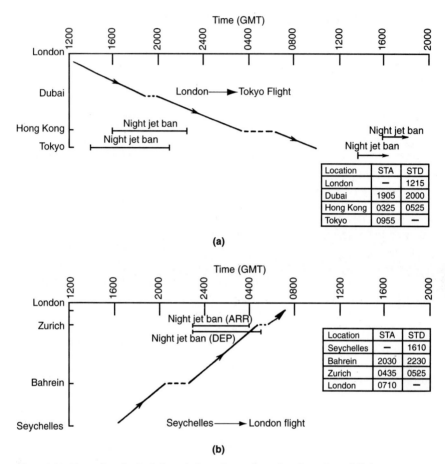

Figure 2.11 Examples of scheduling windows for eastbound and westbound flights.

lems to schedulers who must make assumptions on the likely slots available to them.

Terminal Constraints. Another constraint increasingly faced by schedulers is that of airport passenger apron and terminal capacity. Many airports are running at near to the capacities of these facilities often built 20 or more years ago. In the case of terminals, authorities often limit the number of passengers that can pass through a terminal during a half hourly period, stating that this flow is the "declared capacity." This obviously sets a limit on the number of arrivals, departures or combination of aircraft apron movements that can be scheduled in capacity strained periods, presenting schedulers with yet another constraint.

Figures 2.12a and 2.12b indicate the scheduling limitations on a return flight from London to Sydney, via Hong Kong, in the years 1981 and 1995 re-

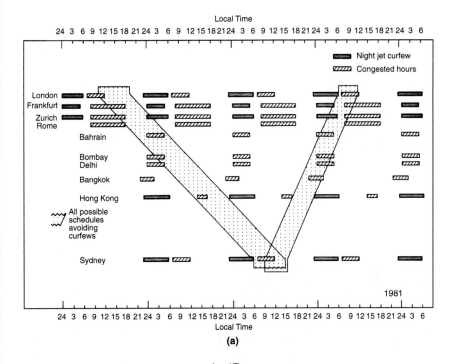

(a)

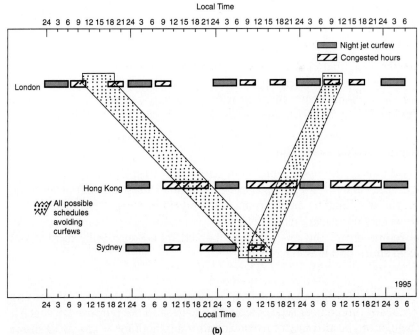

(b)

Figures 2.12a and 2.12b Scheduling limitations for the London-Hong Kong-Sydney route. *(British Airways)*

spectively, showing very clearly how narrow the scheduling window will be when night curfews and runway congested hours are considered. With the introduction of long-range aircraft, technical refueling stops, which were used in 1981 at airports such as Bahrain, Bombay, and Delhi, are no longer necessary. Consequently, the traffic peaks at these airports have radically changed since they no longer provide intermediate stopping points for European-Asian traffic, stops that frequently occurred at strange early morning hours. However, it can be seen that over the 14-year period shown, the runway at Hong Kong has become congested or curfewed for all but three hours of the day. Congested periods have also increased at London Heathrow while the addition of a second runway at Sydney has eased runway congestion at that airport. It is impossible to schedule this route without the aircraft being on the apron of at least one of the three airports during congested hours.

Long-haul crewing constraints

On long-haul flights, crews may not be used continuously. Typically, a maximum tour of duty could be 14 hours, which includes 1½ hours of pre- or post-flight time; there is also a required minimum rest period (usually 12 hours). Therefore, crews are changed at *slip* posts, and timing must be arranged so that fresh crews are available at these posts to relieve incoming flights that will be continuing their journeys.

Short-haul convenience

Because short-haul flights frequently carry large numbers of business travelers, departure and arrival times are critical to marketing the flights. Short-haul flights that cannot provide a one-day return journey suitably scheduled around the business working day are difficult to market.

General crewing availability

In addition to the special problems associated with layovers of long-haul flight and cabin crews at slip ports all schedules must be built around the availability of maintenance, ground, air, and cabin crews. There is clearly a very strong interrelationship between the numbers of various crew personnel required and the operations to be scheduled, especially in terms of mixed short- and long-haul flights.

Aircraft availability

Depending on the fleet type of aircraft, its age, and the purpose for which it is being used, availability of particular aircraft will differ. Figure 2.13 shows the availability of a short-haul fleet. Maintenance requirements vary greatly among different types of aircraft. Typically a B747 might be limited to 120 hours of continuous operation. After this, 8 hours of maintenance is required, including terminal and towing times, which could mean 12 hours downtime. A further 24-hour maintenance break is required every three weeks, and at three-month intervals a major maintenance check is necessary. This might be 2½ days , 5 days, or even a month depending on the aircraft's position in its 20,000-hour maintenance cycle.

Marketability

The scheduled times of departure or arrival must be marketable by the airline. Connections are especially important at major transfer points, such as Atlanta and Dallas-Fort Worth. Whenever possible, passengers avoid long layovers at the airport. Other factors that the airline considers are that departure and arrival times at major generating hubs must be at times when public transport is operating and might have to coincide with hotel check-in and check-out times and room availability. It is also important to have continuity of flight times across the days of the week if the flight operates several times a week.

Summer-winter variations

Where there is a large amount of seasonal traffic, usually vacational, there can be substantial differences in scheduling policies between summer and winter operations. The large variation in demand that can occur at airports serving such areas as the U.S. Gulf Coast and the Mediterranean resort areas is substantial and will affect the schedules of airports with which their services link. In general, however, the difference between summer and winter operations has tended to decline over the last 10 years, with winter providing its own peaks around winter holiday periods.

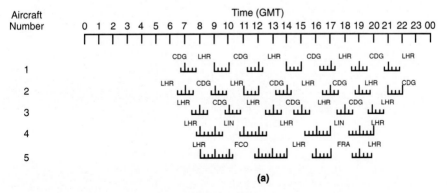

(a)

Figure 2.13a Aircraft fleet utilization—Boeing 767 fleet. *(British Airways)*

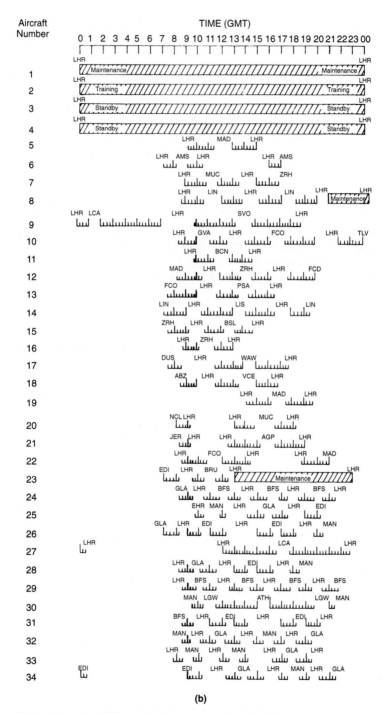

(b)

Figure 2.13b Aircraft fleet utilization—Boeing 757. *(British Airways)*

Landing fee pricing policies

At some airports, an attempt has been made to vary landing and aircraft-related fees in order either to use a pricing policy to spread peaking or to recoup extra finance for operations carried out in the economic night hours. An example of the former policy is that which was used by the former British Airports Authority (BAA), which at a stroke adopted punitive peak hour tariffs at London Heathrow to encourage airlines to transfer operations from Heathrow to Gatwick airports and to move operations from the peak period. Under this policy, a typical turn around of a long-range B747 at Heathrow during the peak period was 2.8 times the cost for an operation outside the peak tariff times and 183 percent of the cost that would have been involved had the operation taken place at the less popular London Gatwick airport at the same peak time. The effect of this peak tariff rate was not large, as can be seen in Table 2.3, which shows the operational impact of this particular differential tariff.

In general, there is little evidence to indicate that airlines do reschedule significantly to avoid such tariffs. Airline operators claim that there are far too many other constraints precluding massive rescheduling outside peak demand periods and that, therefore, such tariffs are almost entirely ineffective in achieving their proclaimed purpose.

The truth would appear to lie somewhere between these two positions. Where there is no differential peak pricing, airlines have no particular incentive, other than congestion-induced delay costs, to move operations from the congested peak period. On the other hand, the commercial viability of a flight and its ability to conform with bans and curfews might necessitate operations in peak hours. High differentials for peak operations might at first appear to be a reasonable step for the operator to take to spread congestion. However, any such action should be evaluated in the light of the impact on the based carriers whose operations inevitably represent a very large proportion of the airport's total movements. The short-term economic gain to the airport could put a long-term economic strain on the finances and the competitiveness of the based carriers. Withdrawal of services, movement of the airline base, or even collapse of the carrier will have a serious financial impact on the airport.

TABLE 2.3 Effect of Peak Tariffs on Traffic

	London Heathrow airport: peak passengers as a percentage of total passengers			
	July	August	September	October
1976 (prepeak tariffs)	30.7	30.8	30.4	30.5
1977 (postpeak tariffs)	29.7	26.3	24.5	24.3

The second type of tariff, that was made to support uneconomic operations during slack night hours, is exemplified by the 15 percent surcharge on handling fees formerly levied at Rome for arrivals and departures between 1900 and 0700, This amounted to a 30-percent surcharge if operations occurred within the period. A tariff of this nature has the bizarre effect that a transiting aircraft arriving and departing at and, respectively, using a period well outside the peak can in fact halve the surcharge by remaining on the stands for five more minutes, consequently utilizing more airport resources.

There is in fact a very wide variation in the manner in which airports structure landing fees. Table 2.4 shows that for the major European airports landing fees are computed from some combination of:

- Aircraft weight
- Apron parking requirement
- Passenger load
- Noise level created
- Security requirement
- Peak surcharge

Table 2.5 indicates the wide variation between airports of handling charges for the same aircraft.

2.5 Scheduling Within the Airline

As a fundamental element in the supply side of air transport, the question of scheduling involves a large number of persons and sections within the structure of the airline itself. A typical functional interaction chart is shown in Figure 2.14. The commercial economist takes advice from market research and interacts with the various route divisions, which control the operations of the various groupings of the airline routes. In some airlines there are no route divisions. The commercial economist and the route divisions are both part of the commercial department in this case. In advising schedules planning, which is concerned with the overall planning of the airlines schedule, the commercial economist will take note of a number of factors that will affect the decision whether or not to attempt to incorporate a service within a schedule. These factors could include some of the following:

- Historical nature of the route
- Currently available route capacity
- Aircraft type
- Fare structure (standby, peak, night, etc.)

TABLE 2.4 Landing Fee Structures at Selected Major International Airports (1993)

Airport	Country	Weight fixed	Weight varying	Parking fixed	Parking varying	Passenger arrival	Passenger departure	Noise scheme	Security charge	Runway charge	Peak surcharge
Atlanta Hartsfield	U.S.A.	X		X					X	X	
Sydney	Australia	X		X					X	X	X
Frankfurt	Germany	X		X		X			X		
Annaba	Algeria		X		X		X	X	X	X	
Linate, Milan	Italy		X		X		X			X	
Malpensa, Milton	Italy		X		X		X			X	
Brussels	Belgium		X		X		X			X	
Toronto	Canada		X		X	X	X	X	X	X	
Dublin	Ireland		X		X		X				
Amsterdam	Netherlands	X		X			X	X	X		
Changi	Singapore		X		X		X				
Bristol	U.K.		X	X	X	X	X		X		
Nairobi	Kenya	X	X								
Dusseldorf	Germany	X	X		X	X		X	X		
Nice	France		X		X	X	X	X			
Boston Logan	U.S.A.		X		X	X					
Shannon	Ireland	X			X		X				
Copenhagen	Denmark		X		X		X			X	
Vienna	Austria		X		X		X			X	
London City	U.K.	X	X		X	X	X	X	X	X	
Stockholm, Arlanda	Sweden		X		X	X	X		X	X	
Birmingham	U.K.		X		X		X				
Bahrain	Bahrain		X		X		X				
Manchester	U.K.		X		X		X	X			
Rio de Janeiro Int.	Brazil		X		X		X			X	
Munich	Germany		X		X		X	X			X
Vancouver	Canada		X	X			X		X	X	

TABLE 2.5 International Airport Charges for Selected Airports (1993–94) US dollars

	737–400	757–200	DC9–30	A320–100	747–400	767–400	DC10–30
Frankfurt	5739	7743	3530	5701	17,008	—	—
Athens	5510	7779	2250	4191	12,258	—	—
Amsterdam	4496	6823	2667	3840	12,461	—	—
Manchester	4000	4912	2163	3328	11,645	—	—
Dusseldorf	3772	6346	3315	3914	15,564	—	—
Brussels	3667	4667	1634	3708	11,691	—	—
Faro	3539	4733	1942	3608	11,630	—	—
Madrid	3924	4558	1273	3174	11,049	—	—
Bilbao	3252	4543	1145	2845	9900	—	—
London Gatwick	3000	3823	1582	2495	8185	—	—
Miami	—	6352	—	—	—	8011	9685
Montego Bay	—	5327	—	—	—	6495	8556
Orlando	—	8187	—	—	—	10,271	12,936
Barbados	—	4396	—	—	—	5608	6796

(Adapted from References 4 and 5)

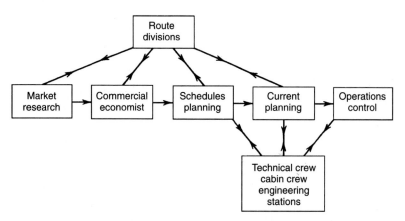

Figure 2.14 Organization of scheduling within a typical airline.

- Social need for route and subsidies
- Political considerations (in the case of the flag airline of a country)
- Competition
- Requirements for special events.

Once the decision has been made that a service should be incorporated into the schedule, the schedules planning section of the airline, which is frequently divided into long-haul and short-haul functions, will examine the scheduling implications of planning the service. At this stage, the factors that affect the planning are:

- Length of haul: long or short
- Availability of aircraft allowing extras for aircraft in maintenance and on standby
- Acceptability of service schedule to the airport
- Availability of technical, flight, cabin, and engineering crews
- Clearances with countries concerned where there are no bilateral agreements to overfly or use airports for technical stops

When schedules planning is satisfied that all overall planning considerations have been satisfactorily resolved, the service is passed to current planning, which is charged with the implementation of the particular service schedule. This is done in conjunction with input from the technical, cabin, engineering, and stations staffs. The final implementation of the service is carried out under operations control, which deals with daily operations and the need to provide service in terms of difficulties from crew sickness, fog, ice, delays, aircraft readiness status, and so on.

2.6 Fleet Utilization

Figures 2.13a and 2.13b present two fleet utilization diagrams for a European airline using relative new aircraft. Several points are noteworthy. First, the aircraft in service are heavily utilized between the hours of 0700 and 2200; there is little use of these aircraft outside these hours because they constitute part of the short- and medium-haul fleet of this carrier. Second, where there is a small young fleet there are no standby aircraft and maintenance is carried out in the night time hours. For older and larger fleets, such as the Viscount and Trident services offered by British Airways in the late 1970s, aircraft are withdrawn from service for maintenance and several aircraft are held back as standby units for use should an aircraft have to be withdrawn for maintenance or repair.

2.7 IATA Policy on Scheduling

The scheduled airlines industry organization, International Air Transport Association (IATA), has developed a general policy on scheduling, which is set out in its Scheduling Procedures Guide (1995). At some airports there are official limitations, and the coordination is likely to be carried out by general governmental authorities. Much more common is the situation where the airlines themselves establish an agreed schedule through the mechanism of the airport coordinator. It is recommended that the airport coordinator be the national carrier or the largest carrier. Coordination compiles a recognized set of priorities that normally produce an agreed schedule with minimum serious disagreement. These priorities include:

1. *Historical precedence.* Carriers have "grandfather rights" to slots in the next equivalent season.

2. *Effective period of movement.* Where two or more airlines are competing for the same movement slot, the airline intending to operate for the largest period of time has priority.

3. *Emergencies.* Short-term emergencies are treated as delays; only long-term emergencies involve rescheduling.

4. *Changes in equipment, routing, and so on.* Application for a schedule of new equipment at different speeds or adjustments to stage flight times to make them more realistic have priority over totally new demands for the same slot (Ashford and Wright 1992).

Aviation is a mixture of a number of segments that can be broadly classified into regular scheduled, programmed charter, irregular general aviation, and military operations. It is the function of the airport to arrange to afford appropriate access to any limited facility after consultation with the representative of the categories. IATA's policy states that the aims of coordination are:

- To resolve problems without recourse to governmental intervention

- To ensure all operators have an equitable opportunity to satisfy their scheduling requirement within existing constraints

- To seek an agreed schedule that minimizes the economic penalties to the operators concerned

- To minimize inconvenience to the traveling public and the trading community

- To arrange for regular appraisal of declared applied limits.

Schedules are set on a worldwide basis at the semiannual IATA scheduling conferences for the summer and winter seasons. More than 100 IATA and non-IATA airlines meet at these huge conferences where, by a process of reiterated presentation of proposed schedules, the airport coordinators are eventually able to set an agreed schedule for the airports that they represent.

2.8   The Airport Viewpoint on Scheduling

Most large airports with peak-capacity problems have strong and declared policies that affect the manner in which scheduling is carried out. The viewpoint of the airport operator is that which represents not only its own needs but also the interests of the air traveler, the airlines as an industry group, and, in some cases, even the nontraveling public. These interests are protected by obtaining a schedule that provides for the safe and orderly movement of traffic to meet the passengers' needs within the economic and environmental constraints of the airport. The viewpoints of the various interested parties differ substantially. The airport operator seeks an economic and efficient operation within the constraints of the facilities available. The air traveler is looking for travel in reasonably uncongested conditions with a minimum of delays and a high frequency of service at desirable times unmarred by unreliability. As an industry group, the airlines are also seeking efficiency of operation and high frequency and reliability of service. However, each airline will quite naturally desire to optimize its own position and will seek to gain its own best competitive situation. In the case of the airlines, the aims of the individual company are not necessarily the same as the interests of the industry group. At some airports, the nontraveler becomes involved where limits have been set for environmental reasons on the number of air transport services that can be scheduled, such as London Heathrow where there was a limit of 275,000 air transport movements per annum. This restraint was lifted in the mid 1980s and by 1995 the number of annual movements had increased to more than 350,000 with very little increase in nighttime movements. The increased capacity was achieved almost entirely by *peak spreading* or infilling the non-peak troughs. Similar aircraft movement constraints exist at Washington, DC, National.

It is the custom of many airports to declare their operational capacities at six monthly or annual intervals. This operational capacity is observed by the scheduling committee, which is composed of the representatives of the scheduled airlines serving the airport. Normally the airport operator is not directly represented on this committee. As already stated, the airport's interests are represented by the major carrier at the airport. At Los Angeles (LAX) this is United Airlines, at Frankfurt it is Lufthansa, and at London Heathrow, British Airways. Therefore, the airport that has capacity limitations is often in an arm's-length relationship with the individual airlines seeking additional services.

2.9 Hubs

There is some ambiguity in the term *hub* when used in the context of air transport. Prior to deregulation of the airlines the FAA used the term to designate large airports serving as the major generator of services, both international and domestic within the United States. With the advent of deregulation airlines were able to control levels of service provision in terms of routes and frequencies. This enabled the establishment of what the airlines designated as *hubs,* which provided services both to other major airports also designated as hubs and to smaller airports providing spoke services. The airline hubbing system was associated with much greater frequency of services between hubs and from the spoke airports, supposedly accompanied by higher load factors on the aircraft. Direct services between smaller non-hub airports were generally abandoned. Some airports operate as hubs for one airline only (e.g., Pittsburgh, which is the USAir hub, where the operations of this one airline account for 80 percent of all activity). Others such as Dallas-Fort Worth are hubs for two or more airlines. Hubbing presents airlines with the opportunity to better use their aircraft and passengers with many more flight combinations, although these combinations almost always require transfer at the hub. Flights from hubs are usually non-stop and those to other hubs are usually in larger more comfortable aircraft than formerly. Flights to spoke airports are often on smaller aircraft with capacities of fewer than 50 persons. The effectiveness of a hub airport depends on:

- its geographic location
- the availability of flights to multiple destinations
- the capacity of the airport system to handle aircraft movements and passenger volumes
- the ability of the terminal layout to accommodate passenger transfers.

Hub airports have very different patterns from airports supporting long-haul flights on predominant sectors. Dallas-Fort Worth with its two-airline hub operation has 12 peaks throughout the day during which aircraft are on the ground providing for transfers. Aircraft therefore arrive and depart in 12 arrival and departure waves, which in FAA terminology are described as *banks*. Hub terminals in the U.S. typically have high terminal usage with peaks occurring at roughly two-hour intervals between 0700 hours and 2200 hours.

References

Ashford, N. and P. H. Wright. 1992. *Airport Engineering, 3rd edition* New York: Wiley Interscience.
Civil Aviation Statistics of the World. 1995. Montreal: International Civil Aviation Organization (ICAO).
Doganis, R., A. Lobbenberg, I. Stockman, R. Echevarne and A. Fragoudakis-Romeos. 1994a. *A Comparative Study of User Costs at Selected European Airports, Research Report No. 1,* College of Aeronautics, Cranfield University.
Doganis, R., A. Lobbenberg, I. Stockman, R. Echevarne and A. Fragoudakis-Romeos. 1994b. *Turnaround Costs at Mediterranean and Caribbean Airports, Research Report No. 2.* College of Aeronautics, Cranfield University.
Scheduling Procedures Guide. 1995. Geneva: International Air Transport Association.
World Traffic Forecasts. 1994, Boeing: Seattle.

3

Airport Noise Control

3.1 Introduction

Airport noise is a worldwide problem. It inhibits the development of new airports and can seriously constrain the efficient and economic operation of existing facilities. In 1968, the Assembly of ICAO, recognizing the seriousness of the problem and that the introduction of new aircraft types could aggravate the situation, instructed its Council to call an international conference on the subject of aircraft noise in the vicinity of airports. This took place in 1969, providing the source document for Annex 16 to the Convention on International Civil Aviation in 1971. Since then Annex 16 has been revised through several editions. This document continues the essential international guidelines for noise control at airports in the form of standardized recommendations. Other governments, such as the United States, have their own standards. In some cases these are more stringent than ICAO recommendations. However, all such standards are designed to combat perhaps the most significant airport problem—noise. Noise, which can be defined as unwanted sound, is a necessary by-product of the operation of transportation vehicles.

Air transport is not the only transport that generates noise. Automobiles generate it from such sources at the engine, the tires, and the gear box. Railroad trains generate noise aerodynamically and from rail wheel contact, suspension, and the traction motors. Aircraft produce noise from their engines and from the aerodynamic flow of air over the fuselage and wings. It is important to bear in mind that airports generate little noise. It is the noise generated by aircraft in and around airports that causes problems. The scale of noise generation of air transport can be seen in Figure 3.1. Whereas the air mode is not

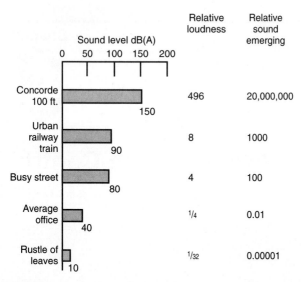

Figure 3.1 A scale of noise and sound.

the only generator of noise, it can be seen that it is the generator of the loudest and most disturbing noise. Consequently, noise at airports is a significant and troublesome problem for nearly all airport authorities, and operators find that they must have some knowledge of the technical terms used by noise experts. A few of these are introduced in this section and the one which follows.

Loudness is the subjective magnitude of sound, which is normally considered to double with an increase in sound intensity to 10 dB. The ear, however, is not equally sensitive to all ranges. The audible spectrum of sound is 20 to 20,000 Hz. Maximum sensitivity to sound is perceived around the middle of this range. It has been recognized for some time that the ear is particularly sensitive to frequencies within the A-range. Sound is therefore normally measured using A-weighted decibels (dBA), which reflect the sensitive ranges.

Since noise is unwanted sound, there is a strong element of subjectivity in its measurement. Obviously intensity of sound alone is not a suitable measure. For example, another factor that strongly affects the reactions of individuals and groups is time. Sound is only rarely constant over time, and time as a variable has been found to influence the subjective evaluation in noise in terms of:

- The length or duration of the sound
- The number of times the sound is repeated
- The time of day at which the noise occurs[1]

---

[1]For a more complete description of measures of transportation noise in general and aircraft noise in particular the reader should consult the bibliography given at the end of this chapter.

3.2 Aircraft Noise

Aircraft noise can be described by measuring the level of noise in terms of sound intensity. Where a noise level measure is required, the simple dBA method is not entirely satisfactory. Following the introduction of jet aircraft, research carried out at JFK Airport, New York, indicated that the ear summed noise in a much more complicated way than the A-scale weighting of dBA. As a result, another noise level measure was devised, the *perceived noise level* (PNL), a D-weighted summation that is sufficiently complex to warrant computer calculation.

Single event measures

Noise intensity by itself is not a complete measure of noise. Intensity requires the factor of duration, which has been found to have a strong influence on the subjective response to noise. The principal measures of single-event noise used are *Effective Perceived Level* (EPNL or L_{EPN}) and *Sound Exposure Level* (SEL, sometimes abbreviated to LSE or LAE).

Annex 16 of the ICAO recommends use of EPNL, which modifies the PNL figure by factors that account for duration and the maximum pure tune at each time increment. This measure of the single event therefore incorporates measures of sound level, frequency distribution, and duration. FAA practice, on the other hand, uses a measure based on the sound exposure levels, weighted on the A-scale, over the time during which the sound is detectable. The accumulation procedure takes note of the logarithmic nature of sound addition. Both the EPNL and the SEL are used as a basis for developing environmental measures of noise exposure.

Cumulative event measures

In the case of noise nuisance generated in the process of airport operation, it is not simply the magnitude of the worst single noise event that gives a measure of environmental impact. Over the operational day of the airport, many noise "events" occur. Therefore, single event indices are not useful methods of measuring aircraft noise disturbance, which is related to annoyance and interference with relaxation, speech, work, and sleep. Quantifying such interference requires noise measurements in terms of instantaneous levels, frequency, duration, time of day, and number of repetitions. Many surveys have been carried out to correlate community response to all these factors.

Day/night average sound levels (United States)

The form of cumulative noise event measure used in the United States is the day/night average sound level (DNL or L_{DN}), which is computed from:

$$L_{DN}\,(i,j) = SEL + 10 \log_{10} (N_D + 10N_N) - 49.4 \qquad (3.1)$$

where N_D = number of operations 0700–2200 hours
N_N = number of operations 2200–0700 hours
SEL = average sound exposure level
i = aircraft class
j = operation mode.

Partial L_{DN} values are computed for each significant type of noise intrusion using (Wright and Ashford 1989). They are then summed on an energy basis to obtain the total L_{DN} due to all aircraft operations:

$$L_{DN} = 10 \log \Sigma_i \, \Sigma_j \, (10) \, \frac{L_{DN}(i,j)}{10} \qquad (3.2)$$

Noise and number index

Another cumulative event measure that is widely quoted in airport noise literature is the noise and number index (NNI). This is a rather simple measure that was used widely in the United Kingdom and had limited use elsewhere. Equation (3.3) is the relevant formula for computation:

$$NNI = \bar{L}_{PN} + 15 \log N - 80 \qquad (3.3)$$

Surprisingly perhaps, the definition of terms within the formula is not completely standardized between users. It is common practice, however, to define N as the number of occurrences of aircraft noise exceeding 80 PNdB, the peak level caused by a Boeing 707 at full power at approximately 13,000 feet (4000 m) height. L_{PN} is the logarithmic average of peak levels. NNI has been largely replaced in the United Kingdom by L_{eq}, but in the mid 1990s it was still in use in Ireland and Switzerland.

Noise exposure forecast (United States)

Prior to the development of the L_{DN} index, the measure of cumulative noise exposure in the United States was the noise exposure forecast (NEF), which still occurs in much FAA literature. It is also used in Canada and Australia. It is computed from:

$$NEF = EPNL + 10 \log_{10} (N_D + 16.7N_N) - 88 \qquad (3.4)$$

where N_D = number of daytime operations (0700–2200)
N_N = number of night operations (2200–0700)

Equivalent continuous sound level, L_{eq}

Usually specified for a relatively long measurement period, L_{eq} is defined as the level of equivalent steady sound that, over the measurement period, contains the same weighted sound energy as the observed varying sound. It is stated in mathematical form as:

$$L_{EQ} = 10 \log_{10} \left\{ (1/T) \int_0^T 10^{L(t)/10} \, dt \right\} \tag{3.5}$$

where $L(t)$ is the instantaneous sound level at time t and T is the measurement period. In practice, this is the same as:

$$L_{EQ} = 10 \log_{10} \{ (1/T) \, \Sigma \, 10^{SEL_1/10} \} \tag{3.6}$$

which is the summation of the individual aircraft sounds over the measurement period T.

Because of shortcomings of the NNI index, which was not entirely applicable to all airports, L_{eq} has largely replaced NNI in the United Kingdom. It has been found at Heathrow that the NNI and L_{eq} equivalences were:

35 NNI   57 L_{EQ}
45 NNI   63 L_{EQ}
55 NNI   69 L_{EQ}

Noise exposure forecast (United States)

Prior to the development of the L_{DN} index, the measure of cumulative noise exposure in the United States was the noise exposure forecast (NEF), which still occurs in much FAA literature. It is computed from:

$$NEF = \bar{L}_{EPN} + 10 \log N - K \tag{3.7}$$

where $\bar{L}_{EPN}$ = average effective perceived noise level which is computed from individual L_{EPN} values. This is the EPNL defined previously.

   K = 88 for daytime periods (0700 to 2200)
   K = 76 for night-time periods (2200 to 0700)

and

$$L_{EPN} = 10 \log \frac{1}{T} \int_0^T 10^{\,0.1 \, L \, (1)} \, dt \tag{3.8}$$

where $L(t)$ = sound level in dB(A) or PNdb
   T = 20 or 30 seconds to avoid including quiet periods between aircraft

The combined 24-hour NEF is computed using (3.9):

$$NEF_{day/night} = \log_{10} \left(\text{antilog} \frac{NEF_{day}}{10} + \text{antilog} \frac{NEF_{night}}{10} \right) \tag{3.9}$$

It can be seen that (3.1), (3.3), and (3.4) for the L_{DN}, NNI, and NEF are all very similar in basic form. There is much evidence to indicate that community response to noise impact can be correlated to any of these measures. The general relationships between noise measures is shown in Figure 3.2.

3.3 Community Response to Aircraft Noise

By definition, noise being unwanted sound, there is a wide range of individual response to noise from aircraft operations in the vicinity of airports. Noise levels that are extremely annoying to some individuals cause little disturbance to others. The reasons for differences are complex and are largely socially based. Research has indicated that unlike individual reaction, community response is more predictable because of the large number of individuals involved. Figures 3.3a and 3.3b show relationships that have been found to exist between levels of noise exposure and community disturbance in terms of the percentage of persons annoyed. It can be seen that below exposure levels of 55 L_{DN}, 57 L_{EQ} and 35 NNI, the percentage of affected individuals who are highly annoyed by aircraft noise is very low. At exposure levels of 65 NNI, 69 L_{EQ} and 80 L_{DN}, more than half the community

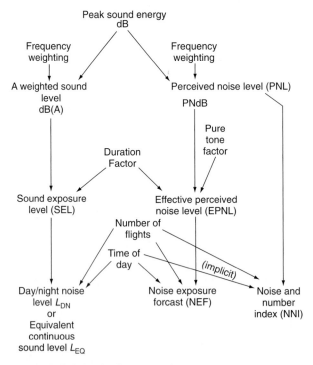

Figure 3.2 Relationship between noise measures.

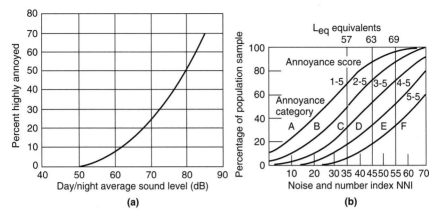

(a)   (b)

Annoyance Category	Feelings About Aircraft Noise
A	*Not annoyed.* Practically unaware of aircraft noise
B	*A liitle annoyed.* Occasionally disturbed
C	*Moderately annoyed.* Disturbed by vibration; interference with conversation and TV/radio sound, may be awakened at night
D	*Very annoyed.* Considers area poor because of aircarft noise; is sometimes startled and awakened at night
E	*Severly annoyed.* Finds rest and relaxation disturbed and is prevented from going to sleep; considers aircraft noise to be the major disadvantage to the area
F	*Finds noise difficult to tolerate.* Suffers severe disturbance; fells like moving away because of aircraft noise and is likely to complain

Figure 3.3 *(a)* Degree of annoyance from noise observed in social surveys; *(b)* distribution of degrees of annoyance due to aircraft noise exposure. *(a) Schultz 1978; (b) Ollerhead 1973.*

is highly annoyed. Figure 3.3b is interesting in that it indicates that even at near tolerable levels of noise exposure about 10 percent of the population is either unaware of the noise or only occasionally disturbed.

3.4 Noise Control Strategies

There are many ways in which operations on and in the vicinity of the airport can be modified to control noise and to decrease its impact on airport neighbors. Several of these are discussed briefly.

Quieter aircraft

Although much noise is generated by aerodynamic flow over the aircraft frame, most of the noise from modern aircraft has the engine as its source. The two principal components of engine noise are high velocity exhaust gas flows and air flows in the compressor fan system. Noise power is highly sensitive to velocity and the early turbojet engine was extremely noisy because of the high velocity of the compressor tip speeds and the jet exhaust. Second generation high bypass ratio engines included a number of noise reducing improvements including:

- Sound absorbing material in the inlet and outlet engine ducts
- A combination of increasing the spacing of turbine blade rows and eliminating inlet guide vanes to minimize noisy flow interactions between stator and rotor
- Most important, the use of low-flow speeds in the fan, compressor, and exhaust areas of the engine

Noise levels of the first generation of wide-bodied aircraft were a considerable improvement upon levels observed from the much smaller narrow-bodied turbojets. As new wide-bodied aircraft replace the older noisier aircraft, noise disturbance at constant levels of operation will decrease. Fleet replacement is being expedited by rising fuel costs, which make the replacement of the old turbojet aircraft imperative. In the 1970s, the possibility of retrofitting existing engines with hush kits was first examined with the aim of converting early turbojets to high-bypass ratio engines. This was carried out but the experience with hush kits has been very variable.

The Environmental Protection Agency in the United States required that all civil subsonic turbojet aircraft operating in the United States should comply with FAR Part 36 limits by retrofitting where necessary. All three- and four-engine jets were required to conform by 1985 but with exemptions for operations into small communities until 1988. Table 3.1 indicates just how important these modifications have been in improving the noise levels of existing aircraft. At some airports, such as Manchester, Frankfurt, Brussels and Amsterdam, the airport has actively encouraged the use of quiet aircraft by offering preferential landing charges to operators using quiet aircraft.

Noise preferential runways

Modern transport aircraft are not particularly sensitive to the crosswind component on landing and takeoff. Consequently, these operations can conveniently be carried out on a less than optimally oriented runway, if that facility will reduce the noise nuisance to the community at large. Amsterdam Schiphol is an example of an airport that might well have abandoned the use of one particular runway were it not for the fact that this runway is well suited to direct noise nuisance away from the heavily populated suburbs of Amsterdam. At Los Angeles, heavier aircraft generally use only one of the two main runways and takeoffs are mainly to the west over the sea.

TABLE 3.1 Noise Levels (in EPNdB) for "Quiet Nacelle" and "Refanned" JT8D Engines Installed in 727, 737, and DC9 Aircraft

Aircraft	Position	FAR 36[a]	Untreated	Quiet nacelle	Refanned
B727	Sideline	104	99	100	92
	Takeoff[b]	99	100	97.5	93
	Takeoff[c]	99	107	106.5	98
	Approach	104	108	100	101
B737	Sideline	103	101	101	86
	Takeoff[b]	95	92	92	82.5
	Takeoff[c]	95	100	100	—
	Approach	103	109	102	101
DC9	Sideline	103	101	101.5	92
	Takeoff	96	97	94.5	85
	Approach	103	108	99	98

SOURCE: Reference 5

[a]US certification requirement

[b]US certification requirement with power cutback

[c]US certification requirement without power cutback

Very much related to the noise preferential runway concept is that of *minimum noise routings* (MNR), or *preferred noise routings* (PNR), which are designed to direct departing aircraft to follow routes over areas of predominantly low population density. Although the size of the noise footprint is not significantly altered, the impact in terms of disturbed population is decreased. The use of MNRs and PNRs has been hotly contested by those adversely affected in terms of the social justice of a few bearing high noise exposure levels in order to protect the many. In the United Kingdom, the Noise Advisory Council has examined the practice of using MNRs and has recommended continuation of this practice as being the best course of action from the community's viewpoint.

Operational noise abatement procedures (NAP)

Several operating techniques are available that can bring about significant and worthwhile reduction in aircraft noise in the takeoff and approach phases in the vicinity of the airport as well as during operations while on the ground.

Takeoff

To reduce noise over a community under the takeoff flight path, power can be cut back once the aircraft has attained a safe operating altitude. Operation continues at reduced power until reaching a depopulated area, when the full power climb is resumed. At the point of cutback, noise levels can be reduced

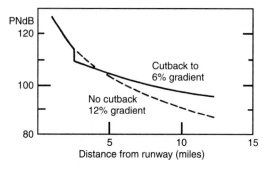

Figure 3.4 Effect of power cutback on ground noise levels.

by 5 PNdB. There will, of course, be reduced benefits further down route, where noise levels will be higher than those produced by a full power climb. However, by carefully planning the NAP on takeoff, the level of noise exposure on the total community can be reduced.

Figure 3.4 shows the easily calculated effect of a power cutback procedure on maximum flyover noise levels for points directly below the flight path. Staged climb NAPs are common at many airports around the world.

Approach

Noise on approach can be reduced by adopting operational NAPs that keep aircraft at increased heights above the ground. If a large turbojet aircraft is operated at 3000 feet (914 m) above ground level rather than 1500 feet (457 m) the noise reduction for those on the ground is in the region of 8 PNdB. Operation at 5000 feet (1524 m) reduces the noise by about 16 PNdB. Several procedures can be used to increase the height of approach operations:

1. Interception of glideslope at higher altitudes when interception is from below the slope. Manchester airport prohibits descent below 2000 feet (610 m) until the glideslope has been intercepted, and requires aircraft making visual approaches to use VASIs to avoid unnecessary low flying.

2. Performing the final descent at a steeper than normal angle. Descents at 4° have been used, but 3° is a more normal angle.

3. Two-segment approaches with the initial descent at 5° or 6° flaring to 3° for final approach and touchdown. Evaluation studies in the United States have indicated that noise can be reduced by this procedure by 10 EPNdB at 5 nm (9.27 km) from the threshold and by 6 EPNdB at 3 nm (5.56 km).

4. Low-grade approaches with reduced flap settings and lower engine power settings demonstrate some reduction of noise. Reducing flap set-

tings on the B727 and B737 from the normal 40° or 30° reduces the noise by 3 and 2 EPNdB, respectively. Slightly lower noise reductions are achieved by decreasing B707 flap settings from the normal 50° to 40°.

5. Use of continuous descent approaches, utilizing secondary surveillance radar for height information. This prevents the use of power in a stepped descent and consequently reduces noise under some parts of the descent path.

A combination of low power and low drag approach procedures has been used with considerable success at Frankfurt Airport, which has severe environmental noise problems due to its position within an urban area. Figures 3.5a and 3.5b show a noise abatement procedure that has been successfully used elsewhere for the reduction of the noise impact for the BAC-11 aircraft, which as a Stage 2 aircraft might still be used into the 21st century.

Runway operations

The most significant improvement in noise impact that can be achieved when aircraft are on runways is the control of the use of thrust reversal. Although thrust reversed is usually about 10 dB below takeoff noise, it is an abrupt noise that occurs with little warning. Aircraft operations should be restrained from the use of thrust reversal on noise nuisance grounds, except in cases where no other adequate means of necessary deceleration is available.

Insulation and land purchase

Some relief to noise nuisance can be attained by the use of sound insulation. In some countries, those adversely affected by defined levels of noise nuisance are eligible for governmental or airport authority grants, which must be used for double glazing and other sound insulation procedures. A scheme of this nature operates, for example, in the noise impacted areas around London Heathrow and Amsterdam Schiphol.

A more direct and more expensive method of reducing noise nuisance was adopted at Los Angeles International Airport, where many homes and businesses in the immediate vicinity of the airport were purchased by the airport authority through mandatory purchasing procedures (eminent domain). In some cases, this type of action is the only recourse open to the airport when continued operation means intolerable living and working conditions for the neighboring population.

3.5 Noise Certification

In the spirit of its resolution of September 1968, the ICAO established international specifications recommending the noise certification of aircraft

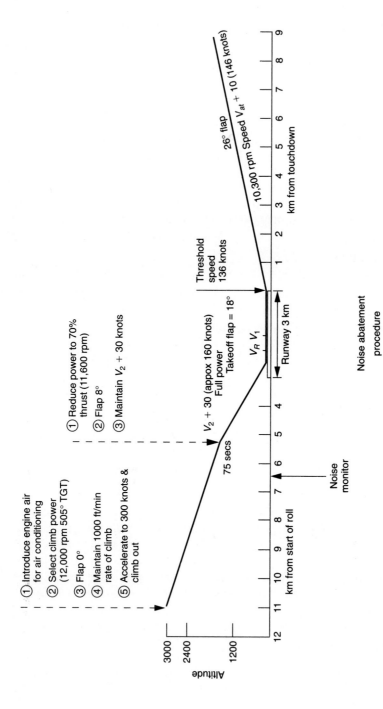

Figure 3.5a Noise abatement procedure for BAC 11.

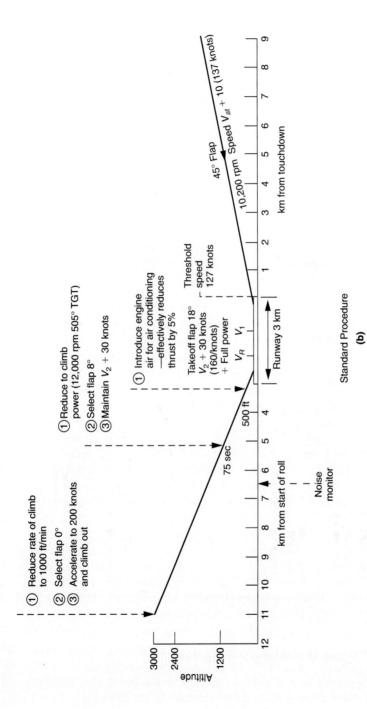

Figure 3.5b Standard procedure.

that have reached acceptable performance limits with respect to noise. Other countries developed their own parallel standards. Most notably the United States developed a set of standards through the FAA; these are published in the Federal Aviation Regulations.

ICAO certification standards principally relate to the noise generated by an aircraft on approach, while on the runway, and on flyover. The general aircraft population is divided into two main categories, Chapter 2 and Chapter 3 (called Stage 2 and 3 in the U.S.) aircraft, and four subcategories:

1. Chapter 2 aircraft: subsonic jet airplanes certificated before 6 October 1977

2. Chapter 3 aircraft:
    a. subsonic jet airplanes certificated on or after 6 October 1976
    b. propeller driven airplanes over 12,500 pounds (5700 kg) on or after 1 January 1985 and before 17 November 1988
    c. propeller driven airplanes over 2000 pounds (900 kg) certificated on or after 17 November 1988

The form of the recommended ICAO standards are shown in Figures 3.6a and 3.6b which relate to Chapter 2 and Chapter 3 aircraft respectively. Chapter 2 aircraft are being gradually phased out in favor of the quieter Chapter 3 aircraft. All Chapter 2 aircraft are scheduled for phase out by the year 2002. Three noise measurement points are defined under the approach and takeoff paths and laterally to the side of the runway. Maximum noise levels are set at these reference noise-measuring points, which are dependent on maximum certificated takeoff weights, presumably with the rationale that small very noisy aircraft are socially undesirable and should therefore not be certificated. Even with very large aircraft absolute maximum noise limits are set. In examining Figures 3.6a and 3.6b, it can be seen that an aircraft with noise characteristics plotting above or to the left of the curves is acceptable; those with plots to the right or below the curve fail to meet certification standards.

Federal Aviation Regulations (FAR) are slightly more demanding than those of ICAO. Although the noise levels are the same, there are minor differences in the location of the measuring points. Figure 3.7 shows the FAR measuring points and the limits. For general reference purposes, Table 3.2 shows the noise levels generated by a number of certificated and noncertificated aircraft in general usage in the mid 1990s.

3.6 Noise Monitoring Procedures

In the whole area of airport noise problems, the airport authority can probably make no greater contribution than by establishing and operating an effective airport noise monitoring program. The most beneficial programs have been those that have honestly and conscientiously monitored the status of noise pollution and encouraged an open exchange of information between the airport operator, the airlines, airline crews, the public, other airport authorities, and researchers in the field. The effectiveness of any such program is measured in terms of the computed reduction in noise due to implementing monitoring procedures.

Noise monitoring at Manchester Airport is a good example of a highly interactive program that has resulted in significant reduction of noise nuisance and aims to obtain continuing improvement. This airport is one of the world leaders in noise monitoring and noise control procedures. Figure 3.8 shows the location of the runway in reference to surrounding urban development,

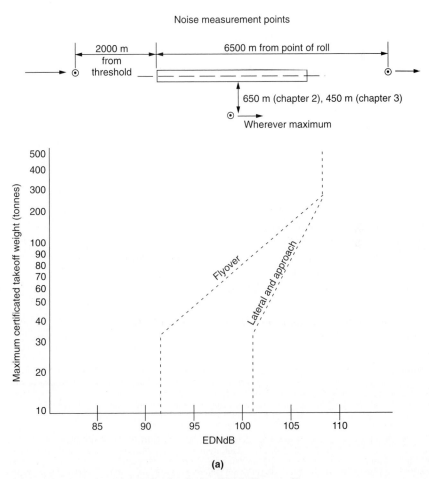

Figure 3.6a Aircraft noise certification limits. *(ICAO Chapter 2)*

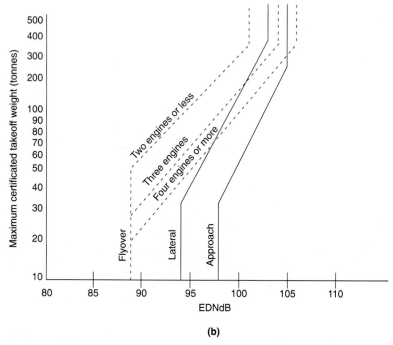

(b)

Figure 3.6b Aircraft noise certification limits. *(ICAO Chapter 3)*

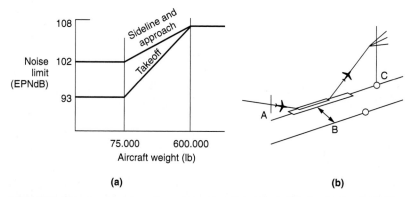

(a)   **(b)**

Figure 3.7 Aircraft noise measuring points. Aircraft noise certification limits—FAA. The noise limits set by ICAO Annex 16 and Federal Aviation Regulations Part 36 are based on fixed measuring points and noise levels calculated in EPNdB, which are allowed to vary on a logarithmic scale with aircraft weight. Although the permitted noise levels under each set of regulations are the same, there are slight differences in the locations of the measuring points—those of the ICAO are slightly less demanding. Between 75,000 and 600,000 lb (see upper diagram) the permitted sideline and approach noise levels increase by 2 EPNdB for each doubling of weight, takeoff noise increasing by 5 EPNdB. The FAR measuring points (lower illustration) are: A—1 nm from touchdown; B—0.25 nm for aircraft with three or fewer engines; 0.35 nm for aircraft with more than three engines; C—35 nm from brake release. ICAO points are: A—120 m below the flight path for a 3° glideslope (this corresponds to 1.08 nm from the threshold); B—650 m. 0.35 nm for all aircraft; C—6,500 nm, 3.5 nm. Under the ICAO regulations, tradeoffs can be made between the different measuring points, although noise levels at any point must not be more than 3 EPNdB above the limit appropriate to that point. *(FAR Part 36)*

TABLE 3.2 Measured Aircraft Noise Levels and FAA/FAR 36 Requirements

	Engine	no.	Thrust (lb)	Gross weight (lb)	Takeoff measured	Takeoff Part 36	Sideline measured	Sideline Part 36	Landing weight (lb)	Approach measured	Approach Part 36
DC-10-10	CF6-D1	3	40,300	440,000	99.0	106.0	96.0	107.0	363,500	106.0	107.0
DC-10-30	CF6-150A	3	48,400	555,000	104.0	107.5	96.0	108.0	403,000	108.0	108.0
DC-10-40	JT9D-20 Dry	3	44,500	530,000	101.0	107.0	94.0	107.5	380,000	105.0	107.0
DC-9-10	JT8D-1"	2	14,000	90,700	90.0	94.5	102.5	102.5	81,700	109.0	102.5
DC-9-30	JT8D-11"	2	15,000	114,000	95.0	96.0	103.0	103.5	102,000	109.0	103.5
DC-8-30	JT4A-9"	4	16,800	315,000	113.0	103.5	109.0	106.0	207,000	111.0	106.0
DC-8-50	JT3D-3B"	4	18,000	315,000	114.0	103.5	106.0	106.0	240,000	118.0	106.0
DC-8-61	JT3D-3B"	4	18,000	325,000	117.0	103.5	103.0	106.5	245,000	117.0	106.5
747-100	JT9D-7 Wet	4	47,000	735,000	110.0	108.0	102.0	108.0	564,000	112.0	108.0
	JT9D-7 Wet (Fixed lip)				107.0	108.0	100.0	108.0	564,000	107.0	108.0
747-200B	JT9D-7 Wet	4	47,000	735,000	107.0	108.0	98.0	108.0	564,000	106.0	108.0
707-320B/C	JT3D-3B"	4	18,000	332,000	114.0	103.5	108.0	106.5	247,000	120.0	106.5
720B	JT3D-1"	4	17,000	235,000	104.0	101.5	108.0	105.5	175,500	117.0	105.5
727-200	JT8D-9"	3	14,500	190,000	102.0	100.0	102.0	105.0	154,500	109.5	105.0
	JT8D-9" Quiet nacelle			175,500	97.0	99.0	102.0	104.5	142,500	103.0	104.5
737-100	JT8D-9"	2	14,500	111,000	96.0	96.0	100.0	103.0	101,000	111.0	103.0
737-200	JT8D-9"	2	14,500	111,000	98.0	96.0	101.0	103.0	103,000	112.0	103.0
737-200 (Adv)	JT8D-15"	2	15,500	115,500	96.0	96.0	104.0	103.5	100,800	108.0	103.5
TriStar	RB.211-22C	3	41,030	430,000	97.0	105.5	95.0	107.0	358,000	103.0	107.0
One-Eleven (500)	Spey Mk 512-14"	2	12,550	100,000	103.0	95.0	108.5	103.0	87,000	102.5	103.0
Caravelle 10B	JT8D-9"	2	14,500	123,500	99.0	96.5	102.0	103.5	109,200	107.0	103.5
F28 Mk 2000	Spey Mk 555-15	2	9850	65,000	90.0	93.0	99.5	102.0	59,000	101.8	102.0

ᵃUncertificated noise levels, a combination of data from the manufacturer, ICAO, and FAA research.

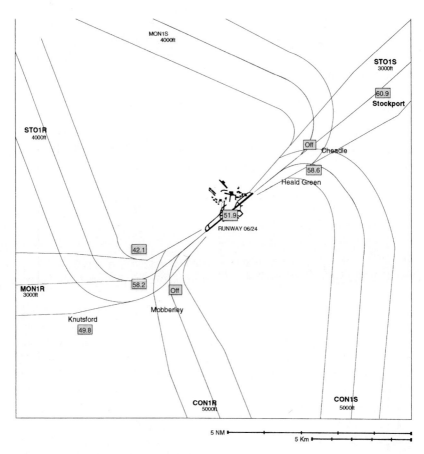

Report Generated 18/07/95 05:22 pm
MANTIS Manchester Airport Noise and Track Information System

Figure 3.8 Manchester Airport noise and location map, preferred noise routes, fixed monitoring terminals.

designated preferred noise routings (PNR) and the fixed noise monitoring terminals. For special purposes, mobile noise monitoring units are also used. These are moved to various locations according to need. Following United Kingdom practice, only departing operations are monitored. The microphones are located

- To conform as closely as possible with FAR Part 36/ICAO Annex 16 requirements of being 3.5 nautical miles (6.5 km) from start of roll and under the flight path.

- To provide measurement in the vicinity of the nearest possible built-up area where a noise abatement procedure can safely be carried out.

- To be easily accessible for maintenance.

- To provide one sideline monitor that allows correlation of air traffic control records with those of the monitoring system, thus making it possible to identify specific flights.

In common with other United Kingdom airports, departure noise levels at Manchester Airport are limited to 105 PNdB by day (0700 to 2300) and 100 PNdB by night (2300 to 0700). Monitoring is geared around detecting and assigning responsibility for violations to these limits by combining the techniques of noise monitoring with aircraft tracking that determines deviations from the acceptable swathes either side of the PNRs.

The automatic system in use at Manchester records each departure from the ATC radar track and records the noise level for each at the fixed monitoring points. Deviations from the PNRs are flagged.

Noisy operations at Manchester are penalized in a number of ways. Jet aircraft not meeting the noise certification standards set out in ICAO Annex 16 Chapter 2 are subject to a 50-percent surcharge on runway use charges. (Chapter 2 aircraft themselves are scheduled to be phased out by the year 2002.) Furthermore, when any aircraft exceeds noise limits as measured at the noise monitoring terminals there is a penalty of (US) $800 (£500) plus $230 (£150) for each full PNdB above the limit. Manchester has also indicated its intention to move to introducing penalties for aircraft that deviate unacceptably from the published preferred noise routes.

Each month the airport publishes a report on the monitored noise levels, the number of daytime and nighttime infringements and the deviations from the PNRs. Figure 3.9 is an example of the reported average monthly daytime and nighttime peak noise levels over a year. Table 3.3 shows the published deviations from PNRs and a breakdown of noise infringements by airline for a particular month. Although associated with penalties to airlines for noisy operations, the chief purpose of the Manchester noise monitoring control is positive rather than negative. The system is used as a continuous

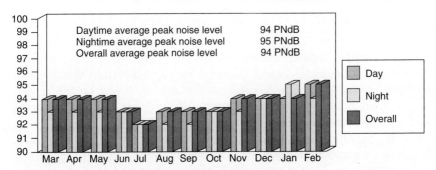

Figure 3.9 Average monthly noise levels, Manchester Airport.

TABLE 3.3 Example of Monthly Infringement Table,
Manchester Airport Environmental Monitoring System

Tracking Infringements

Airline	Departures	Extreme deviations from PNR	%
1. Egyptair	1	1	100
2. Sabre	2	1	50
3. Gulf Air	8	3	37.5
4. Caledonian Airways	30	5	16.6
5. Air Ops of Europe	9	1	11.1
6. Sabena	94	7	7.4
7. Air France	81	4	5.0
8. Airtours	175	8	4.5
9. Swissair	25	1	4.0
10. Delta Air Lines	26	1	3.8
11. Iberia Airlines	28	1	3.5
12. British Airways	1540	39	2.5
13. Brittania Airways	223	5	2.2
14. Air UK	109	2	1.8
15. UK Leisure	57	1	1.7
16. Lufthansa	184	3	1.6
17. Monarch	93	1	1.0
18. Air 2000	143	1	0.7
19. Aer Lingus	207	1	0.4
Total extreme deviations from PNR tracks		86	

Noise Infringements

Airline	No	Type
Pakistan Int	2	747-200
British Airways	2	747-200
Uzbeckistan	1	IL62
Total infringements	5	

Total no of movements = 11,007

SOURCE: Manchester Airport Plc Summary Report

feedback to the airlines, to airport management and to the airport's Consultative Committee, which includes membership from the environmentally affected communities around the airport. Based on the results documented in *MANTIS*, the publication of the environment monitoring system, Manchester Airport maintains a close working relationship with its surrounding communities to achieve an operation that minimizes avoidable negative environmental impacts.

Similar systems of noise monitoring exist widely in countries where noise impact creates severe problems with communities that surround busy urban airports. Not surprisingly Chicago O'Hare has a noise compatibility program that closely monitors noise impact according to FAR Part 150 requirements. Figure 3.10 shows the flight tracks for the six-runway airport and the associated noise contours that are generated using the FAA Integrated Noise Model. The validity of the model contours is verified using mobile noise recording units, unlike the static recorders used at Manchester. Between 1979 and 1993, the number of homes adversely affected by noise in excess of 65 L_{DN} had decreased from 87,000 to 45,000, largely due to the introduction of less noisy Chapter 3 (Stage 3) aircraft. This is a reduction of 48 percent. By the turn of the century that reduction from 1979 levels is expected to be well over 80 percent. The shrinkage of the noise impact contours between 1979 and 1993 is shown in Figures 3.11a and 3.11b.

3.7 Night Curfews

The operation of any aircraft at night, especially large air transport aircraft, can be the cause of considerable noise annoyance. The factor of sleep disturbance is so troublesome that the standard measures of cumulative noise events such as L_{DN}, L_{eq}, CNR, and NNI weight each night operation by a factor between 10 and 15 in an attempt to describe its real importance in nuisance value. Many governments and authorities claim that even such a heavy weighting will not lead to adequate safeguards with respect to night operations. Even very low levels of operation of noisy air transport aircraft (e.g., one operation each half hour) could cause intolerable night conditions for many neighboring residents.

Consequently, night curfews on aircraft operations exist at many airports throughout the world (e.g., Zurich and Sydney). The nature of the curfews, however, varies substantially between airports. At some facilities, there is a complete ban on all operations, and the runways are effectively closed. Other airports permit the operation of propeller aircraft that have low noise characteristics. Frequently night freight movements are made by such "quiet" aircraft. Some airports, such as London Heathrow, allow for a quota of night movements that permits a heavily reduced level of operation. In Amsterdam, London, Frankfurt, and Hong Kong, some curfew exemptions are granted under certain operational and scheduling circumstances to permit operations by noise-certificated aircraft. These normally include all wide-bodied aircraft. Hong Kong, London, Tokyo, and Paris bend the curfew to allow delayed flights to come in, whereas Sydney operates a very restrictive seven-hour curfew allowing none of these exemptions.

The nature of the curfew depends greatly on the local political atmosphere, the location, and physical climate of the city involved and the nature and volume of air transport through the airport. Curfews can be very effective in limiting night disturbance. However, before activating a curfew, the airport must examine very carefully the effect that this constraint will have

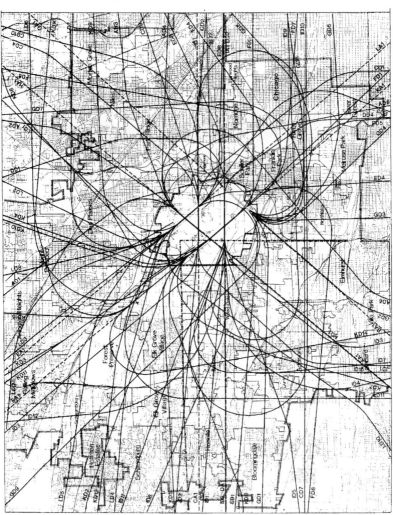

Legend:
— Study area
--- County line
---- Community
···· Airport
—··— Arrival track
—— Departure track

Figure 3.10 Flight tracks for Chicago O'Hare International Airport.

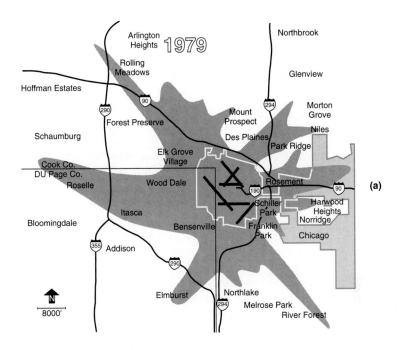

(a)

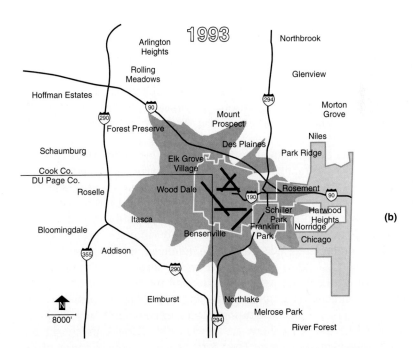

(b)

Figures 3.11a and 3.11b 65 L_{DN} contours due to operations at Chicago O'Hare International Airport for the years 1979 and 1993.

on airlines. Curfews increase the problem of peaking, and stringent curfews that accept no delayed aircraft, such as that in operation at Sydney, can, when located at the end of long flight sectors, produce alarming "scheduling window" constraints (see Chapter 2).

3.8 Noise Compatibility and Land Use

Set in an empty environment, such as farming land or forest, airports present no real noise problems. It is the interaction of the noise from aircraft operations and land utilized for residential, commercial, industrial, and other urban uses that creates the undesirable noise impact of airports so familiar to airport operations and planners. Because airports are work places and terminal points for a mode of transport, left to themselves they will generate urban development in their vicinity. This is likely to be in the form of residential areas for those working on an airport and commercial and industrial development located near the airport either because of convenience to the transport mode or because of some commercial connection with activities on the airport. Directly associated land use changes themselves generate secondary growth in the form of residences for the industrial and commercial workers, schools, shops, and a variety of other developments necessary for an expanding community. Because it is a large employer and consequently a generator of urban activity, without land use control an airport will very rapidly find that it has developments in its immediate vicinity that are incompatible with its own function.

However, not all types of land use are equally incompatible with airports. Residential areas are recognized as being highly sensitive to aircraft noise. Therefore, every effort must be made to discourage residential land use in the vicinity of airports. Some types of commercial area are less sensitive; uses such as manufacturing and resource extraction, where the internally generated noise levels might be very high are usually reasonably compatible with a large modern airport.

Recognizing the peculiar ability of airports to choke themselves environmentally, several governments around the world have developed land use planning controls that apply specifically to airports to minimize the degree of incompatibility with their surrounding land uses. In the United States, the FAA has developed standards of airport land use compatibility planning for use in the development of U.S. airports. Land surrounding airports is classified into four categories of noise exposure: minimal, moderate, significant, and severe. Their locations relative to a typical airport configuration are shown in Figure 3.12. Each category is defined by a range of one of four noise exposure indices: day/night average sound level (L_{DN}), noise exposure forecast (NEF), composite noise rating (CNR), or community noise equivalent level (CNEL). Table 3.4 shows that areas within Zone A are considered to be minimally affected by noise. Therefore, no special consideration of air-

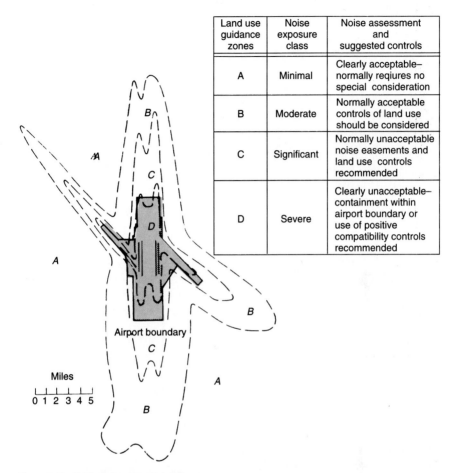

Land use guidance zones	Noise exposure class	Noise assessment and suggested controls
A	Minimal	Clearly acceptable– normally reqiures no special consideration
B	Moderate	Normally acceptable controls of land use should be considered
C	Significant	Normally unacceptable noise easements and land use controls recommended
D	Severe	Clearly unacceptable– containment within airport boundary or use of positive compatibility controls recommended

Figure 3.12 Typical airport noise patterns.

port noise need enter into the designation of land use within that zone. At the other extreme, Zone D is severely affected and land in this zone either should fall within the airport boundary or must be subject to positive compatibility controls. Table 3.5 is a land use guidance chart of 9 major land use classifications and 80 subcategories with recommendations for zone location. This chart should be used not only for initial development of land, but also for evaluating potential land use changes. In the United Kingdom, the Department of the Environment (DOE) has drawn up a similar guideline for use in conjunction with the British Noise and Number Index. This has been reinterpreted in terms of L_{eq}. This is shown in Table 3.6. Zone D in the FAA chart correlates reasonably well with the zone where the NNI exceeds 60.

Many airports that were originally put down in virtual "greenfield" sites have found themselves severely constrained within 20 to 30 years of opera-

TABLE 3.4 Major Land Use Guidance Zone Classifications

| Land use guidance zone (LUG) | Noise exposure class | Land Use Guidance Chart I: Airport Noise Interpolation | | | | | | Suggested noise controls |
| | | Inputs: aircraft noise estimating methodologies | | | | HUD noise assessment guidelines | |
		LDN	NEF	CNR	CNEL		
A	Minimal exposure	0 to 55	0 to 20	0 to 90	0 to 55	"Clearly acceptable"	Normally requires no special considerations
B	Moderate exposure	55 to 65	20 to 30	90 to 100	55 to 65	"Normally acceptable"	Land use controls should be considered
C	Significant exposure	65 to 75	30 to 40	100 to 115	65 to 75	"Normally unacceptable"	Noise easements, land use, and other compatibility controls recommended
D	Severe exposure	75 and higher	40 and higher	115 and higher	75 and higher	"Clearly unacceptable"	Containment within an airport boundary or use of positive compatibility controls recommended

SOURCE: AC 150/5050-6

TABLE 3.5 Noise Sensitivity

			Land Use Guidance Chart II: Land Use Noise Sensitivity Interpolation					
SLUCM no[b]	Land use name	LUG zone[a] suggested study	SLUCM no[b]	Land use name	LUG zone[a] suggested study	SLUCM no[b]	Land use name	LUG zone[a] suggested study
10	Residential.	A-B	30	manufacturing (continued)[c]		60	Services.	
11	Household units	A-B	31	Rubber and miscellaneous plastic products—manufacturing	C-D	61	Finance, insurance and real estate services	B
11,11	Single units—detached	A	32	Stone, clay, and glass products—manufacturing	C-D	62	Personal services	B
11,12	Single units—semiattached	A	33	Primary metal industries	D	63	Business services	B
11,13	Single units—attached row	B	34	Fabricated metal products—manufacturing	D	64	Repair services	C
11,21	Two units—side by-side	A	35	Professional, scientific and controlling instruments: photographic and optical goods; watches and clocks—manufacturing	B	65	Professional services	B-C
11,22	Two units—one above the other	A	39	Miscellaneous manufacturing	C-D	66	Contract construction	C

TABLE 3.5 Continued.

No.	Category	Class	No.	Category	Class	No.	Category	Class
11,31	Apartments—walk up	B	40	Transportation, communication, and utilities		67	Governmental services	B
11,32	Apartments—elevator	B-C	41	Railroad, rapid rail transit, and street railway transportation	D	69	Miscellaneous services	A-C
12	Group quarters	A-B	42	Motor vehicle transportation	D	70	Cultural, entertainment and recreational	
13	Residential hotels	B	43	Aircraft transportation	D	71	Cultural activities and nature exhibitions	A
14	Mobile home parks or courts	A	44	Marine craft transportation	D	72	Public assembly	A
15	Transient lodgings	C	45	Highway and street right-of-way	D	73	Amusements	C
19	Other residential	A-C	46	Automobile parking	A-D	74	Recreational activities^f	B-C
20	Manufacturing^c	C-D	47	Communication	A-D	75	Resorts and group camps	A
21	Food and kindred products—manufacturing	C-D	48	Utilities	D	76	Parks	A-C
22	Textile mill products—manufacturing	C-D	49	Other transportation, communication and utilities	A-D	79	Other cultural, entertainment and recreational^f	A-B
23	Apparel and other finished products made from fabrics,	C-D	50	Trade^e	C-D	80	Resource production and extraction	

No.	Land use	Code
24	leather, and other similar materials—manufacturing	C-D
25	Lumber and wood products (except furniture)—manufacturing	C-D
26	Paper and allied products—manufacturing	C-D
27	Printing, publishing, and allied industries	C-D
28	Chemicals and allied products—manufacturing	C-D
29	Petroleum, refining and related industries[a]	C-D
51	Wholesale trade	C-D
52	Retail trade—building materials, hardware, and farm equipment	C
53	Retail trade—general merchandise	C
54	Retail trade—food	C
55	Retail trade—automotive, marine craft, aircraft and accessories	C
56	Retail trade—apparel and accessories	C
57	Retail trade—furniture, home furnishings, and equipment	C
81	Agriculture	C-D
82	Agricultural related activities	C-D
83	Forestry activities and related services	D
84	Fishing activities and related services	D
85	Mining activities and related services	D
89	Other resource production and extraction	C-D
90	Undeveloped land and water areas	

TABLE 3.5 Continued.

59	Retail trade—eating and drinking. Other retail trade	C-D	91	Undeveloped land and unused land area (excluding non-commercial forest development)	D
			92	Non-commercial forest development	D
			93	Water areas	A-D
			94	Vacant floor area	A-D
			95	Under construction	A-D
			99	Other undeveloped land and water areas	A-D

SOURCE: AC 150/5050-6

a Refer to Table 3.5
b SLUCM: *Standard Land Use Coding Manual*, paragraph 21.
c Zone C suggested maximum except where exceeded by self-generated noise.
d Zone D for noise purposes; observe normal hazard conditions.
e If activity is not substantial, air-conditioned building, go to next higher zone.
f Requirements likely to vary—individual appraisal recommended.

TABLE 3.6 British Recommended Criteria for the Control of Development in Areas Affected by Aircraft Noise

Level of aircraft noise to which site is, or is expected to be exposed	>60NNI (>72 dBAL$_{eq}$)	50–60 NNI (66–72 dBAL$_{eq}$)	40–50 NNI (60–66 dBAL$_{eq}$)	35–40 NNI (57–60 dBAL$_{eq}$)
Dwellings	Refuse	No major developments. In-filling only permitted with sound insulation.		Permission to develop should not be refused on noise grounds alone.
Schools	Refuse	Most undesirable. Development permitted only under exceptional circumstances	Undesirable	Permission not to be refused on noise grounds alone.
		◄— Insulation required —►		
Hospitals	Refuse	Undesirable	Each case to be considered on its own merit.	Permission not to be refused on noise grounds alone.
		◄— Insulation required —►		
Offices	Undesirable	Permit	Permit but require insulation in special rooms (eg., conference rooms)	Permit
	Full ◄— insulation required —►			
Factories and warehouses	Permit with occupier taking all necessary precautions.			

SOURCE: Department of Energy Circular 10/73, as reinterpreted, and ICAO.

tion. The airport operation, therefore, has a strong and legitimate interest in ensuring that the future viable operation of the airport is not constrained by piecemeal development of incompatible land uses. Adoption of standards such as those set out by the FAA in the United States and the DOE in the United Kingdom will, if adhered to, provide a reasonable basis for the continued compatible operation of the facility with its environment.

References

Aitchison T. W. and L.I.C. Davies. 1980. *The Low Power/Low Drag Approach Procedure at Frankfurt Airport.* CAA Paper 76032. London: Civil Aviation Authority.

Ashford N. J. and P. H. Wright. 1992. *Airport Engineering,* 3rd edition. NY: John Wiley & Sons.

Federal Aviation Administration (FAA). 1977. *Airport Land Use Compatibility Planning, AC150/5050-6.* Washington DC: Department of Transportation.

Federal Aviation Administration (FAA) *Federal Air Regulation Part 36.* Washington DC: Department of Transportation (as amended).

Her Majesty's Stationery Office. 1973. *Department of the Environment Circular 10/73.* London.

International Civil Aviation Organization (ICAO). 1993. *Annex 16 Volume I Aircraft Noise,* 2nd edition. July. Montreal.

Ollerhead J.B. 1973. "Noise: How Can It Be Controlled?" *Applied Ergonomics.* September: 130–138.

Schultz T. J. 1978. "Synthesis of Social Surveys on Noise Annoyance." *Journal of the Acoustical Society of America* 64:August.

U.S. Environmental Protection Agency (EPA). 1974. *Noise Standards for Civil Subsonic Turbojet Engine Powered Airplanes (Retrofit and Fleet Noise Level).* Project report. December.

Wright P. H. and N. J. Ashford. 1989. *Transportation Engineering,* 3rd edition. NY: John Wiley & Sons.

Aircraft Operating Characteristics

ROBERT CAVES

4.1 General Considerations

The operations of airports are closely linked to the aircraft they serve. The linkage is ultimately of an economic nature based on the premise that public transport safety norms must never be degraded. Therefore, the function of the design and operation of runways and their approaches must be to allow safe transition between flight and ground maneuvering over the complete spectrum of air transport operations.

Operations of aircraft close to the ground are usually at slow speed and therefore at the extremes of the lift and control envelopes. At the same time, they are exposing the aircraft to the risk of inadvertent contact with the ground. This risk is inescapable twice on every flight, so special emphasis must be placed on the safety of these operations. In particular, the aircraft performance must be matched carefully to the infrastructure in terms of ground maneuvering of the runway length and of the aircraft's ability to climb and descend over obstacles.

The relationship between aircraft performance and economics is indicated diagrammatically in Figure 4.1, where the average load is taken as a measure of yield per kilometer. It can be seen that a given reduction of load-carrying capability causes a disproportionately large reduction in profit potential. Aircraft performance is primarily a function of the excess of thrust

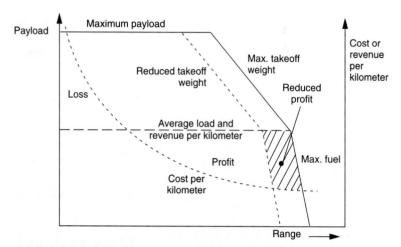

Figure 4.1   Effect of field performance on aircraft economics.

to drag and the excess of lift to weight, as shown in Figure 4.2. At a given speed, drag is primarily a function of lift and therefore weight, so it is clear that aircraft weight is the variable most able to influence obstacle clearance. A similar, but more complex, analysis would show that weight is also the principal variable in determining field length. Thus the most critical variable from the performance point of view is also the most critical economically.

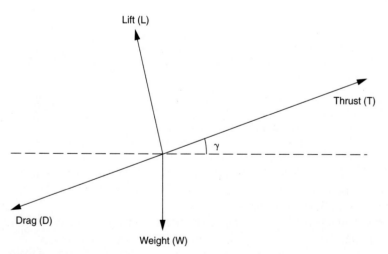

Figure 4.2   Forces in the climb. Resultant climb force $\simeq (L - W) + (T - D)\ \alpha$; resultant accelerating force $\simeq (T - D) - L\alpha$, where α is a small angle of climb and the angle between the thrust vector and the flight path is negligible.

Because of the economic impact of load reduction, operators will not use airfields where a significant reduction in payload is required. Equally, there is an economic limit to the airfield operator's ability to extend available field lengths or protect obstacle clearance surfaces. The importance of matching the infrastructure to the aircraft is clear, but it is not always possible to achieve this economically under all conditions of operation. The protection of safety standards then calls for an additional balance, this time between the regulation of performance and the regulation of operations.

There are, in fact, several principles involved in matching aircraft to infrastructure:

- The demonstrated performance of the aircraft
- The use of net performance
- The assessment of the probability of failure
- The regulation of operations as well as performance

Table 4.1 shows a sample of aircraft data indicating the airfield design requirements for each category of operation and gives some pertinent data for a representative set of aircraft that might provide a focus for the discussions which follow. The weights and dimensions are firm data, but care must be taken with the use of the other values. The limiting field lengths given (which are always takeoff values for these aircraft) are related to maximum available weight and might therefore be longer than necessary. The cruise fuel consumption is an average figure given to allow some feel for the trade-off between weight and range. The values given are only indicative. Reference should always be made to the relevant flight manuals.

Regulations[1] require that each new aircraft type demonstrate the distance required to land and take off under closely controlled conditions, with defined limitations on the pilot's reactions and carefully constructed safeguards to obviate any actions that might be inherently unsafe. Thus, for example, in the United States, in demonstrating takeoff distance, the pilot may not rotate the aircraft nose up before the speed has reached the highest of:

- Decision speed (V_1)[2]
- 1.05 times the minimum control speed (V_{mc}) in the air

---

[1]The regulations govern the aircraft in the nation where they are registered: in the United States, they are embodied in the Code of Federal Regulations. Title 14, Chapter 1, Part 25 (14CFR 1.25), and in the United Kingdom formerly by British Civil Airworthiness Requirements (BCAR) and now within Europe in the Joint Airworthiness Requirements (JAR).

[2] V_1 is the decision speed, such that the takeoff should normally be continued if an engine fails at or above V_1.

TABLE 4-1. Aircraft Data

Aircraft	Weight (lbx 1000)		Max Seats	Span (m)	Length (m)	Limiting FAR Field Length (m)		Cruise Fuel Consumption (lb/hr)	Track (m)
	Max TO	Empty				ISA.SL	ISA+20°		
Long-range									
747-400	869	389	660	64.3	70.7	3340	5000	21,912	11.0
MD-11	625	295	405	51.7	61.2	2926	4031	15,532	10.6
DC10-30	572	261	380	50.4	55.2	2996	4225[a]	16,440	10.7
A340	558	270	375	60.3	59.4	2765	—	11,594	10.7
L-1011-500	496	246	330	50.1	50.0	2636	4038[a]	15,300	11.0
DC-8-72	350	148	189	45.3	48.0	2900	—	9870	6.3
Medium-range									
DC-10-10	440	237	380	47.3	55.3	2575	4118	15,260	10.6
A300B4-200	364	195	336	44.8	53.8	2850	4040[a]	13,800	9.6
A310-300	328	178	265	43.9	46.7	2515	3750	8500	9.6
767-300	350	192	290	47.6	59.4	2545	—	8382	9.3
757-200	240	128	239	38.0	47.3	1880	3410	6570	7.3
727-200	210	102	189	32.9	46.7	3033	4176[a]	10,000	5.7
Tu-154B	216	112	164	37.5	48.0	2100	3240	10,100	11.5
Short-range									
A300B2-200	313	192	336	44.8	53.8	1676	2923[a]	12,600	9.6
A320-200	162	90	179	34.1	37.6	2190	3980	4510	7.5
DC-9 Super 82	147	80	172	32.9	45.0	2170	4365	6240	5.1
737-400	138	73	168	28.9	36.4	2222	3709	5229	5.25
146-200	89.5	48.5	106	26.3	28.6	1612	3429[a]	4270	4.7
F28 Mk 4000	73.0	37.1	85	25.1	24.4	1676	2930[a]	4780	5.0

a Weight limited by climb performance, tire speed, or brake energy.

- A speed that will not allow lift-off at less than 1.10 times the minimum un-stick (V_{mu}) or lift-off speed[3]

A speed that allows the greater of 1.2 times stall speed V_S, or the speed necessary to meet second segment climb requirements to be achieved by the screen or obstacle clearance height of 35 feet with one engine inoperative. The screen may be considered to be a virtual obstacle at the end of the available takeoff distance.

Similarly, all other certificable performance measures must be demonstrated for all applicable configurations of power and geometry, with all engines operating and with the critical engine inoperative.

The performance as demonstrated is referred to as the *gross performance*. For the purpose of dimensioning the geometry of the environment within which it is considered safe to operate, the gross performance is factored down to take account of in-service variables. The variation might be caused by pilot skill, instrument inaccuracies, weight growth, or thrust reduction between overhauls. Thus, for example, the demonstrated landing distance is factored by 1.67 under some regulations, including those of the United States, to derive the schedule landing field length[4], and the gross climb performance is reduced by 0.9 percent in order to derive the *net* performance that can be guaranteed. This information is published in the aircraft flight manual[5].

For the matching of field length requirements to the distances available, the airfield is required to publish for each runway the following distances:

- Takeoff run (TOR) available

- Takeoff distance (TOD) available

- Emergency distance (ED) available

- Landing distance (LD) available

in the nation's *Aeronautical Information Publication (AIP)*, together with the airfield reference temperature, the runway elevation, and the runway slope. The distances declared should take into account displaced thresholds, stopways, clearways, and *starter strips,* as shown in Figure 4.3.

It is the responsibility of the airport to notify, by means of Notices to Airmen (NOTAM), any changes in these distances caused by, for example,

---

[3]Unless the aircraft is limited in its ability to rotate by lack of elevator power or by tail dragging, in which cases the factor may be 1.05 $V_{mu.}$

[4]See Part 121.195 of Title 14 of the Code of Federal Regulations.

[5]The flight manual is part of the aircraft certification purposes, and is specific to the aircraft model and operator. The more limited information published in airport manuals is not adequate for day-to-day operational decisions on allowable takeoff weight and/or thrust settings.

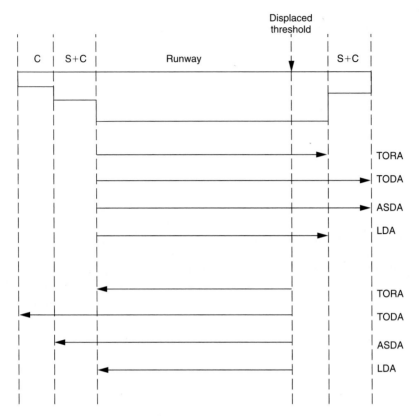

Figure 4.3 Declared distances. C = clearway; S = stopway.

work in progress or accidents. When the aircraft scheduled performance has been corrected to the appropriate altitude, temperature, slope, wind and runway surface condition at the required takeoff weight, the resultant required distances for takeoff, aborted takeoff, landing and landing at an alternative airfield may be compared with the declared distances available. An excess of Takeoff Run Available (TORA) and Takeoff Distance Available (TODA) may allow a reduced thrust takeoff to preserve engine life, or a lower V_1 to reduce the number of rejected takeoffs and hence off-base engine changes or an increased speed over V_2[6] at the screen in order to improve climb performance. An excess of EDA may allow the choice of a high V_1 to compensate for obstacles in the flight path; an excess of LDA may allow flexibility in planning for bad weather at the destination or may allow tanking of fuel. A more detailed discussion of field length requirements is given elsewhere (Ashford and Wright 1992).

---

[6]V_2 is the minimum allowable speed at the screen height, usually 1.2 V_2.

The factors placed on demonstrated field length and gross performance are part of a general treatment of accident probability that has developed historically on the basis of the empirical evidence of risk. This approach is being increasingly supplemented by the use of statistically based modeling. The underlying philosophy is based on the premise that the present risk of an accident per flight is acceptable (see Chapter 12) and that risk can be categorized and quantified, as depicted in Table 4.2.

A pertinent example of empirical risk determination and consequent policy is given by the CAA's analysis of aircraft undershoot and overshoot accidents (Monk 1981). The results of the analysis are given in Table 4.3 for all public carrier accidents involving undershoots or overshoots. The original purpose of the work was to assess the need to protect the public by means of public safety zones (PSZ), but it is also a valuable empirical vindication of the existing relationship between the aircraft performance and the field length and protected surface requirements. The analysis shows a risk of 1 in 10^7 of a ground contact between 7000 feet (2100 m) and 2 miles (3.2 km) of a runway and a 1 in 10^6 chance of contact anywhere between 200 feet (61 m) and 2 miles. The results are relevant to the current public safety zone (PSZ) criteria which are set at 4500 feet (1372 m) to give protection to a level of 1.7×10^7 for jet operations, and also to the determination of reasonable distances for runway end safety areas (RESA), about which there is no clear agreement within ICAO. The indications are that a RESA of 1000 feet (300 m) beyond the runway strip would contain half of all the under- and overshoot accidents; however, there seem to be few near undershoots by jets. There is a fairly even distribution between undershoots and overshoots and a significant number of takeoff accidents, by no means all of which resulted from rejected takeoffs. This compares with data collected in the United States between 1965 and 1969, showing 186 undershoots and 315 overshoots by general aviation, compared with only 4 undershoots and 2 overshoots by air carriers. This indicates better missed approach decision making but also probably indicates the importance of vertical guidance.

Observation of the statistical distribution of deviations from a prescribed flight path allow models to be built from which predictions can be made of

TABLE 4.2   Categorization of Accident Probability

Category	Probability (per flight)	Example
Frequent	10^{-3}	
Reasonable probability	10^{-5}	Engine failure
		Stall on approach
Remote	10^{-7}	Ground contact off runway
		Engine failure in critical second
		Ground contact more than 2 km from runway
Extremely remote	10^{-9}	Double engine failure on turn

TABLE 4.3 Public Service Under or Overshoot Accidents (1971–1977)

Aircraft Type	Jet Aircraft			Nonjet Aircraft		
Flight Sector	Landing		Takeoff	Landing		Takeoff
Relation to runway	Before	Beyond	Beyond	Before	Beyond	Before
Distance from threshold						
200–1000 ft	4	11	9	6	6	4
All known	20	12	13	18	7	9
All accidents	29	31	21	31	21	13
Third party fatalities	18	0	120	0	12	6

the likelihood that deviations will exceed a given distance depending on the extent of the guidance received. In this way it has been possible to construct improved funnels of protected surfaces to limit the probability of aircraft deviating outside them to the *remote risk* category. The funnels, as illustrated in Figure 4.4, have to protect the aircraft during approach, missed approach, and takeoff following engine failure.

There are occasions when it is not possible to meet full safety requirements and permit economic operation. Examples of this are the requirement to climb after engine failure at the critical speed in light twin-engined aircraft and an airport with a mountain off the end of the runway. In such situations, the required category risk is achieved by invoking progressively severe operational limitations as the intrinsic risk increases. This concept of protection is illustrated in Figure 4.5. In the examples quoted, the light-twin regulations accept zero climb gradient initially; the airport use would be restricted to conditions where landing and takeoff operations could be conducted safely from the one runway direction.

The general philosophy of matching the safe operation of aircraft and airports has been reviewed in this section. The details of the requirements for takeoff and landing are considered, after which some attention is given to operations in inclement weather and other special operations.

4.2 Departure Performance

The takeoff portion of the departure procedure until the screen height is achieved is dealt with in detail elsewhere (Ashford and Wright 1992). The essentials are restated here in order to introduce the operational choices open to the pilot.

The aircraft flight manual contains the following required distances:

- To 35 feet (10 m) with all engines
- To 35 feet with the loss of one engine at the critical speed (V_1)[7]
- To stop after loss of one engine at the critical speed

---

[7]JAR allowed a 15-feet (4.6m) screen for this continued takeoff case because the risk of engine failure in jets at the critical speed is remote (Table 4.2).

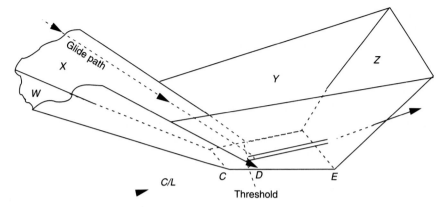

Figure 4.4 Obstacle assessment surfaces. *W* and *N* are approach surfaces; *CDE* is the footprint. *Y* is a transitional surface; *Z* is the missed approach surface.

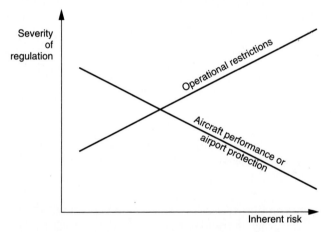

Figure 4.5 Operational and performance regulation trade-off.

These distances will have been demonstrated in conformance with constraints on minimum speeds for rotation and for crossing the screen (the takeoff safety speed) and within the criteria for reacting to the engine failure. The FAA has recently required an extra 2 seconds to be added to this reaction time. The all-engine case is then factored by 1.5 in order to bring the probability of exceeding the resultant distance into the *remote risk* category.

Over the course of time, the regulations have evolved from having no specific knowledge of engine failure (early British Civil Airworthiness Requirements, BCAR) through the use of a balanced field length[8] (Civil

---

[8]Balanced field is derived by choosing V_1 so the distance to screen equals distance to stop.

Airworthiness Requirements, CAR) to the present situation where ICAO allows the use of unbalanced field lengths and BCAR and the European Joint Airworthiness Requirements (JAR) specify demonstration of wet-runway performance but allows a 15-foot (4.7 m) screen and a V_1 up to 4 seconds earlier than the dry runway case.

The TOD required is then given directly by the greater of the factored distance to the screen with all engines and the unfactored distance with the engine out. The TOR required is the distance to a point equidistant between lift-off and the screen, factored for all engines and unfactored with engine out. The ED required is the unfactored rejected takeoff distance.[9] These distances may then be corrected to specific conditions and compared with the TORA, TODA, and EDA.

There is still a certain amount of disagreement about the adequacy of the regulations for the rejected takeoff case, stemming from the number of accidents that have either overrun the runway after aborting or where the continued takeoff has been unsuccessful. Frequently these have been caused by substandard acceleration rather than a hard and noticeable engine failure. On a DC 8-63 at Anchorage (November 27, 1970), the brakes locked after taxi, resulting in blowouts and a distance to reach V_1, which was 171 percent of the normal 4500 feet (1372 m). At Washington National (January 13, 1982), a Boeing 737 elected to continue takeoff in blowing snow and with wing ice despite the loss of acceleration resulting from erroneous engine pressure ratio settings due to iced up sensors. In this case lift-off was extended to 150 percent of the normal 30 seconds and 159 percent of the normal 3400 feet (1000 m). Both accidents might have been prevented by distance markers against which to check acceleration, but the DC 8 accident was at night, and the snow in the second case would have lessened their effectiveness.

The most appropriate solution appears to be a ground speed indication inside the cockpit, but the growing trend to the certification of rolling takeoffs will make it difficult to use this accurately, as will the trend to using reduced thrust takeoffs to conserve engine life. The onus is moving to the airport operator for the provision of RESA in addition to the trip and prepared stopway.

The situation is potentially much worse for aircraft of less than 12,500-pound (6700 kg) all-up-weight, because there is no engine failure accountability for these aircraft below 200 feet (61 m). This is compensated to some extent by the all-engine requirement to reach a 50-foot (15 m), rather than a 35-foot (10 m), screen, by a 1.25 factor on the demonstrated distance and by the TODR being not greater than the EDR. However, it should

---

[9]JAR allows reverse thrust in the rejected takeoff demonstration, but then requires a factor of 1.1.

be noted that, while the official figures for a popular light twin are 2500 feet (762 m) with all engines and 5000 feet (1500 m) with one engine out, the more likely engine-out distance would be 6000 feet (1829 m) with average piloting and atmospheric conditions. This would extend the distance to reach a 500-foot (150 m) altitude to nearly 4 miles (6.4 km), while until that height it would be very unwise to perform any sort of turn. The implications for obstacle clearance are severe.

In fact, the only engine-out climb gradient that must be demonstrated by light aircraft used for air transport purposes is 0.8 percent climb in the clean configuration, at the best angle-of-climb speed. In contrast, the engine-out requirements for large aircraft are shown for each segment of the takeoff climb in Table 4.4, where the segment configurations are defined in Figure 4.6. It should be noted that twin-engined aircraft are not required to demonstrate a positive rate of climb with engine out until the gear is up. They will, however, still tend to have a positive climb during the first segment because of residual ground effect even above 35 feet (11 m), but this does mean that where clearway is declared by virtue of a fall away of ground beyond the runway, the climb away will tend to be more sluggish.

The most severe gradient requirement occurs in the second segment, that is, with takeoff power, gear up, takeoff safety speed, and the flaps still at the takeoff setting. The maximum weights at which the gradients in Table 4.4 can be met reduced the increased altitude and temperature. It is unusual for aircraft to be able to meet the gradient at maximum takeoff weight much above sea level and the temperature of the international standard atmosphere (ISA). The maximum weights at which the requirements can be met are given in the flight manual by WAT (weight, altitude, temperature) charts for each takeoff flap setting. It is possible to optimize the climb performance by choice of flap setting, but this is usually at the expense of field performance. In hot and high conditions, the second segment normally limits the maximum takeoff weight, while at sea level it is usually field length that is limiting.

It should be noted that aircraft must also demonstrate gross climb gradients in the second and fourth segments with all engines at maximum continuous rating. In JAR, these are 5.2 percent and 4.1 percent, respectively, regardless of the number of engines.

In order to assess the ability to overfly obstacles, the flight paths implied by these gross gradients are reduced by 3.2 percent in the all-engine cases

TABLE 4.4 Gross Climb Gradient Requirements (%)

Climb segment	1	2	4
Twin engines	0	2.4	1.2
Three engines	0.3	2.7	1.5
Four engines	0.5	3.0	1.7

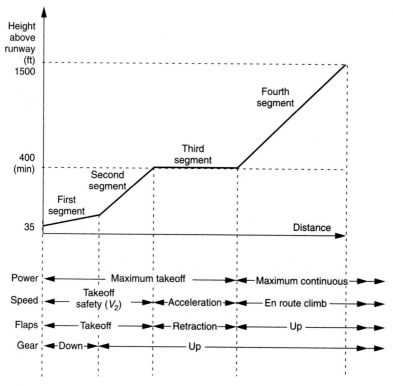

Figure 4.6 Climb path segments.

and as shown in Table 4.5 with one engine out, to give net flight paths that provide *remote risk* of being underflown. All obstacles within a defined fan of the end of the TODA must be cleared by 35 feet (11 m) by the net flight path or by 50 feet (15 m) if any turn included in the flight path is greater than 15°. The turn must not, in any case, be greater than defined in the flight manual and the start of the turn must be identifiable to the pilot while flying on instruments. The fan is defined symmetrically about the planned flight path, the initial width of 390 feet (120 m) plus the span of the relevant aircraft) increasing at a rate of 12.5 percent until it reaches a maximum of 9843 feet (3000 m). The regulations for light aircraft assume operations on a see-and-avoid basis, but with a similar expanding funnel under their own defined net engine-out flight path from the height of any cloud base that calls for IFR flight. Thus, for light aircraft, the maximum allowable takeoff weight may vary with the cloud base.

The preceding review of performance requirements shows how the maximum allowable takeoff weight may be limited by field length or by climb performance to meet either WAT limits or dominant obstacle clearances. It is clear that frequently there is a margin available even at maximum structural weight over all these requirements. There are in fact other requirements that a dispatcher must check associated with en route climb performance, landing performance, and limits on tires and brakes, but these are seldom critical. The pilot therefore has the discretion to perform the takeoff with less than takeoff thrust, provided that not less than 90 percent of the available power is selected. The amount of thrust required to stay within field length and climb requirements can normally be selected accurately for specific airfield conditions by using the thrust computer, with consequent advantages to engine life and fuel consumption.[10] On the other hand, there is not only increased tire wear to be considered, but also the possible overall increase if risk compared with the historical statistical base for accidents per takeoff, which will have had a predominant percentage of safe takeoffs (i.e., those with a significant margin between actual and allowable takeoff weight).

Although the climb-out technique shown in Figure 4.6 is the one that must be used for performance certification, it is frequently modified in practice by considerations of noise abatement or fuel economy. Efforts to reduce noise range from the early U.S. Air Transport Association (ATA) *get them high early* technique,[11] which is recommended by the International Air Transport Association (IATA), to specific combinations of thrust and heading to avoid noise-sensitive areas. The successful completion of these techniques depends on maintenance of dominant obstacle clearances and very careful handling of the aircraft. The accuracy of the flying depends on the monitoring of pitch angle, as it does during the initial period from lift-off to screen. In each case, small errors in pitch can lead to large changes in flight path or to increased risk of stalling.

Now that fuel economy has become important, other procedures have been developed by operators to minimize fuel consumption in the climb, as instanced by the ATA *full clean-up* technique. By retaining full takeoff thrust until the flaps are fully retracted, up to 80 pounds (36 kg) fuel might be saved on a 737 and up to 350 pounds (159 kg) on a 747 at the cost of

TABLE 4.5 Gradient Reductions to Give "Net" Climb Performance

Number of engines	Gradient reduction (%)
2	0.8
3	0.9
4	1.0

---

[10]Up to 300 pounds (136 kg) of fuel can be saved per takeoff on a BAC 1-11 (i.e., 10 percent of the fuel consumed on a 200-mile [321. km] sector).

[11]The technique was essentially to continue to climb at V_2 + 10 knots with takeoff flaps to between 1500 and 3000 feet (457 and 914 m) at maximum continuous power.

rather higher noise levels. However, the technique is expensive in terms of engine life, and, when all the cost and environmental variables are considered, particular compromise solutions have to be chosen depending, among other things, on the field length and obstacle clearance margins discussed earlier.

The required takeoff weight is a function of the payload and fuel requirements. Ultimately, the payload will, if necessary, be tailored to the takeoff weight available by off-loading that part of the payload that will produce least revenue. The primary fuel requirement may be reduced by fuel management techniques such as those described earlier, but a considerable proportion of the fuel uplift is to allow reserves for en route winds, holding, and diversion to alternative landing fields. Thus on short-haul flights, the reserves can exceed the primary fuel requirements, and this leads to an increase in the takeoff-weight requirement, which is particularly significant when it is realized that a long-haul jet might burn a quantity of fuel equal to a quarter of the reserve fuel simply to carry the reserves. It is open to question whether aircraft really need to carry the reserves traditionally required, with modern improvements in fuel flow management, navigational accuracy, and weather forecasting. On short-haul flights, with these improvements and with excellent destination weather at departure time, the need to carry reserves for an alternative destination is particularly questionable if the destination has two independent runways. Until there is a change in the regulations, operators will continue to minimize fuel use and/or maximize payload by filing for closer destinations and then, if possible, refiling in flight for the original destination.

In summary, departure performance is dominated by the allowable takeoff weight, which is determined as the lowest of:

- Maximum structural takeoff weight
- Climb performance limited by the WAT curve
- Takeoff field lengths: TOR, TOD, and ED available
- Obstacle clearance
- En route climb requirements
- Maximum structural landing weight
- Landing field length, WAT curve and diversion requirements
- Tire and brake limits

Any resulting margins between these limits and the required takeoff weight may then be used for tankering, to ease other limits, or to alleviate economic or environmental considerations.

4.3 Approach Performance

Landing performance is seldom the limiting factor in fight planning, with the exception of short takeoff and landing (STOL) aircraft, yet this phase of flight still attracts more than its share of accidents both near the airfield (Table 4.3) and in the descent and missed-approach operations.

Most regulations allow the manufacturer a choice of demonstrating the landing performance. One option is the earlier method, landing on a dry hard runway with conservative assumptions as to height at the threshold, a large factor of safety and no credit for reverse thrust. The second option is to demonstrate landing on a wet hard surface from a lower height and higher speed at threshold and using all forms of retardation for which a practical procedure has been evolved. The latter case is referred to as the *Reference Landing Distance* in JARs and uses a much smaller factor of safety. A demonstration with a critical engine out is also usually required.

Specific flight plan calculations must take account of forecast runway conditions and wind, with the limiting field length being the lower of the no wind and forecast wind cases. The diversion airport must also be taken into account, but in this case the wet runway factor is allowed to be 0.95.

Although the regulations are easy to state, it should not be inferred that the operation is similarly easy to carry out. There are all the problems of accurate alignment and speed control on the approach; adjustment of speed and heading for crosswinds, gusts and windshear, and maintenance of direction on the runway, as well as the primary problem of arresting the descent rate without inducing an extended float. The aircraft design is severely tested in this phase of the flight, yet the pilot can seldom compromise in favor of sparing the structure. Indeed, the latest short-haul aircraft specifications are very concerned that brake cooling should not affect turnaround times, that crosswinds should not affect regularity, and that the autopilot should be able to cope with nonlinear and decelerating approaches.

Performance and handling on the approach are just as important as the ground phase of landing in producing a safe completion of a flight. The most vital consideration is the accurate achievement of the correct conditions at the threshold (i.e., height, speed, descent rate, track, and power). In order to attain these conditions consistently, the ground aids must be satisfactory and the aircraft must have adequate performance on the approach to correct discrepancies in the flight path and to respond to emergencies.[12]

---

[12]For example, an extra knot of airspeed at the threshold converts to an extra 100 feet (30.5 m) of runway. Speed control is harder on modern jet aircraft because the minimum drag speed is higher relative to the stall speed and so the aircraft can arrive at a condition of zero speed stability.

The high vertical momentum of modern jets combined with wind gradients, gusts, and windshear, make it essential to provide slope guidance in the form of *Visual Approach Slope Indications* or *Precision Approach Path Indicators*. The use of radio markers and arithmetic is better than nothing, but is of little benefit in the crucial final phase of the approach. Full precision approach is, of course, the best aid to accurate flying. NASA, with their Boeing 737 terminal-configured vehicle program, has shown that the tracking capability to two standard deviations improves for Category II with MLS compared with ILS from 63.9 feet (19.5 m) to 22.3 feet (6.8 m) laterally and from 12.1 feet (3.7 m) to 8.5 feet (2.6 m) vertically.

In the limit, the approach can be flown completely automatically, the various categories being defined in Chapter 5. The original purpose in developing automatic landings was to increase regularity and save the costs of diversions. There are, however, many other advantages in the form of pilot workload, control of touchdown dispersion, softer landings, and the maintenance of flight path even in difficult windshear conditions. Experience with the equipment is showing increased reliability and a greater harmony between pilot and autopilot. It appears that the monitoring and takeover function of the pilot is sufficiently undemanding for fail-passive equipment to be used even for Category IIIA. Most of the benefits therefore accrue to the airline, while only full use of the system can justify the expense of the ground equipment, particularly for Category III. Also, different regulations must be brought into effect[13], the equipped aircraft must be allowed to bypass other stacked aircraft, and it must be provided with positive control on the ground in minimum visibility.

Safe automatic landing operations depend on the same factors that have been considered for visual landings, but with much greater emphasis on the concept of decision height. This depends on the method of operation, the specification of the equipment, and the runway visual range (RVR) as shown in Table 4.6, as well as on the obstacle-clearance criteria. The later must take account of the demonstrated height loss during a missed approach.[14]

The required RVR is a function of the pilot's angle of view cutoff, the intensity and beam spread of the lighting system, the vertical and horizontal structure of the fog as well as the location of the pilot's eyes relative to the aids and his intrinsic visual reference needs. It must be measured at least at three positions along the runway. The touchdown zone reading must be passed to the pilot within 15 seconds of reading, followed by the other readings if they are lower than the first and less than 2625 feet (800 m).

---

[13]For example, more strict procedures must be in force when RVR is less than 1300 feet (400 m), and a Category III landing is declared, including a minimum separation between the landing aircraft and the preceding departure of 4 miles (6.5 km).

[14]This ranges from 45 feet (13 m) on a BAC 125 to 80 feet (24 m)on a B747.

**TABLE 4.6 Decision Heights
(DH) and Runway Visual
Ranges (RVR) for Precision
Approach Runways**

Category	DH (ft)	RVR (m)
I	200	800
II	100	400
IIIA	--[a]	200
IIIB	--[a]	50

[a] No decision height applicable

Any proposal to operate automatic landings must show the feasibility of the proposed minima for each runway, including the adequacy of the facilities and the obstacle clearance capability. An airport wishing to declare a runway as suitable to receive automatic landings must consider:

- Obstacle clearances
- Glide path angle[15]
- Terrain on approach (should be essentially flat for 1000 feet (300 m) before threshold over a 200 foot- (60 m) wide strip)
- Runway length, width[16], and profile
- Conformity and integrity of the ILS[17]
- The visual aids and their integrity
- The level of ATC equipment and its monitoring meteorological requirements[18]

The most critical aspect of performance is the ability to climb after a missed approach has been declared. It must be possible to demonstrate adequate climb performance in each of the three following flight conditions:

1. Positive net gradient at 1500 feet (457 m) above airfield in the cruise configuration with one engine out

---

[15]Should be less than 3° normally.

[16]24.6-feet (7.5-m) shoulders are needed if less than 147 feet (45 m) wide.

[17]The critical areas need special protection—see ICAO Annex 10, Part 1.

[18]Including the need for cloud reports at the middle marker—see ICAO PANSMET and PANSRAC documents.

2. A gross gradient not less than 3.2 percent at the airfield altitude with all engines at maximum takeoff power[19] in the final landing configuration— this is to allow a safe overshoot (balked landing)

3. A minimum-gross climb gradient at the airfield altitude with the critical engine out and all others at maximum takeoff power in the final landing configuration but with gear up (the gradients should be not less than 2.1 percent for twin-engined aircraft, 2.4 percent for three-engined aircraft, 2.7 percent for four-engined aircraft)

Landing WAT charts indicate the maximum landing weights at which these gradients can be achieved as functions of altitude and temperature.

These performance criteria will allow safe operations in the vicinity of airfields only if used in conjunction with minimum descent altitudes and that take account of the handling of each aircraft type, the ground aids available, and the local terrain conditions. In setting minimum descent altitudes and associated decision heights, it is important to realize that there is an inevitable height loss between the decision to declare a missed approach and the establishment of a positive climb gradient, even with the demonstrated performance quoted earlier. This is due to delays in responding to pitch and power inputs and to the initial downward momentum of the aircraft. In the extreme case of a light twin-engined aircraft with one engine out, the height loss can be several hundred feet, and it is essential that the speed on the approach does not drop below the speed for best rate of climb.

The protected-surface funnel shown in Figure 4.4 includes surfaces to protect the missed-approach situation. These surfaces are designed to give clearance below the climb paths guaranteed by the landing WAT curves to the *remote risk* level.

The approach path is governed by the need to maintain clearance over obstacles both on the expected flight path and in the general vicinity of the airfield. The former are protected by the surfaces in Figure 4.4 or by those in Annex 14, while the latter (and any obstacles that break the protected surfaces) are cleared by the impositions of margins over the declared obstacle clearance altitudes (OCA) or obstacle clearance heights (OCH). The derivation of these is fully explained in Reference 3, from which Figure 4.7 is taken to illustrate the particular case of nonprecision approaches. Within the limits imposed by obstacle clearance and the required conditions at the threshold, the method of approach may be varied to achieve secondary objectives concerned with noise abatement or fuel saving. Some of these are shown relative to the standard procedure in Figure 4.8.

In essence, reductions in noise and fuel use are obtained by avoiding until the latest time the drag and hence the thrust that comes with the selec-

---

[19]Or the power that can be generated within an 8-second spool-up time.

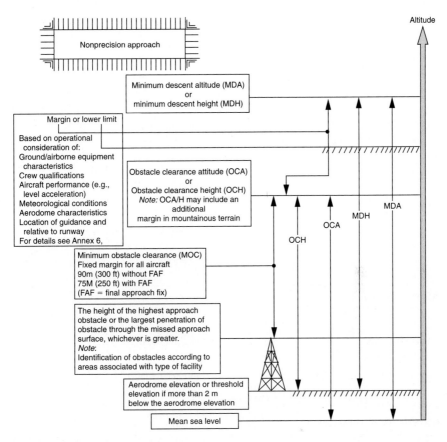

Figure 4.7 Relationship of obstacle clearance altitude/height to minimum descent altitude/height for nonprecision approaches. *(ICAO 1979)*

tion of full landing flap: the time at which the speed is stabilized is thus also delayed so as to maintain a safe margin over the stall speed. Lufthansa first developed the technique to decrease noise over a sensitive area 8 miles (13 km) out on the approach to Frankfurt. A more general version of the procedure has been adopted by IATA. Fuel consumption is reduced, as are internal and external noise levels (by 5 or 6 dBA), but there are some disadvantages. It is not possible to operate in this way without affecting decision heights, ATC must be able to allow 210 knots in the early stages of the approach, and either all operations must adopt the technique or separations on the approach must increase.

A more radical technique, the two-segment approach, was examined in depth in the middle 1970s. In common with the preceding technique, thrust was reduced until the final stages of the approach, but by increasing the initial descent angle to 6° rather than by increasing speed. *High and fast* tech-

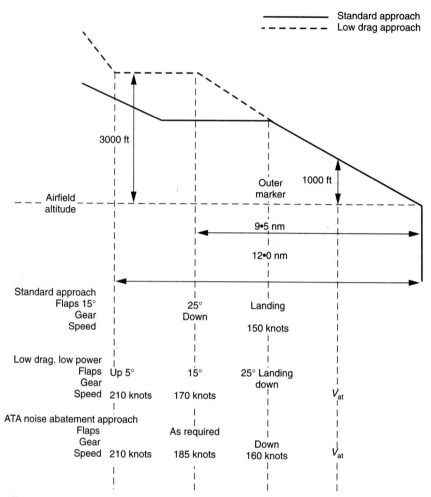

Figure 4.8 ILS approach options.

niques had been used by Air Traffic Control (ATC) at noise-sensitive airports on an ad hoc basis, and research at NASA and in the United Kingdom was aimed at rationalizing this. Both sets of investigations found the technique impractical because of the descent rates and the aids required, even though it used less fuel and did reduce noise significantly between 3 and 8 miles (4.8 and 13 km) from the threshold. It was necessary to use a descent rate in excess of 1500 feet (457 m) /minute, while pilots are not generally comfortable with more than 1000 feet (305 m) /minute.

More recently, profile descents have been developed to save much more fuel by close local management of the complete descent phase of the flight. The arrival rate into the TMA is matched to the acceptance rate of the airport, thus minimizing radar vectors and path stretching. This allows the use of a near-idle thrust descent from cruise altitude and minimized maneuvering time at low altitude. Thus a 737 might save 441 pounds (200 kg) of fuel or perhaps 23 percent of its short-haul fuel burn by using this technique rather than taking a 5-minute hold at 25,000 feet (7620 m). As developed by NASA, the descent is first at constant Mach number, changing to a constant indicated air speed (IAS) before decelerating in level flight at 15,000 feet (457 m) to a metering fix at approach speed.

A final consideration in approach performance is the problem of wake turbulence. It is necessary to pay attention to the order in which aircraft of different weights are allowed to approach or takeoff, so that smaller aircraft do not follow larger aircraft when runway capacity must be maximized. At Heathrow, the arrival runway capacity was found to vary from 34.7 to 31.8 movements per hour as the percentage of heavy aircraft varies from 10 to 50 percent, because the approach separations are more critical than runway occupancy. Although quick solutions to wake vortices in terms of aircraft design are unlikely, there is a valid suggestion to improve the situation operationally. Because smaller aircraft generally need shorter runways and can accept higher approach angles due to their slower speed, they could be brought in over the sinking vortices of preceding larger aircraft by positioning a second set of VASIs farther down the runway and with a higher approach slope. Unfortunately, there still remain the problems of mixed takeoffs and landings and also the sideways dislocation of vortices from operations on parallel runways in crosswind conditions.

4.4 Operations in Inclement Weather

As with vortices generated from preceding aircraft and with the use of automatic landings, one of the prime consequences of operations in inclement weather is reduced runway capacity, in this case due to increased runway occupancy time. Not only will average times increase by 33 percent as traction varies from excellent to nil, but the standard deviation on occupancy time will also increase. This is a function of the decreasing value of high speed exits as well as the increased braking distances, in that a radius of 1000 feet (304.8 m), which would be acceptable for 50 mph (88 kph) in good weather, will be usable at only 20 mph (32 kph) in slippery conditions. On the other hand, the aiming point for touchdown is a function of the distance from threshold to the most likely exit, but it does not vary significantly with weather conditions.[20] Since the runway occupancy is likely to be the critical capacity parameter when inclement weather forces the occupancy time up toward 90 seconds, a case can be made for designing exit location and angles in relation to an aircraft's poor weather landing performance.

---

[20]The major piloting adjustment for weather is in approach speed.

Crosswinds affect operations in inclement weather in that it is usual to put a limit of 10 knots on them for wet runway conditions. However, they are also a form of inclement weather per se, particularly for STOL aircraft. In fact, these aircraft can land satisfactorily in high crosswinds despite the excessive vector caused by their own low approach speeds, by judicious use of a combination of crab, slip, and the use of the runway width. The main problem comes on roll-out as the aerodynamic controls lose effectiveness, but this can be countered by the use of spoilers to dump lift, by improvements in nose-gear steering, and by improving thrust response.

Wet runways do not really qualify as inclement weather in that, as described earlier, regulations exist to cover this rather normal situation and the condition can normally be controlled by good runway grooving and drainage. The real difficulty comes when the runway is contaminated with standing water, slush, snow, or ice. There are some regulations to cover these cases. JARs require that data be established for aborting or continuing takeoff on runways with very low friction and with significant precipitation as well as simply wet conditions. They also require that 150 percent of the average depth of precipitation be used in subsequent calculations based on these data.

Flight or operations manuals contain data on performance in slush and ice[21] and on aquaplaning. Slush is perhaps the worst offender in that it affects both acceleration and braking. It can thus increase the emergency distance required on takeoff by 50 percent and can be particularly dangerous on an exploratory touch-and-go. When the friction coefficient is less than 0.05, flight manuals must advise a V_{stop}. This is similar to the normal decision speed V_1, but it is concerned only with the speed from which the aircraft at a given weight may be stopped within the runway length available. It might be less than the minimum control speed and does not imply any ability to continue takeoff if an engine fails. Aquaplaning is a function of speed and the depth of the standing water, the friction reducing to levels for an icy runway as the water fails to disperse beneath the tire.

All of these contaminated runway conditions give rise to two specific problems. First, there is the difficulty, already discussed, of the consistent evaluation of the contamination on aircraft performance. The variation is not confined to variations in contamination, but includes factors such as tire wear and pressure, pilot technique, and the extent to which the initial dynamic aquaplaning gives way to a sustained aquaplaning even at much reduced depths of precipitation.

The second group of problems concerns the method of informing the pilot correctly of the prevailing runway conditions. For some years, the two methods in use have been feedback from pilots who just used the runway

---

[21]The Citation operating manual suggests 50- and 100-percent factors on landing distance for wet and icy runways, respectively.

and the measurement of the depth of precipitation (the depth and type of precipitation being calibrated to give a percentage increase in takeoff and landing distance required). There is now a move to measure the runway friction directly by towing a device down the runway. An early British system, called the *Skidometer*, consisted of a wheel towed at 7° to the runway direction, with the side force being calibrated. Later systems (Mu Meter, Friction Tester) calibrate an aircraft wheel fitted with a conventional anti-skid system set to a 15-percent slip-braking ratio. The later systems appear superior because the equipment and the speed at which it is used approximate to real aircraft conditions; it can read in turns, and it has a low runway occupancy time. On the other hand, there are still many variables concerned with the tire and runway surface to bear in mind when attempting to calibrate their friction readouts to predict roll-out distance. Some airlines already use them in this way, but the FAA recommends them only for checking of surfaces after repair, and ICAO still recommends pilot reporting of three levels of braking. This simple method has advantages in that it copes with the rapid changes, the airport does not have to invest in the equipment, and there is no legal liability problem associated with the airport's reporting of the readings.

There are clearly doubts about the reporting of contamination and the performance in known contamination. When considered together with other problems of *soft*[22] failures on takeoff, there will always be occasions when the pilot is in doubt in the region of V_1. In these circumstances, it should always be safer to reject rather than to risk a nonflying takeoff, even if it means leaving the runway at 40 knots. It is also safer to overrun in a straight line rather than to risk losing directional control, because the gear is stronger in a fore-and-aft sense. This points to a real need for RESA, because the implied safety in these pilot decisions is false if the terrain at the end of the runway is difficult or nonexistent.

A remaining item of inclement weather is windshear. This is now suspected of causing several accidents previously attributed to pilot error. The most serious form of windshear is associated with the cold air gust front preceding a thunderstorm. It is the horizontal shears produced by turbulence in the cold sublayer and by the reversed direction of the warm inflow moving up and over it that causes the worst problems. The associated changes in airspeed can produce very strong tailwinds very soon after the pilot has taken remedial action for a gusting headwind, causing a stall. Systems of low level anemometers have been developed that compare horizontal wind strength and direction around the periphery of an airport with the normal central reading. Windshear warning is given when the discrepancy exceeds a preset tolerance. It is hoped that aircraft-based systems will be developed using

---

[22]*Soft* failures are instanced by tire bursts and partial loss of power in contrast to the *hard* complete engine failures around which the regulations are drawn up.

pulse Doppler radar with the requisite response fed directly to modify the flight director control laws. When ready, this system would be able to compensate for vertical windshear between two horizontal planes and also the shearing of vertical wind on the same horizontal plane, as well as the horizontal windshear. As such, it should be effective against microbursts, which are the other main suspected windshear source.

4.5 Implications of the Development of the New Large Aircraft

In recent years there has been considerable discussion concerning the introduction of new large aircraft (NLA) capable of carrying between 800 and 1000 passengers. The call for such vehicles comes from the severe shortage of runway slots at some very busy airports and the existence of a number of heavily traveled routes that already have enough traffic to require aircraft larger than the 747-400. The design and construction of such aircraft present a number of formidable problems such that many aeronautical engineers believe that a new generation of technology is required to allow designs that will carry up to 1000 seats, conventional technology and configurations becoming untenable for cost, safety and compatibility reasons above 800 seats.

There are many facets to the issue of airport compatibility. These are summarized in Table 4.7. In terms of capacity, strong vortex wakes and increased runway occupancy times might negate most of the gains in passengers per aircraft movement. Increased runway, taxiway and apron separation requirements due to larger dimensions will put great pressure on airside space. Stand requirements might be greatly increased unless onerous new turnaround time targets can be met; particular difficulties are foreseen with cabin cleaning and servicing of fully double decked aircraft. Loading and unloading passengers from such aircraft presents major interface problems with existing passenger terminals.

In mid 1996 the FAA and Boeing announced a program to develop a full-scale runway at the FAA Technical Center to determine the technological needs of runways for the accommodation of the new large aircraft. Simultaneously, airlines and airport operators are also urgently examining the implications for airports of the use of these aircraft. In 1995, British Airways carried out in-depth studies of the effect of the NLA geometry on existing airports including London Heathrow and Tokyo Narita.

In terms of environmental compatibility, the problems relate primarily to noise. Current certification requirements are hard to meet on the approach to landing due to the cut-off at high weight, even the self-noise from the large airframe being substantial. Additional environmental concerns relate to the need to eliminate the use of toxic materials and fluids and to control waste products and their disposal. Economically, the rebuilding that would be required at airports, or the loss of operational flexibility caused by ac-

TABLE 4.7 Impact of New Large Aircraft (NLA) on Airports

	Aircraft Characteristics							
	Overall Length	Wingspan	Overall Height	Weight	Wheeltrack	Wheelbase	Door Height	Service Points
Terminal interface - air bridges, mobile lounges, steps	●						●	●
Ground handling - loading and unloading equipment, ground traffic requirements							●	●
Apron parking positions - fixed hydrants, clearances, size, docking systems, ground traffic requirement	●	●			●	●		●
Runways - length, width, clearances, strength	●	●		●	●			
Taxiways - length, width, clearances, strength	●	●		●	●	●		
Aprons - area, clearances, strength	●	●		●	●	●		
Intersections - radii, clearances	●	●	●					
Maintenance facilities - size, weight	●							

cepting the NLA will be acceptable at only the few major airports whose future traffic growth is strongly tied to its introduction.

The main development not shown in Table 4.1 is the stretched 747, which might take any of several forms. The span could increase to 250 feet (76 m), while a full-length upper deck would give a capacity of 800 passengers and perhaps 992,250 pounds (450,000 kg) takeoff weight.

References

Ashford, N. J. and P. H. Wright. 1992. *Airport Engineering, 3rd edition,* New York: Wiley-Interscience.

Monk, K. A., March 1981. *A Review of Aircraft Accidents Between 1971 and 1977 Relating to Public Safety Zones,* UK Civil Aviation Authority DORA Communications 8103.

Procedures for Air Navigation Services: Aircraft Operations, Volume II, Doc. 8168-Ops/611, 1979. Montreal: ICAO, 1st edition.

Chapter 5

Operational Readiness

5.1 Introduction

One of the criteria for judging the efficiency of an airport is the availability of operational facilities: runways, instrument approach aids, lighting, fire and rescue services, mechanical and electrical systems, people movers, baggage handling systems, airbridges, and so on; in short, the "readiness" state of the airport to provide the operational facilities appropriate to the types of airlines and aircraft using the airport. All of this involves a considerable commitment to maintenance on the part of airport management. Increasingly, managements are documenting this aspect of their responsibilities. Table 5.1 shows The British Airport's Heathrow management report on "Passenger Sensitive" equipment as an example of a detailed analysis of monthly performance for a year. The availability target for all five items of equipment was 98.5 percent.

5.2 Airport Licensing

The principle of central government responsibility for safety aspects of public transportation extends to the licensing of airports. In the U.S. the FAA uses the term *certificate*, not license, although most states will require the airport to obtain a license. Public concern is primarily for the safety of aircraft and passengers and therefore with those elements of the total airport operating system related to this. In common with most countries, the United States makes certification compulsory for airports used by air carriers, both scheduled and charter. In Great Britain, the Civil Aviation Authority (CAA)

TABLE 5.1 Heathrow Passenger Sensitive Equipment Performance Results

Monthly	Jan 92	Feb 92	Mar 92	Apr 92	May 92	Jun 92	Jul 92	Aug 92	Sep 92	Oct 92	Nov 92	Dec 92
Loading bridges	99.03	98.89	98.36	97.95	98.54	97.14	96.54	96.35	97.02	96.85	97.98	97.21
Passenger conveyors	97.67	98.69	97.78	98.74	94.97	96.24	98.47	98.24	98.84	98.69	99.13	98.54
Baggage conveyors	98.85	99.20	98.80	98.64	98.64	94.07	98.76	98.24	99.06	98.59	99.17	99.09
Lifts	99.50	99.25	99.24	98.95	99.17	98.33	98.73	98.34	99.30	99.35	99.10	98.36
Escalators	98.97	99.09	99.45	99.48	99.25	99.26	98.55	99.62	99.48	98.98	97.52	97.65
Aggregate result	98.25	99.07	98.73	98.64	98.41	97.03	97.95	97.63	98.46	98.27	98.53	98.03
Cululative												
Loading bridges	98.73	98.74	98.71	97.95	98.23	97.97	97.52	97.29	97.25	97.19	97.28	97.27
Passenger conveyors	98.00	98.10	98.07	98.74	96.86	96.65	97.10	97.05	97.35	97.53	97.73	97.82
Baggage conveyors	99.05	99.07	99.04	98.87	98.75	97.19	97.58	97.71	97.94	98.03	98.17	98.27
Lifts	99.16	99.17	99.17	98.95	99.06	98.78	98.77	98.68	98.79	98.88	98.90	98.84
Escalators	98.80	98.83	98.88	99.48	99.37	99.33	99.14	99.23	99.28	99.24	99.03	98.86
Aggregate result	98.82	98.85	98.84	98.64	98.52	98.01	97.99	97.92	98.01	98.05	98.11	98.10
Moving annual												
Loading bridges	98.78	98.76	98.71	98.63	98.61	98.47	98.30	98.11	97.98	97.83	97.78	97.56
Passenger conveyors	98.22	98.18	98.07	98.03	97.72	97.60	97.61	97.47	97.47	97.51	97.72	97.90
Baggage conveyors	99.09	99.08	99.04	99.03	98.99	98.65	98.53	98.47	98.48	98.44	98.44	98.44
Lifts	99.20	99.20	99.17	99.14	99.15	99.05	99.03	98.97	98.99	99.02	99.02	98.96
Escalators	98.89	98.89	98.88	98.97	98.95	99.85	99.85	98.98	98.98	98.93	98.93	98.94
Aggregate result	98.89	98.88	98.84	98.80	98.74	98.58	98.51	98.42	98.39	98.33	98.34	98.30

requires, in addition, that any airport being used for flight training must also be licensed. The CAA issues two classes of license, *public use* and *ordinary*. The essential difference between them is that a public use airport must be available to all would-be operators without discrimination, while the ordinary category may be restricted if the owner so desires. Typically, an aircraft manufacturer's airfield falls into the latter category.

The requirements for the licensing or certification of an airport are set out in national regulations. In the case of the United States, they are to be found in Part 139 of the *Code of Federal Regulations* which specifies certain criteria that must be met in relation to pavement areas (runways, taxiways, and apron), safety areas (overrun areas), marking and lighting of runways, thresholds and taxiways, airport fire and rescue services, handling and storage of hazardous articles and materials, emergency plan, self-inspection program, ground vehicles, obstructions, protection of navaids, public protection, bird hazard reduction, and the assessment and reporting of airport conditions, including areas where work is in progress and other unserviceable areas. In the case of Great Britain the legal requirements are set out in the *Air Navigation Order and Regulations CAP 393, Articles 76–79 (CAA 1995)*.

Although the requirements are very similar in the United States and Great Britain, U.S. regulations contain additional rules dealing with maintaining the operational readiness of the airport (e.g., pavement repairs, clearance of snow, ice, etc., and lighting maintenance). In addition, the FAA issues very comprehensive guidelines regarding implementation in the 150 and 139 series of Advisory Circulars. In general, the certificate holder has to satisfy the regulating authority that:

1. Airport operating areas on the airport and in its immediate vicinity are safe

2. Airport facilities are appropriate to the types of operations taking place

3. The management organization and key staff are competent and suitably qualified to manage the aircraft flight safety aspects of the airport.

5.3 Operating Constraints

Visibility

Air traffic moves either under visual flight rules (VFR) or instrument flight rules (IFR), depending on weather conditions and prevailing traffic densities. VFR operations are possible where weather conditions are good enough for the aircraft to operate by the pilot's visual reference to the ground and to other aircraft (see Section 11.2). Operational runways are classified according to the weather conditions in which they can operate. The worse the condition in which a runway is to operate, the greater the amount of visual and instrument navigational equipment that must be provided. Runways can be classified according to their ability to accept aircraft at different degrees of visibility (ICAO 1995).

Noninstrument runway. A runway intended for the operation of aircraft using visual approach procedures only.

Instrument approach runway. A runway served by visual aids and a nonvisual aid providing at least directional guidance for a straight-in approach.

Precision approach runway—Category I. An instrument runway served by Instrument Landing System (ILS) and visual aids, intended for use in operations down to a decision height of 200 feet (60 m) and a RVR (Runway Visual Range) of 2600 feet (800 m).

Precision approach runway—Category II. An instrument runway served by ILS and visual aids, intended for use in operations down to a decision height of 100 feet (30 m) and a RVR of 1200 feet (366 m).

Precision approach runway—Category III. An instrument runway served by ILS to and along the runway with further subcategories.

Category IIIA. Intended for operations down to a RVR of 700 feet (213 m) and zero decision height, using visual aids during the final phase of landing.

Category IIIB. Intended for operations down to a RVR of 150 feet (46 m) and zero decision height, using visual aids for taxiing.

Category IIIC. Intended for operations without reliance on visual reference for landing or taxiing.

Runway visual range is the distance over which the pilot of an aircraft on the centerline of the runway can see the runway surface markings or the lights delineating the runway or its centerline. This range is now frequently determined automatically by RVR sensors, such as those shown in Figure 5.1, which are set just off the runway shoulders. Decision height is defined as the minimum height at which the pilot will make the decision either to land or to abort the attempt to land.

Figure 5.2 shows the sort of record that should be available to help determine the economic viability of high category operations. Simply recording the number of hours that the RVR is below 2600 feet (792 m), the limit for Category I, is not particularly helpful. At airports where low RVRs occur at night or in the very early morning when there is little traffic; the number of hours of poor visibility overestimates the level of traffic disruption it would cause. At other airports, however, severe and prolonged morning mist or haze could affect peak-hour operations, and without Category II or III capability the development of the airport might be made difficult. The figure shows graphically the results of a computation done by the BAA on

Figure 5.1 RVR-Skopograph. *(Hagebuk GmbH)*

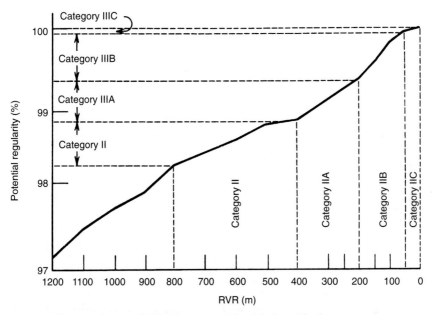

Figure 5.2 Impact of reduced visibility on potential regularity at Heathrow.

the effect of reduced visibility in terms of potential regularity (i.e., operational impact). It can be seen that the proportion of operations requiring Category II and III operations is less than 2 percent. Category IIIC conditions affect less than 0.05 percent of operations. Nevertheless, the principal operator at Heathrow, British Airways, has decided that a "blind landing" capability on its Concorde, and Boeing fleet is economically justifiable, because the airline finds itself able to operate when its competitors are grounded. Completely automatic "hands-off" landings of BA aircraft at London Heathrow are now routine.

Crosswind effects

Regulating bodies such as the ICAO and FAA require that an airport have sufficient runways, both in number and orientation, to permit use by the aircraft for which it is designed, with a usability factor of at least 95 percent with reference to wind conditions. Modern heavy transport aircraft are able to operate in crosswind components of up to 30 knots without too much difficulty, but for operational purposes, runway layouts are designed more conservatively. Annex 14 of the ICAO requires an orientation of runways that permits operations at least 95 percent of the time with crosswind components of 20 knots (37 kmh) for Category A and B runways, 15 knots (27.8 kmh) for Category C runways, and 10 knots (18 kmh) for Category D and E runways (ICAO 1995). FAA regulations differ slightly. Runways must be oriented so that aircraft can be landed at least 95 percent of the time with crosswind components not exceeding 15 mph (24 kmh) for all but utility airports and 11.5 mph (18.5 kmh) for utility airports.

The usability factor should be based on reliable wind distribution statistics collected over as long a period as possible, preferably not less than ten consecutive years. As aircraft have become heavier, the provision of crosswind runways has become less important at large hubs, where there is a generally prevailing direction of wind. However, crosswind runways are still operated at many airports when winds vary strongly from the prevailing direction or where light aircraft are operated.

The usability of a runway, or a combination of runways, is most easily determined by the use of a *wind rose,* which is compiled from a tabular record of the percentage incidence of wind by direction and strength as shown in Table 5.2. For clarity of presentation, the table shows a record of the percentage of time the wind falls within certain speed ranges (in knots) with the direction recorded to the nearest of 16 compass points.

A wind rose is drawn to scale with rings at 4, 7, 12, 18, 24, 31, and 38 knots, as shown in Figure 5.3. The percentage of time that a crosswind component occurs in excess of 15 knots can be determined using the following example with a runway (36-18) that runs due north-south. (For the purposes of this example, it is assumed that true north and magnetic north are identical; in practice, runway bearings are magnetic and wind data are referred to true north. Therefore, runway bearings must be corrected prior to plotting.) The direction of the main runway 36-18 is plotted through the center of the rose and 15-knot crosswind component lines are plotted to scale parallel on either side of this centerline. The sum of percentages of wind components falling outside the parallel component lines is the total amount of time that there is a crosswind component in excess of 15 knots. Table 5.3 indicates that this occurs for a total of 12.22 percent of the time for this particular runway direction. Therefore, this runway would not conform to FAA standards if proposed as the only runway of a U.S. airport. The reader is invited to check that it would also fail to meet ICAO standards. Note that estimates of part areas of the rose must be made in compiling Table 5.3. The effect of using a crosswind runway in addition to the main runway can be seen by adding a runway (14-32) that can cope with the

TABLE 5.2 Wind Table: Wind Direction and Percentage Incidence

Wind Strength (knots)	N	NNE	NE	ENE	E	ESE	SE	SSE	S	SSW	SW	WSW	W	WNW	NW	NNW
0–4							3.4 for all directions									
4–7	1.1	0.7	0.3	0.4	0.5	0.7	0.9	1.3	1.2	0.7	0.5	0.1	—	—	0.3	0.8
7–12	2.9	2.0	0.2	0.3	0.4	2.0	0.5	8.0	5.0	2.1	0.2	—	—	—	0.1	2.0
12–18	3.2	2.0	0.2	0.2	—	9.0	0.5	9.8	6.0	3.2	0.1	—	—	—	—	2.2
18–24	1.5	0.5	0.2	0.1	—	6.0	0.2	2.1	2.7	1.9	—	—	—	—	—	1.6
24–31	0.6	0.2	0.1	0.1	—	2.1	—	0.4	1.8	0.9	—	—	—	—	—	0.7
31–38	0.2	0.1	—	—	—	0.1	—	0.1	0.2	0.1	—	—	—	—	—	0.2
over 38	—	—	—	—	—	—	—	—	0.1	—	—	—	—	—	—	0.1

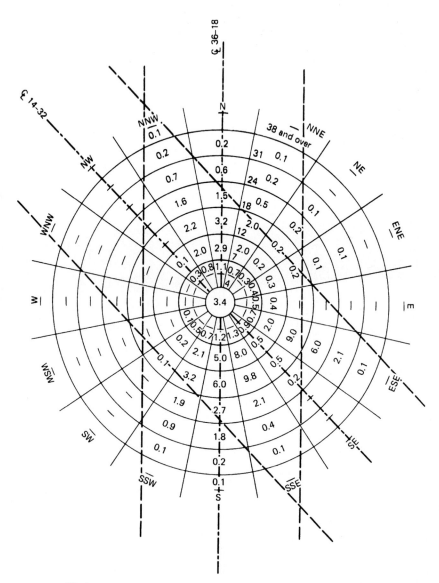

Figure 5.3 Wind rose.

strong components accruing from the east-southeast direction. To show that crosswind runways can still be very effective even if not exactly aligned in the major direction of troublesome crosswinds, the direction chosen for this example is 14-32 rather than 11-29. By now examining the components that fall outside the shaded area it can be seen from Table 5.4 that all but 0.5 percent of crosswind components in excess of 15 knots would be ac-

TABLE 5.3 Percentage of Time Crosswind Component Exceeds 15 knots on Runway 36-18

Wind Strength (knots)	Direction																
	N	NNE	NE	ENE	E	ESE	SE	SSE	S	SSW	SW	WSW	W	WNW	NW	NNW	Total
0–4	—	—	—	—	—	—	—	—	—	—	—	—	—	—	—		
4–7	—	—	—	—	—	—	—	—	—	—	—	—	—	—	—	—	
7–15	—	—	—	0.07	—	—	—	—	—	—	—	—	—	—	—		—
15–18	—	—	—	0.1	—	3.0	—	—	—	—	—	—	—	—	—		
18–24	—	0.05	0.1	0.1	—	6.0	0.1	—	—	—	—	—	—	—	—		
24–31	—	0.1	0.1	—	—	2.1	—	—	—	—	—	—	—	—	—		
31–38	—	—	—	—	—	0.1	—	0.05	—	0.15	—	—	—	—	—		
over 38	—	—	—	—	—	—	—	—	—	0.05	—	—	—	—	—	0.05	
Total	—	0.15	0.2	0.27	—	11.2	0.1	0.05	—	0.20	—	—	—	—	—	0.05	12.22

TABLE 5.4 Percentage of Time Crosswind Component Exceeds 15 knots Using Both Runways 36-18 and 14-32

Wind Strength (knots)	Direction																	
	N	NNE	NE	ENE	E	ESE	SE	SSE	S	SSW	SW	WSW	W	WNW	NW	NNW	Total
0–4	—	—	—	—	—	—	—	—	—	—	—	—	—	—	—	—	
4–7	—	—	—	—	—	—	—	—	—	—	—	—	—	—	—	—	
7–15	—	—	—	—	—	—	—	—	—	—	—	—	—	—	—	—	
15–18	—	—	0.1	0.1	—	—	—	—	—	—	—	—	—	—	—	—	
18–24	—	—	0.1	0.1	—	—	—	—	—	—	—	—	—	—	—	—	
24–31	—	—	—	—	—	—	—	—	—	—	—	—	—	—	—	—	
31–38	—	0.05	—	—	—	0.05	—	—	—	—	—	—	—	—	—	—	
over 38	—	—	—	—	—	—	—	—	—	—	—	—	—	—	—	—	
Total	—	0.05	0.2	0.2	—	0.05	—	—	—	—	—	—	—	—	—	—	0.5

counted for by this configuration, which would conform to FAA and ICAO standards.

Bird strike control

Since the beginning of aviation, birds have been recognized as a hazard to aviation. In the early days, damages tended to be minor, such as cracked windshields, dented wing edges, and minor fuselage damage. However, fatal accidents due to bird strikes occurred as far back as 1912 when Cal Rogers, the first man to fly coast to coast across the United States, was subsequently killed in a bird strike. As aircraft have become faster, birds have become less able to maneuver out of the way, and the relative speed at impact has increased. Damage also increased when turbine-engined aircraft were introduced. Ingestion of birds into the engine can cause a blocking or distortion of an airflow into the engines, severe damage to the compressor or turbine, and an uncontrollable loss of power. Loss of life through bird strike is unusual, but the airport operator must be aware that the potential for a disaster can exist in the vicinity of an airport where aircraft are operating at the low altitudes at which they are likely to come in contact with birds. International and national aviation regulating bodies have therefore prepared advisory documents that guide airport operators in methods of reducing the risk of bird hazards through programs of bird strike control (ICAO 1991b).

Fundamental to a successful bird control program is an understanding of bird types and their habits. Any bird, if present in sufficient quantities, can present a hazard to aviation at an airport. However, because different birds exhibit remarkably different behavior patterns, only a few are likely to create hazards. Past accidents involving large passenger aircraft on the U.S. east coast indicate the particular hazard associated with gulls. Birds present on the airport are there because the facility provides a desirable environment for such natural requirements as food, shelter, safety, nesting, rest, and passage for migratory routes. Successful bird strike control largely depends not on driving birds off, but in creating an environment on the airport and in its immediate vicinity that is not attractive to birds in the first instance. ICAO recommends that a control program should:

- Identify problem species
- Determine bird behavior patterns
- Study the ecology of the airport environment
- Determine specifically what encourages the problem species to the area

Typical measures will:

1. Control garbage, especially the location of garbage dumps near the airport. It is recommend that garbage disposal dumps should not be located within 8 miles (13 kms) of an airport reference point.

2. Control of other food sources such as insects, earthworms, and small mammals by a variety of measures such as poisons, insecticides, cultivation, and hunting to ensure that the open space at and near the airport does not encourage a food supply likely to be attractive to the troublesome species.

3. Eliminate as much as possible the occurrence of surface water that can form suitable habitat for water birds. Control measures include filling, draining areas or netting open water areas.

4. Where possible, control farming in close proximity to operations. If open areas held by the airport for future expansion are leased as farmland, a ban on cereal cropping, for example, should be written into the leases.

5. Promote vegetation that discourages the presence of birds and avoid vegetation, such as trees, hedgerows, and berry-bearing shrubbery, that attracts birds.

6. Ensure that buildings in the airport area do not provide suitable nesting places for birds such as swallows, starlings, and sparrows that have become used to living in a man-made environment.

Even where habitat control measures have been taken to discourage some birds from being attached to the airport, other birds might appear in significant numbers. It might become necessary to disperse and drive off birds using more active measures. These measures should not be used instead of habitat control, because, given an attractive environment, when one bird is driven off, another is likely to take its place. Dispersing and expulsion techniques include:

- Pyrotechnic devices (firecrackers, rockets, flares, shell crackers, live ammunition, gas cannons)
- Recorded distress calls
- Dead or model birds
- Model aircraft and kites
- Light and sounds of a disturbing nature
- Trapping
- Falcons
- Narcotics and poisons

If the presence of birds is a serious problem that threatens to disrupt the safe operation of the airport, the operator has no choice but to initiate a control program that will reduce the hazard to an acceptable level. On occasions airport management can then be faced with the intervention of wildlife conservation organizations.

5.4 Operational Areas

Pavement surface conditions

It is essential that the surfaces of pavements, especially runways, be kept as free as possible of contaminants and debris to ensure safe aircraft operations. A *contaminant* is defined as "a deposit (such as snow, ice, standing water, mud, dust, sand, oil, and rubber) on an airport pavement, the effect of which is detrimental to pavement braking conditions" (FAA 1981). *Debris*, on the other hand, refers to loose material such as sand, stone, paper, wood, metal, and pavement fragments that could be detrimental to operation by damaging aircraft structures or engines or by interfering with the operation of aircraft systems.

Especially since the introduction of jet aircraft, airport operators find that they must pay increasing attention to the available friction between runways and tires on landing, and to the precipitant drag effect on takeoff. The danger of damage from debris has also increased due to the higher speeds at takeoff and landing, the nature of the jet engine, and the danger of ingestion, especially for underslung engines. The seriousness of the problems involved is recognized by national airworthiness authorities, which routinely recommend that landing distances on wet runways be increased over those for dry conditions. Jet aircraft are also highly susceptible to the effect of precipitant drag, which occurs on slush- or water-covered runways, seriously affecting the ability of aircraft to obtain flying speed safely on takeoff. ICAO publishes special recommendations on operational measures dealing with the problem of taking off from slush- or water-covered runways. Ideally, the airport operator, although unable to keep a runway dry, would like to be in a position to keep the runway clear of contaminants and debris. However, snow, slush, and blowing sand might present conditions where operation will continue with less than optimum pavement surface conditions during a continuous clearing process. Therefore, procedures are set up to measure runway surface friction and the precipitant drag effect so that the pilots can adjust their techniques to existing conditions. In summary, the occasions under which assessment of the runway surface condition might be required are:

- The dry runway case—infrequent measurement to monitor texture, wear and tear through the normal life of the runway

- The wet runway case—taking care to note the dramatic interaction of wet conditions with rubber deposits, which can result in a serious deterioration of the friction coefficient

- The presence of a significant depth of water and the possibility of aquaplaning

- The slippery runway case due to the presence of ice, dry snow, wet snow, compacted snow, or slush, which reduce the coefficient of surface friction

- A significant depth and extent of slush, wet snow, or dry snow that can produce a significant level of precipitant drag

At a very busy airport that frequently experiences conditions where braking might be impaired by contaminants, an adequate level of runway cleaning equipment must be maintained. Equipment must also be available to check the results of cleaning by measuring friction and drag. At a less busy airport where conditions of impaired braking are only infrequently experienced, but where operations must continue in spite of inadequate cleaning equipment, assessment of runway friction and drag potential is essential, and equipment for measuring these effects must be available to enable pilots to adjust their operations to the existing conditions. At an even less important airport where operations can be suspended, it is essential to have equipment to assess runway friction and precipitant drag in order to be able to make a decision on when conditions have reached the point where suspension of activities is necessary. It is important to remember that even where the removal of snow and ice is given high priority, there is frequently a significant loss of friction on an apparently dry, cleared runway. At airports that regularly experience heavy snowfalls, for example, in northern Europe and North America, clearance might have to be discontinued for a short while during a storm to permit some operations to continue. Runways are unlikely, in such conditions, to be completely clean. There are also likely to be local slippery patches. The airport authority will need to measure and assess surface conditions to inform pilots of the overall condition and to determine those areas requiring more cleaning treatment.

Various types of friction testing equipment are available. Several versions are small trailers with a measuring device (Mu Meter) that is attached behind a towing vehicle. Another type developed by SAAB utilizes a standard sedan car with an integral measuring unit within the car that is lowered underneath the car as it proceeds along the runway. An illuminated control panel by the driver gives readings, at the same time as a record is made, of the coefficient of friction with a corresponding estimated braking action. (Figures 5.4a and 5.4b). For further descriptions of such equipment, the reader is referred to FAA 1991a. Table 5.5 indicates the relationship between the coefficient of friction and the subjective estimate of braking efficiency. It is quite possible for a thin film of ice to reduce the coefficient of

(a)

Figure 5.4a SAAB auto outfitted to measure runway friction.

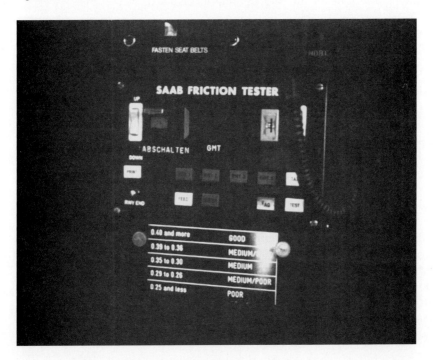

(b)

Figure 5.4b Digital readout of the SAAB friction tester.

**TABLE 5.5 Relationship
Between Coefficient of
Friction and Braking Efficiency**

Measured/ Calculated Coefficient	Estimated Braking Action
0.40 and above	Good
0.39 to 0.36	Medium to good
0.35 to 0.30	Medium
0.29 to 0.26	Medium to poor
0.25 and below	Poor

SOURCE: FAA 1991

friction on an aircraft pavement from 0.50 to 0.15, reducing the braking efficiency to less than a third of that in the dry condition.

Recognizing that an airport authority must be in a position to evaluate the level of runway friction, assessment of pavement condition should never take precedence over the clearance operations themselves. Within the operational areas, safety and efficiency require observing the following clearance priorities for the various areas involved:

- Runways

- Taxiways

- Aprons

- Holding bays

- Other areas

Snow clearance is frequently coordinated through the operation of the *snow committee* consisting of members from the airline operators, meteorology, ATS services, and the airport administration. Clearance is laid down in a snow plan that ensures that agreed procedures exist for the provision and maintenance of equipment; for clearance according to stated priorities; for installation of runway markers, snow fencing and obstruction marking; and for providing for the maneuvering aircraft. For further details, on snow clearance the reader is referred to FAA 1991a and FAA 1992, which indicate the availability and uses of such specialized equipment as the snow blower shown in Figure 5.5. While snow conditions occur for only a limited period during the year, airports might turn to outside contractors to provide snow clearance services, having the equipment moved onto site only prior to the snow season.

Figure 5.5 A snowblower.

Debris presents a separate and different problem at airports. Jet turbine engines are extremely susceptible to damage from ingestion of solid particles of debris picked up from the pavement surfaces. Tire life is also reduced by wear and cuts induced by sharp objects on the pavements, deteriorating pavement surfaces and edges, and poor, untreated pavement joints. Damage can also occur to the skin of aircraft from objects thrown up from the pavements. Problems arising from debris can be reduced by regular inspections of pavement surface condition of all operational areas and by establishing a sweeping and cleaning program setting up priorities and frequencies. In order to help designate the particular location of debris, a plan of the paved areas should be divided into manageable paved segments or approximately 1640 feet (500 m) (ICAO 1984). In apron areas, the cleanliness of the airline operators and other users determine the amount of litter and debris that is present. Cargo areas are particularly susceptible to the presence of fragments of strapping, nails, container, and pallet debris. The problem can be reduced by careful adherence to a disciplined program of maintaining a litter-free pavement. Runways, taxiways, and holding bays are subject to debris eroded from the pavement shoulders. This kind of debris can be reduced by shoulder-sealing treatments, which might be necessary on highly trafficked facilities. Removal of debris is achieved by powered mobile brooms, vacuum and compressed air sweepers, and magnetic cleaners. As a further incentive, some airports provide brightly painted litter bins adjacent to the aircraft gates/parking bays in which any litter or debris, found in the aircraft parking areas, can be deposited.

5.5 Approach/Landing Aids

Instrument Landing System

The most commonly used navigation system for landings in conditions requiring instrument guidance is the *Instrument Landing System* (ILS), shown diagrammatically in Figure 5.6. It is designed to identify for the pilot an approach path that is exact in both alignment and descent. In its extreme form, coupled with airborne cockpit equipment, it provides a completely automatic hands-off approach and landing. There are three component parts:

1. *Guidance.* Given by the VHF localizer and UHF glideslope signals

2. *Range.* Provided by marker beacons along the approach

3. *Visual reference.* Provided by approach lights, touchdown zone, and centerline lights and runway threshold and edge lights.

The figure shows the nonvisual elements of the FAA system. The ground equipment consists of two highly directional transmitting systems and at least two marker beacons. Navigational information is interpreted in the cockpit by an adaptation of the very high frequency omnidirectional radio range (VOR) equipment.

Alignment with the runway centerline is controlled by the directional *localizer* transmitter, which typically is set 1000 feet (300 m) beyond the end of the runway. Deviation either to the right or left of the extended runway centerline is displayed on the combined ILS/VOR instrument, as shown in Figure 5.6. The UHF glideslope transmitter sends out a directional beam along a plane at right angles to the localizer, at a nominal slope of 3° to the horizontal; deviation above or below this slope is also displayed on the VOR receiver in the cockpit. The pilot therefore receives a continuous precise indication of position relative to the correct azimuth and position on the glidepath. Additional information is provided to the pilot in the form of two low-power fan markers, over which the aircraft passes as it progresses down the approach path. The outer marker (OM), is located at approximately 5 miles (8 km) from the threshold, at which point the glidepath is at approximately 1400 feet (430 m) altitude, and the middle marker (MM) is sited approximately 3500 feet (1070 m) from the threshold, where the glidepath is at approximately 200 feet (60 m). Visual indications are given in the cockpit as the aircraft passes first over the outer marker and then over the middle marker.

On Category II ILS systems, there is a further positional indication given at an inner marker (IM) that indicates the position on the glidepath corresponding to a designated decision height at which the landing should be aborted if visibility prevents continuing the approach with the use of the appropriate and necessary visual aids. Safe operation within the ranges of RVR and decision height that correspond to Categories I, II, and III require

ILS

[FAA INSTRUMENT LANDING SYSTEM]

STANDARD CHARACTERISTICS AND TERMINOLOGY

ILS approach charts should be consulted
to obtain variations of individuals systems.

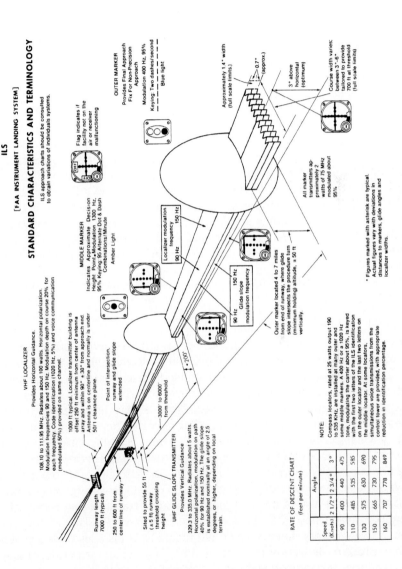

Flag indicates if
facility not on the
air or receiver
malfunctioning

OUTER MARKER
Provides Final Approach
Fix For Non-Precision
Approach
Modulation 400 Hz, 95%
Keying: Two dashes/second
Blue light

MIDDLE MARKER
Indicates Approximate Decision
Height Point. Modulation 1300 Hz,
95% Keying: 90 Alternate Dot & Dash
Combinations/Minute
Amber Light

VHF LOCALIZER
Provides Horizontal Guidance.
108.10 to 111.95 MHz. Radiates about 100 watts. Horizontal polarization.
Modulation frequencies 90 and 150 Hz. Modulation depth on course 20%, for
each frequency. Code identification (1020 Hz, 5%) and voice communication
(modulated 50%) provided on same channel.

1000 ft typical. Localizer transmitter building is
offset 250 ft minimum from center of antenna
array and within 90° ± 30° from approach end.
Antenna is on centerline and normally is under
50/1 clearance plane.

Runway length
7000 ft (typical)

250 to 600 ft from
centerline of runway

Point of intersection,
runway and glide slope
extended

Sited to provide 55 ft
(± 5 ft) runway
threshold crossing
height

UHF GLIDE SLOPE TRANSMITTER
Provides Vertical Guidance
329.3 to 335.0 MHz. Radiates about 5 watts.
Horizontal polarization, modulation on path
40% for 90 Hz and 150 Hz. The glide slope
is established nominally at an angle of 2.5
degrees, or higher, depending on local
terrain.

3000' to 6000'
from threshold

Localizer modulation
frequency

90 Hz   150 Hz

90 Hz   150 Hz
Glide slope
modulation frequency

Outer marker located 4 to 7 miles
from end of runway, where glide
slope intersects the procedure turn
(minimum holding) altitude, ± 50 ft
vertically.

Approximately 1.4° width
(full scale limits.)

0.7°
(approx.)

3° above
horizontal
(optimum)

Course width varies;
between 3°–6°
tailored to provide
700 ft at threshold
(full scale limits)

All marker
transmitters ap-
proximately 2
watts of 75 MHz
modulated about
95%

* Figures marked with asterisk are typical.
Actual figures vary with deviations in
distances to markers, glide angles and
localizer widths.

NOTE:
Compass locators, rated at 25 watts output 190
to 535 KHz, are installed at many outer and
some middle markers. A 400 Hz or a 1020 Hz
tone, modulating the carrier about 95%, is keyed
with the first two letters of the ILS identification
on the outer locator and the last two letters on
the middle locator. At some locators,
simultaneous voice transmissions from the
control tower are provided, with appropriate
reduction in identification percentage.

RATE OF DESCENT CHART
(feet per minute)

Speed (Knots)	Angle			
	2 1/2 °	2 3/4 °	3 °	
90	400	440	475	
110	485	535	585	
130	575	630	690	
150	665	730	795	
160	707	778	849	

Figure 5.6 Instrument Landing System—ILS. *(Federal Aviation Administration)*

increasingly sophisticated ILS equipment as the operating conditions become worse. Although the principle of the ILS system remains the same, to go from Category I operations to Category II and from Category II to Categories IIIA, IIIB, and IIIC requires more operational precision. The equipment itself must be more precise and the conditions of its operation more controlled. As the final approach and landing becomes increasingly blind, the aircraft must receive more accurate signals from the locator and the glideslope antennas. Extreme precision is required for all Category III operations. The provision of a sufficiently accurate glideslope signal for such operations is not always economically possible because current ILS systems depend on the reflection of the signal from the ground in front of the antenna. Frequently, an ILS system is installed and tested to determine the final category to which it can operate. To attempt to achieve ab initio a high category such as IIIB or IIIC might be prohibitively expensive. At most air transport airports, weather would significantly hamper operations if no instrument runway were provided. However, a decision must be made on the category of runway to be installed. To move up from a Category I to a Category II operation requires a significant additional expense in both visual and instrument landing and approach aids. An even more substantial investment is required to move up from Category II to Category III. In these circumstances, the operator should perform an economic evaluation comparing the extra cost of providing higher category operations against the cost and impact of closing down during severe weather.

Microwave Landing Systems

Although ILS gives a substantial increase in airport serviceability in poor weather, the system is not without drawbacks. Introduced as the international standard instrument approach aid in 1947, and based on initial military systems, it requires substantial antennas to radiate sufficiently narrow beam signals at the wavelengths used. These signals are affected by the presence of buildings, vehicles, and taxiing aircraft. The best signals are obtained when the beams are reflected from a smooth and featureless terrain in front of the threshold. In areas of steep topography, high performance ILS become difficult and even impossible to install. The exact category that an ILS will achieve can often be determined only by in-situ testing in difficult terrain. The use of much higher microwave transmission frequencies would overcome most of the problems associated with ILS. Transmitting antennas become very much smaller and more easily installed. The signal is not sensitive to deflection from surrounding objects and is not dependent on terrain for the forming and propagation of the signal beams. Unlike the two-beam ILS signal, *Microwave Landing System* (MLS) guidance can be multidirectional, allowing multiple approach paths as shown in Figure 5.7. The system can also give continuous distance information to the pilot (at

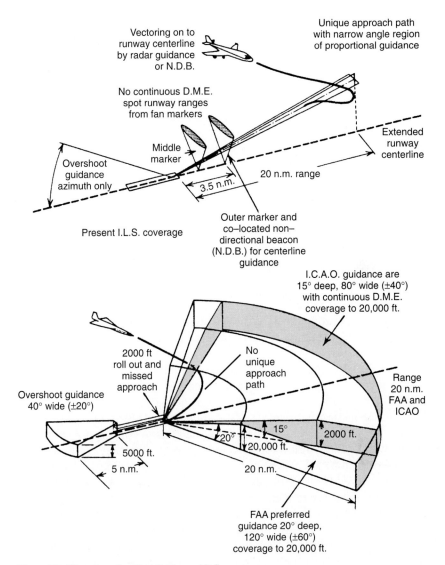

Figure 5.7 Microwave Landing System—MLS.

the moment, only the fan markers on an ILS approach give two- or three-point positional information with reference to distance from the threshold).

Where there was every expectation that the MLS would be the system to replace ILS by the year 2000, there is now considerable doubt on this score. As a result of a meeting of the ICAO Communications and Operations Division in April 1995, the original plan for the transition from ILS to MLS has been replaced by a new global strategy that allows for different areas of the world proceeding at their own pace. The United States, having already developed a satellite navigation system, the *Global Positioning System* (GPS), is working to refine the system further to enable its use as a precision navigation aid for approaches and landings, to replace ILS and MLS. Funding for MLS Research and Development has been withdrawn by the Federal Aviation Administration in the United States, in favor of GPS systems. Some states might wish to keep ILS in service until such time as they decide GPS (GNSS—Global Navigation Satellite Systems in ICAO terminology) have been developed to the stage where the transition can be made directly from ILS to GNSS. Yet a third option, particularly for European countries, might be to install MLS in order to safeguard Category II operations. The possibility of these three variations in implementation of a new precision approach and landing system on a worldwide basis has implications for airline international operations since they will need the availability of an airborne multi-mode receiver.

Satellite navigation systems

Although the operational use of these systems will be the concern only of governments, telecommunications organizations, airlines, and general aviation in terms of equipment installation and maintenance, there are clear implications for airports in so far as the life of their currently installed navigation and landing aids are concerned. At present only two Global Navigation Satellite Systems have been notified to the International Frequency Registration Board; the Global Positioning System (GPS) developed by the United States, and the *Global Orbiting Navigation Satellite System* (GLONASS) under development by the Federation of Russia. The Global Positioning System is comprised of 24 Navstar satellites operating on six orbital planes (four satellites per orbit) at a height of 10,900 nautical miles. GPS utilizes range measurements from the satellites to determine a position anywhere in the world (Figure 5.8). However, it is a U.S. military system rather than a civilian system and this has raised issues of national sovereignty and security, although the U.S. has promised to keep the system available except in the most dire circumstances. Although the system is capable of producing highly accurate position information, including the position of an aircraft taxiing at an airport, there is still some uncertainty regarding the time necessary to determine the feasibility of GNSS in terms of integrity and continuity and understanding the interference and failure mechanisms. There are varying estimates of the time scale for certification of GNSS for Cat II/III operations, and as a result, of the probable date for the replacement of ILS and MLS.

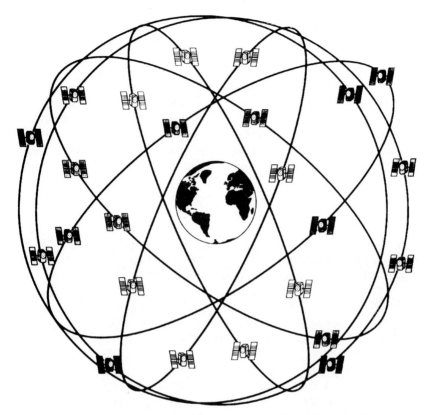

Figure 5.8 The Global Positioning System is composed of 24 satellites operating on six orbital planes (four satellites per orbit) at a height of 10,900 nm.

Radar

Although there has always been a preference on the part of pilots for a pilot-interpreted approach and landing aid, it has become generally acknowledged that a ground-based, controller-interpreted aid—such as radar—has a vital role to play in airport operations. In the role of approach and landing guidance for precision approaches, its application as *Precision Approach Radar* (PAR) is on the decline.

There is no doubt, however, of its usefulness in terms of facilitating arrivals and departures in busy terminal areas and in terms of the consequent increase in airspace capacity. Several forms of radar exist. The air traffic control authority decides which of those available will be employed at a particular airport. There are two basic types of radar in use at airports: *primary* and *secondary* radar. The FAA (1993) gives the following definitions:

Primary radar. A radar system in which a minute portion of a radio pulse transmitted from a site is reflected by an object and then received back at that site for processing and display at an air traffic control facility.

Secondary radar/radar beacon/ATCRBS. Known in Europe as Secondary Surveillance Radar (SSR), this is a radar system in which the object to be detected is fitted with cooperative equipment in the form of a radio receiver/transmitter (transponder). Radar pulses transmitted from the searching transmitter/receiver (interrogator) site are received in the cooperative equipment and used to trigger a distinctive transmission from the transponder. The reply transmission, rather than a reflected signal, is then received back at the transmitter/receiver site for processing and display at an air traffic control facility.

Primary radar. A narrow radar beam that sweeps through 360° of azimuth in much the same fashion as an invisible searchlight beam, primary radar "illuminates" targets (aircraft) with radar energy that is reflected back and displayed as a blip of light at its appropriate bearing on a controller's radar scope, as illustrated in Figure 5.9. The vertical extent of the beam's coverage, its lobe, varies according to the equipment, especially the antenna characteristics. For airport and terminal area uses, the requirement is for a lobe that gives both low coverage (as near to the ground as possible for the immediate vicinity of the airport) and a good solid coverage to an altitude of around 10,000 to 12,000 feet (3000 to 3650 m) out to a range of 20 to 30 miles (32 to 48 k). This type of "search" radar will pick up all aircraft within coverage but indicate plan position only, without any height indication. Traditionally, radar displays have been viewed in a darkened environment. However, new bright displays have been developed that can be viewed in normal, even strong, natural light. Such displays are to be found in some busy airport towers—*Bright Radar Indicator Tower Equipment* (BRITE). They are not used for control purposes (e.g., to give aircraft headings), but merely to provide information on arriving aircraft to airport controllers.

A somewhat specialized form of primary radar is used at some airports to provide a radar picture (map) of the airfield surface. This is especially useful when operations are taking place in very low visibility and controllers can give taxiing instructions even though the aircraft might not be visible in the normal sense from the tower. In especially low visibility, when Category II or III landings are taking place, it is an essential aid for keeping track of taxiing aircraft and vehicles that might be in the operating areas. In Europe, such radars are known as *Airfield Surface Movement Indicators* (ASMI), and in the United States, as *Airport Surface Detection Equipment* (ASDE).

Figure 5.9 Primary radar—Approach Control.

At a relatively small number of airports where local terrain or other conditions might make it impossible to install ILS but where precision instrument approach capability is required, *Precision Approach Radar* (PAR) might be installed. This comprises two elements: a normal-type primary search radar of very short range but high definition and an associated height-finding radar element scanning a vertical glideslope segment. By simultaneous use of the two radar pictures, plan and vertical, a radar controller can pass to a landing aircraft precise heading and height instructions to bring the aircraft to the touchdown point, hence the now largely outdated term "talkdown" radar. In this case, there is no associated cockpit instrumentation; the pilot relies on following the precise guidance given by a controller over radio. The antenna for normal primary surveillance radar is frequently mounted beneath a secondary radar antenna in the array illustrated in Figure 5.10.

Secondary radar. Although the use of the term "radar" is somewhat misleading, this interrogator/responder system is generally classified as a radar system. But unlike primary radar, which merely "illuminates" the aircraft by radar pulses that are reflected back (passive response), the secondary transmission triggers a response from the aircraft transmitter (transponder). This equipment has now become so vital to air traffic control in busy terminal areas and other types of controlled airspace that its carriage by

Figure 5.10 Antennae for normal and secondary radar.

transport aircraft has become mandatory in these areas. The actual response signals from the aircraft are processed at the ground-receiver end by computers, so that the results appear on the controllers radar screen as alphanumerics. The amount of information varies with the "mode" employed.

- **Mode A.** Identifies aircraft and its position
- **Mode A/C.** Identifies aircraft and its position and also provides height read out

In the case of Mode A/C, an automatic tracking facility incorporated in the computer will also provide groundspeed data on the radar screen. It can also be used to predict track and provide a "conflict warning" to the controller. The later SSR systems can automatically address or interrogate a specific aircraft. In Great Britain such systems are described as *Address Selection* (ADSEL) systems; in the United States, *Discrete Address Beacon Systems* (DABS). A common specification that removes the slight differences between the two has now been agreed on. This is now known as *Mode S* and provides a data link by means of which ATC can pass long messages to the aircraft, which in turn can also send long messages back. (128 data bits are available.) In the earlier stages of the development of SSR, the ICAO assigned reserved Modes B and D against future developments, but these will probably not be utilized since they have been overtaken by events, that is, by the development of Mode S. A typical data block of information as it appears on the radar screen is illustrated in Figure 5.11 (A/C Mode). However, the future of radar, whether primary or secondary, will no doubt be assessed in light of the development of satellite position information and of future methods for processing and utilizing this source of information.

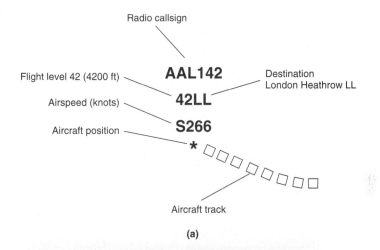

(a)

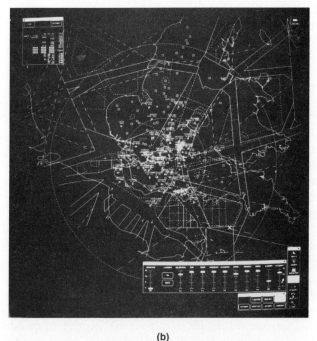

(b)

Figure 5.11 Secondary radar display—data block, London radar

5.6 Approach/Runway Lighting

Operating conditions

Visual guidance for Categories I, II, IIIA, and IIIB is achieved by a combination of approach, threshold, touchdown zone, runway edge, and runway end lighting used in conjunction with taxiway edge and centerline lights. The amount of lighting provided increases significantly from Category I through Category III. The systems are more fully described in References 1, 7, and 10. Figure 5.12 shows the FAA high- and medium-intensity lighting configurations, which conform to ICAO recommendations. There are three basic categories of *Approach Light Systems* (ALS) with adaptations of each category:

- **ALSF-1**—Approach Light System with sequenced flashing lights in ILS Cat I configuration. This configuration is no longer used by the FAA

- **ALSF-2**—Approach Light System with sequenced flashing lights in ILS Cat II configuration. The ALSF-2 may operate as a SSALR when weather conditions permit

- **SSALF**—Simplified Short Approach Light System with sequenced flashing lights

- **SSALR**—Simplified Short Approach Light System with runway alignment indicator lights

- **MALSF**—Medium Intensity Approach Light System with sequenced flashing lights

- **MALSR**—Medium Intensity Approach Light System with runway alignment indicator lights (FAA 1994)

It is important however to realize that especially where conditions are in the Category II and III range, the approach lighting system might be visible only on the initial part of the approach or from directly overhead. Slant visibility and weather conditions near the ground causing severely limited visibility are often so bad that no approach lighting is visible to the pilot on the final stages of approach, the first visual cues coming from lights in the touchdown zone. This effect is shown in Figure 5.13. Aircraft A is in the first position in which a full visual segment can be seen. As the aircraft continues to approach, the visual segment remains angularly constant and the length on the ground shrinks. In conditions of very poor visibility, the visual range of the approach light might always be less than the height of the aircraft from the ground. In such conditions, lights in the touchdown zone will provide the first visual cue to the pilot. Various stages of light intensity are usually controlled from the Air Traffic Control (ATC) tower to avoid the possibility of blinding the pilot on a night approach and to provide sufficient brilliance during a daylight approach in poor visibility.

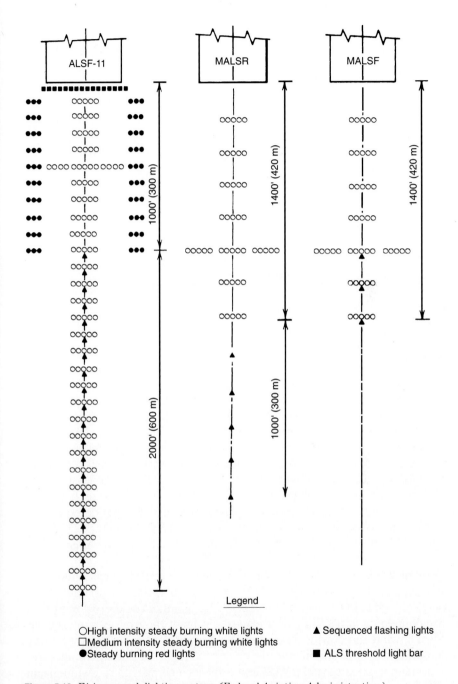

Figure 5.12 FAA approach lighting system. *(Federal Aviation Administration)*

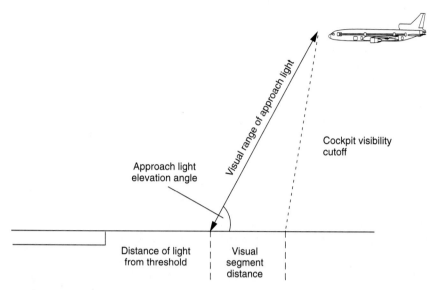

Figure 5.13 Visibility of approach lights.

Other visual approach aids

Approach lighting systems give the pilot information on the azimuth of the runway centerline and the location of the threshold. When large portions of the approach system are visible, the pilot is given a reference for checking yaw and roll. Other visual systems are available to provide information on the altitude of the aircraft with reference to the glidepath. The VASI (or VASIS) was the most commonly used visual aid designed to ensure minimum safe wheel clearances over the threshold and minimum safe clearances over obstacles in the final approach. The aim of approach visual aid systems is at least to provide information on the glideslope down to 200 feet (60 m) and preferably to give a reference throughout the entire final approach. Several different systems are in use, all of which use the same basic principles: two-bar VASIs, two-bar AVASIs, three-bar VASIs, three-bar AVASIs, TVASIs, and ATVASIs.

Two-bar VASI. This system consists of two symmetrically positioned pairs of wing bars of lights bracketing the origin point of the glidepath on the runway centerline in the touchdown zone. The limits of the correct approach are defined by upper and lower planes that meet the runway centerline at ideal upwind and downwind points. The upwind limit is defined by runway length and configuration; the downwind point by obstacle-clearance requirements. The individual VASI unit projects a composite red-white beam with the white portion of the beam above and the red below. Figure 5.14 shows the plan layout of a two- or three-bar VASI. It can be seen from the section that a pilot on the correct approach path of a two-bar system sees the upwind bars showing red and the downwind white. If on an overshoot path, both upwind and downwind bars show white, whereas if the approach is too low upwind and downwind indications are both red. Where the approach is much too low, the signal merges into a broad red band. The standard two-bar VASI is 12 lights, 3 at each of the four-bar positions.

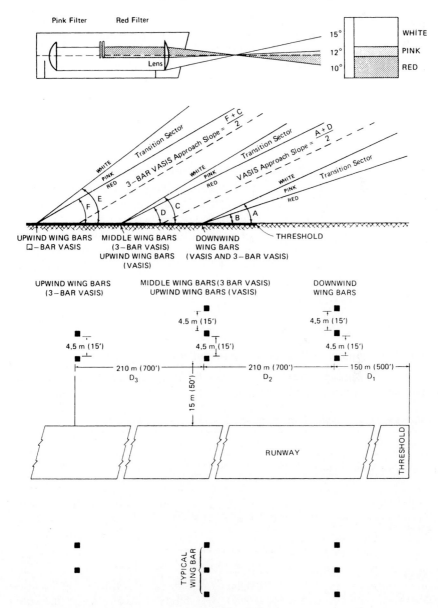

Figure 5.14 Typical VASIs lighting arrangement. *(ICAO Annex 14)*

Two-bar AVASI. Any system with less than the standard number of lights is an *Abbreviated Visual Approach Slope Indicator System* (AVASIS or AVASI). Abbreviation can be in the form of using only two lights at each of the four positions or more usually in an asymmetrical arrangement on one side of the runway centerline with two 3-light or two 4-light units.

Three-bar VASI. The introduction of very heavy wide-bodied aircraft meant that VASIs suitable for smaller aircraft were giving insufficient clearances for larger aircraft. When the vertical distance between the pilot's eye level and the landing gear in the flare attitude exceed 15 feet (4.5 m) on a nominal 3° approach, a three-bar VASI is required. As Figure 5.14 indicates, the upwind bar of the two-bar VASI becomes the middle bar, and a third bar is added upwind. Large aircraft use the upwind and middle bars; smaller aircraft use the middle and downwind bars. Table 5.6 indicates the visual display for various positions of the aircraft with reference to the correct glidepath.

TVASI. This system consists of 20 light units symmetrically arranged around the runway centerline, made up of a wingbar of 4 units bisected by 6 symmetrically placed longitudinal lights. Those units placed downwind of the wingbar are known as *fly-up* lights and those on the upwind side are *fly down* lights. The arrangement of lights, which are also composite beam units, is shown in Figure 5.15. When above the glidepath, the pilot sees the white wingbar and one, two, or three fly-up lights, depending on how far the aircraft is above the correct approach path. Below the glidepath, the wingbar also appears white with one, two, or three fly-up lights. When well below the glidepath, the wingbar and fly-up lights all appear red. The ATVASI is an abbreviated TVASI system with lights on one side of the runway only.

Precision approach path indicator system (MPI)

Although VASIs and TVASIs offer considerable visual assistance to pilots on final approach, they have since their inception been subject to a number of

TABLE 5.6 Visual Display from Three-Bar VASIS to Pilot on Approach

Positive Relative to Glide Path:	Downwind	Middle	Upwind
Small aircraft on downwind and middle-bar approach			
Far above	White	White	(White)
Above	White	White	(Red)
On	White	Red	(Red)
Below	Red	Red	(Red)
Large aircraft on upwind and middle-bar approach			
Above	(White)	White	White
On	(White)	White	Red
Below	(White)	Red	Red
Well below	(Red)	Red	Red

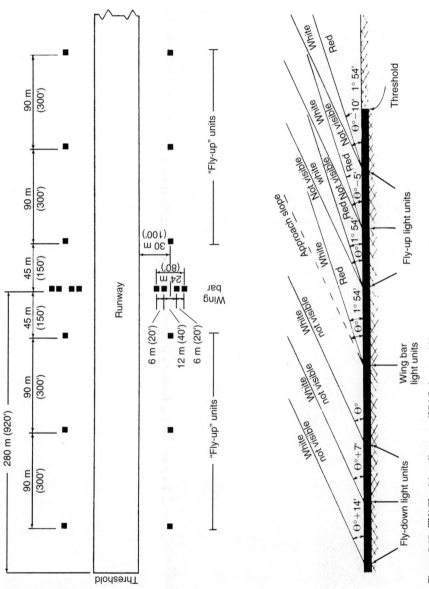

Figure 5.15 TVASIs siting diagram. *(ICAO Annex 14)*

criticisms. VASIs tend to give an oscillatory approach as the pilot moves between the upper and lower limiting approach planes, which results in a large touchdown scatter. The systems are imprecise below 200 feet (60 m), are of no help if a nonstandard approach is used, and in daytime haze, the red is difficult to determine from the pink zone, which can be up to ¾°. The three-bar VASI has an upper approach corridor that is 20 min steeper than the lower corridor. Both two- and three-bar configurations require considerable flight checking and maintenance to keep them in operation.

TVASI, which does not rely on color change except in a severe case of underflying error, avoids some problems of VASIs; it is also more suited to multipath approaches, and therefore more flexible. However, the systems are more complex than VASIs and require more difficult sitting procedures and more precise maintenance. They also fail to give a fail-safe reading if the downwind "fly-up" lights go out.

The PAPI system shown in Figure 5.16 was introduced to overcome these disadvantages. It is a two-color light system using sealed units that give a bi-colored beam, white in the upper part, red in the lower. The units are visible for up to 4.5 miles (7 km) from the threshold. The sealed units are much easier to set up and maintain, are capable of multipath interpretation, and are therefore more flexible. The approach is more precise than the VASI approach and is not subject to the same oscillation. Consequently, it has replaced VASI systems at most transport airports.

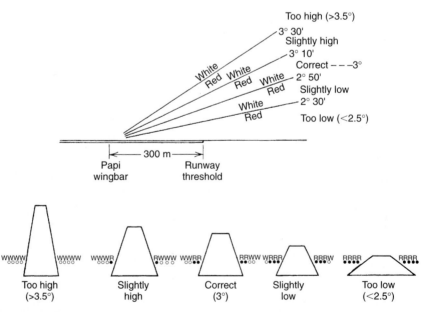

Figure 5.16 PAPI—visual indication to pilot.

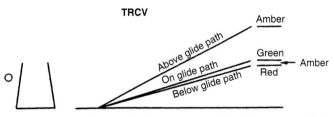

CAUTION: When the aircraft descends from green to red, the pilot may see a dark amber color during the transition from green to red.

Figure 5.17 Tri-color Visual Approach Slope Indicator. *(FAA Aeronautical Information Manual)*

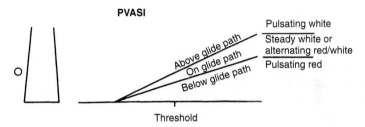

CAUTION: When viewing the pulsating visual approach slope indicators in the pulsating white or pulsating red sectors, it is possible to mistake this lighting aid for another aircraft or a ground vehicle. Pilots should exercise caution when using this type of system.

Figure 5.18 Pulsating Visual Approach Slope Indicator. *(ICAO 1977)*

U.S. VASI variations

The United States has some additional VASI variations and although not in widespread use their existence should be noted. The first of these projects is a three-color visual approach path into the final approach area. The tri-color visual approach slope indicator normally consists of a single light unit. The below glidepath indication is red, the above glidepath indication is amber and the on-glidepath indication is green, as shown in Figure 5.17. There is also a pulsating visual approach slope indicator which again is normally provided by a single light unit projecting a two-color visual approach path (red and white) with a steady white light to indicate on-glidepath. Any abnormal deviations either above or below the glidepath are indicated by pulsating white or pulsating red respectively (Figures 5.18). Pilots are, however, cautioned not to confuse the pulsating light indications with the lights of another aircraft or of a ground vehicle.

5.7 Airfield Inspections

First signs of airfield unserviceability might well arise as a result of the routine daily inspections, which are usually the responsibility of the airport owner. In some cases, the inspections are carried out by air traffic control personnel; in others by airport authority personnel.

Whoever carries out the inspections, however, the objectives will be the same:

- To check the movement area for any surface defects, obstructions, or debris
- To check that aircraft operating are serviceable

In most cases this will not require the person carrying out the inspection to leave the airport.

In some cases, the regulatory authorities might specify a minimum frequency of airport inspection (FAA 1993). The frequency of inspection at most airports would, in any case, be far in excess of any minimum, although it would depend very much on local conditions. At some busy airports, there might be an inspection every two hours while a less busy airport might be inspected only in the morning before commencing operations and shortly before darkness. The first inspection at any airport usually is carried out early in the day before flying operations have commenced, as soon as there is sufficient daylight to ensure that nothing will be overlooked. Special inspections of activities and facilities are usually carried out after receipt of a complaint, or after the occurrence of an unusual condition or event, e.g., aircraft accident.

It is absolutely vital that radiotelephone (RT) contact and liaison between ATC and the vehicle being used for inspection be faultfree since the inspection vehicle will be moving on the operating area, which includes runways and taxiways, where there might be a danger of collision with aircraft. The vehicle should be driven slowly enough for a thorough visual inspection to be made and, if necessary, stopped for a closer inspection of a particular area or for the removal of debris. During the inspection particular attention should be given to:

- Surface condition of runways, taxiways, runway holding points, stopways, and clearways
- Presence of standing water, snow, ice, sand, rubber deposits from aircraft tires, oil and fuel spillage
- Presence of debris or any part of the movement area
- Status of any work in progress on the airport and the presence of any associated materials, obstructions, ditches, and so on together with the correct ground hazard warnings
- Condition of daylight movement/indicator boards and signs including boundary markers

- Damage to light fittings, broken glass, and so on
- Growth of grass/plants or any other causes of lights being obscured
- Any congregation of birds or presence of animals or unauthorized persons likely to interfere with operations

In the event of any objects being discovered that are identified as coming from an aircraft, immediate steps would need to be taken to check recent departures with ATC, who will decide if it is necessary to send a message. Prior to darkness, an inspection to check the operation of all lighting systems should be carried out if night operations are to take place. This inspection will be particularly aimed at an examination of:

- Runway/taxiway lighting
- Obstruction lights
- Airport rotating beacon/identification beacon
- Traffic lights guarding the operating/movement area
- Visual approach slope indicators (VASIs)
- Approach lights visible from within the airport boundaries

A full approach lighting inspection can be carried out only by means of a flight check. With the larger lighting systems, some form of photographic record has to be made.

In view of the variety and complexity of all those aspects of an airport that need to be inspected, it is advisable to use a checklist, together with a map of the airport, so that a systematic record can be made of the results of the inspection. An example of a suggested checklist for an airfield inspection is given in Figure 5.19. Each airport will need to develop lists appropriate to its own particular circumstances. If any unserviceability is discovered as a result of an inspection, and the condition cannot be rectified immediately, then suitable NOTAM action has to be taken (FAA 1988; FAA 1993).

5.8 Maintaining Readiness

Maintenance management

The primary concern of operations management is to ensure the continuous availability of all operational services. To achieve this, a systematic approach to maintenance management is called for, the extent of which will depend on the type of operations at a particular airport. Clearly, a major air carrier hub will require a vastly more complex maintenance program than an airport dealing only with general aviation (GA) type operations. There

REGULARLY SCHEDULED INSPECTION CHECKLIST

DATE: _____   DAY: _____   ✓ Satisfactory
                                                   ✗ Unsatisfactory
TIME: _____   INSPECTOR : _____

FACILITES	CONDITIONS	✓	REMARKS
Pavement areas	Pavement lip over 3"		
	Hole 5" diam. 3" deep		
	Cracks/spalling/bumps		
	FOD: Gravel/debris/etc.		
	Rudder deposits		
	Ponding/edge dams		
Safety areas	Ruts/humps/erosion		
	Drainage/construction		
	Objects/frangible bases		
Markings and signs	Visible/standard		
	Hold lines/signs		
	Frangible signs		
Lighting	Obscured/dirty/faded		
	Damaged/missing		
	Inoperative		
	Faulty aim/adjustment		
Navigational aids	Rotating beacon		
	Wind indicators		
	REILs/VGSI systems		
Obstructions	Obstruction lights		
	Cranes/trees		
Fueling operations	Fencing/gates/signs		
	Fuel marking/labeling		
	Fire extinguishers		
	Grounding clips		
	Fuel leaks/Vegetation		
Snow & ice	Surface conditions		
	Snowbank clearance		
	Lights & signs obscured		
	NAVAIDS/fire access		
Construction	Barricades/lights		
	Equipment parking		
ARFF	Equipment/crew avail.		
	Communications/alarm		
Public protection	Fencing/gates		
	Signs		
Wildlife hazards	Dead birds		
	Flocks of birds/animals		

Figure 5.19 A scheduled inspection checklist.

are two essential aspects, however, that will be common to any airport maintenance program:

- A documented schedule of routine maintenance
- A comprehensive system of maintenance records, including costs

Many of the airport facilities such as radio communications, radio and radar approach aids, and airfield lighting are of such critical importance to flight safety that every effort has to be made to ensure that failures do not occur. Among the elements involved are:

- Radio communications (air/ground) transmitters and receivers
- Aeronautical fixed telecommunications
- Telephones
- Approach and landing aids
  ~Radio/radar
  ~Lighting
  ~Fire and rescue service
- Aircraft movement areas
- Power plant and distribution system

Preventive maintenance

The process of preventive maintenance is concerned mainly with regular inspection of a system and all its component parts with the objective of detecting anything likely to lead to a component or system failure, and taking appropriate action to prevent this happening. Such action might involve cleaning or replacement of parts on a predetermined schedule. Whatever the action called for, this cannot in the first place be determined unless a planned inspection schedule is established. An example of a preventive maintenance schedule for medium-intensity approach lighting is given in Table 5.8. Runway lighting is somewhat simpler to maintain, but the same systematic inspection is required (Table 5.9). Centerline and touchdown zone lighting will, of course, be much more vulnerable to damage as a result of being run over by aircraft. The fact that they are located below ground level also makes them vulnerable to water infiltration.

Electrical maintenance

There are few systems on an airport that do not depend in one way or another on electrical power. Indeed, the power requirements of modern airports are equal to those of small towns. Operational facilities, especially those concerned with aviation technical services, make heavy demands on

TABLE 5.7 Preventive Maintenance Inspection Schedule for Medium-Intensity Approach Lighting

Maintenance Requirement	Daily	Weekly	Monthly	Bimonthly	Semiannually	Annually	Unscheduled
1. Check for burned-out lamps	X						
2. Check system operation		X					
3. Replace burned-out lamps		X					
4. Check semiflush lights for cleanliness		X					
5. Record input and output voltages of control cabinet			X				
6. Clear any vegetation obstructing the lights			X				
7. Check angle of elevation of lights				X			
8. Check structures for integrity				X			
9. Check approach area for new obstructions				X			
10. Check photoelectric controls (if used)				X			
11. Check electrical distribution equipment					X		
12. Check insulation resistance of cable					X		
13. Check fuse holders, breakers, and contacts					X		
14. Replace all PAR 38 lamps after 1800 hours on maximum intensity							X

the public main supply. Standby power must be available to provide a secondary power supply to these essential services in the event of breakdown of the main supply. The arrangements for a secondary power supply depend, in part, on the switchover time requirements, that is, the time inter-

TABLE 5.8 Preventive Maintenance Inspection Schedule for Centerline and Touchdown Zone Lighting Systems.

Maintenance Requirement	Daily	Weekly	Monthly	Bimonthly	Semiannually	Annually	Unscheduled
1. Check for burned-out or dimly burning lights	X						
2. Repair or replace defective lights		X					
3. Clean lights with dirty lenses			X				
4. Check the intensity of selected lights			X				
5. Check the torque of mounting bolts				X			
6. Clean and service light; check electrical connections					X		
7. Check for water in the light base					X		
8. Remove lamps after 80 percent of service life							X
9. Remove snow from around fixtures							X
10. Check wires in saw kerfs							X

SOURCE: FAA 1982

TABLE 5.9 Recommended Switchover Times in the Event of Power Failure

Runway	Lighting	
	Aids requiring power	Maximum switchover time
Noninstrument	Visual approach slope indicators	2 minutes
	Runway edge	2 minutes
	Runway threshold	2 minutes
	Runway end	2 minutes
	Obstacle	2 minutes
Instrument approach	Approach lighting system	15 seconds
	Visual approach slope indicators	15 seconds
	Runway edge	15 seconds
	Runway threshold	15 seconds
	Runway end	15 seconds
	Obstacle	15 seconds
Precision approach category I	Approach lighting system	15 seconds
	Runway edge	15 seconds
	Runway threshold	15 seconds
	Runway end	15 seconds
	Essential taxiway	15 seconds
	Obstacle	15 seconds
Precision approach category II	Approach lighting system	1 second
	Runway edge	15 seconds
	Runway threshold	1 second
	Runway end	1 second
	Runway centerline	1 second
	Runway touchdown zone	1 second
	Essential taxiway	15 seconds
	Obstacle	15 seconds
Precision approach category III	(Same as category II)	

val between loss of power and the availability of a secondary supply. This could be critical in the case of precision approach aids or lighting as indicated in Tables 5.9 (FAA 1982b) and 5.10 (ICAO 1991a) respectively. The demands range from a maximum permitted interruption of 15 seconds to zero or completely uninterrupted supply as in the case of ILS localizer and glideslope for Category II and III approaches. The source of secondary supply usually is one or more diesel-driven generators. In the case of a zero switchover time requirement, the arrangement is to have these facilities supplied by a generator with a coupled energy storage flywheel.

The generator is driven by an electric motor. In the event of main failure, the generator derives the required driving power from the flywheel until the coupled standby diesel generator takes over the full load. To further safe-

TABLE 5.10 Recommended Switchover Times in the Event of
Power Failure—Ground-Based Radio Aids

Type of runway	Aids requiring power	Maximum switchover times
Instrument approach	SRE	15 seconds
	VOR	15 seconds
	NDB	15 seconds
	D/F facility	15 seconds
Precision approach category I	ILS localizer	10 seconds
	ILS glide path	10 seconds
	ILS middle marker	10 seconds
	ILS outer marker	10 seconds
	PAR	10 seconds
Precision approach category II	ILS localizer	0 second
	ILS glide path	0 second
	ILS inner marker	1 second
	ILS middle marker	1 second
	ILS outer marker	10 seconds
Precision approach category III	(Same as category II)	

guard the remote possibility of the generator failing, a second generator is coupled in parallel with the first.

Such stringent requirements as these call for an appropriate level of maintenance and alongside this a suitable level of workforce. The requirements for maintenance personnel in a typical airport electrical shop (Category II airport) are listed in Table 5.11.

Operational readiness—aircraft rescue and fire fighting

There is one essential element of the operating system, the aircraft rescue and fire fighting (ARFF) service, where maintaining readiness applies as much to personnel as to machines. Opportunities for ARFF personnel to carry out their assigned tasks under "real" conditions are fortunately rare; as a result, they can maintain readiness to deal with an aircraft accident only by constant practice. It is especially difficult to maintain a peak of performance under these circumstances, and it is for this reason that facilities should be provided for "hot fire" practices and also, if possible, for practices in smoke-filled confined spaces, preferably simulated aircraft interiors.

Some airports employ independent aircraft and fire fighting services. It will be important in these cases to carry out occasional tests of the ability of the personnel and equipment to meet the required performance criteria (see Chapter 12), and in particular to test their communications and coordination procedures.

TABLE 5.11   Manpower Requirements in a Typical Electrical Shop of a Category II Airport

Workshop office	High-voltage plants	Secondary power supply	Lighting systems	Low-voltage 24-hour service
1 leader of workshop 3 clerks	1 foreman 5 electricians	1 foreman 5 fitters	1 foreman 5 electricians 3 fitters 1 driver	1 foreman 41 electricians 15 fitters

SOURCE: Frankfurt Airport

Safety aspects

Whether maintaining electrical or mechanical systems, the nature of maintenance work exposes those who carry it out to certain risks, including natural phenomena such as lightning strikes while working out on the airfield. A comprehensive set of guidelines on dealing with risks should be drawn up by management. The nature of these risks can be indicated by an examination of some of the common causes of accidents:

- Working on equipment without adequate coordination with equipment users
- Working on equipment without sufficient experience on that equipment
- Failure to follow instructions in equipment manuals
- Failure to follow safety precautions
- Using unsafe equipment
- Failure to use safety devices
- Working at unsafe speed
- Poor housekeeping of work areas

Conclusion

Most air carrier airports will have a multiplicity of electrical and mechanical systems with large amounts of associated equipment requiring continuous checking and servicing by skilled maintenance personnel. The extent of this work will very much depend on the amount and type of aircraft operations, the weather categories in which the airport operates, and the number of passengers, visitors, and staff using the airport.

With the increasing sophistication (and complication) of the equipment used at airports, the maintenance requirement has escalated in recent years. As the effects of equipment failures have become more acute and more rapidly felt throughout the whole of the airport operating system, the need has grown for swift response from maintenance sections.

In response to this need, there is a growing use of automatic equipment/system monitoring to provide prompt warning of equipment failures and also to provide comprehensive performance records. Although automatic monitoring of certain airport operational facilities has long been accepted as a need (e.g., ILS, lighting), there is a growing awareness of the wider capabilities of this technique. The increasing availability and adaptability of micro computers will serve to further extend the use of automatic monitoring as part of a real time maintenance management information system. But a vital element for achieving the exacting state of readiness needed in air transport is for airports to have available the resources—workers and materials—to enable a rapid and effective response to any unserviceablities in the airport's operating infrastructure.

References

National Commission to Ensure a Strong Airline Industry. 1993. *Change, Challenge and Competition, A Report to the President and Congress.* Washington, DC.

Civil Aviation Authority (CAA). 1995. *Air Navigation: The Order and the Regulation, CAP3930.* London.

Code of Federal Regulations. 1994. *Certification and Operations: Land Airports Serving Certain Air Carriers, Part 139.* Washington, DC: Office of the Federal Register National Archives and Records Administration.

Federal Aviation Administration (FAA). 1981. *Debris Hazards at Civil Airports, AC150/5380-5A.* Washington, DC: Department of Transportation.

FAA. 1982a. *Guidelines and Procedures for Maintenance of Airport Pavements, AC150/5380-6.* Washington, DC: Department of Transportation.

FAA. 1982b. *Maintenance of Airport Visual Aid Facilities, AC150/5340-26.* Washington, DC: Department of Transportation.

FAA. 1986. *Standby Power for Non-FAA Airport Lighting Systems, AC150/5340-17B.* Washington, DC: Department of Transportation.

FAA. 1988. *Airport Safety Self-Inspection, AC150/5200-18B.* Washington, DC: Department of Transportation.

FAA. 1989. *Airport Design, AC150/5300-13.* Washington, DC: Department of Transportation.

FAA. 1991a. *Airport Winter Safety and Operations, AC150/5200-30A.* Washington, DC: Department of Transportation.

FAA. 1991b. *Runway Surface Condition Sensor Specification Guide, AC150/5220-13B.* Washington, DC: Department of Transportation.

FAA. 1992. *Airport Snow and Ice Control Equipment, AC150/5220-20.* Washington, DC: Department of Transportation.

FAA. 1993. *Notices to Airmen (NOTAMs) for Airport Operators, AC150/5200-28A.* Washington, DC: Department of Transportation.

FAA. 1994. *Airman's Information Manual (AIM).* Washington, DC: Department of Transportation.

International Civil Aviation Organization (ICAO). 1984. *Airport Maintenance Practices, Doc 9137 Part 9, 1st edition.* Montreal.

ICAO. 1990. *Annex 14—Aerodromes, Vol. 1.* Montreal.

ICAO. 1991a. *Annex 10—Aeronautical Telecommunications, Vol 1.* Montreal.

ICAO. 1991b. *Bird Control and Reduction, Doc 9137 Part 3, 3rd edition.* Montreal.

Chapter

6

Ground Handling

6.1 Introduction

The passenger and cargo terminals have been described as interface points between the air and ground modes (Ashford and Wright 1992). The position of the terminals within the general system has been shown conceptually in Figure 1.3, and the actual flow within the terminals in the more detailed system diagrams of the passenger and cargo terminals, Figures 8.1 and 10.7, respectively. Within the context of these diagrams, the movement of passengers, baggage, and cargo through the terminals and the turnaround of the aircraft on the apron are achieved with the help of those involved in the ground handling activities at the airport. These activities are carried out by some mix of the airport authority, the airlines, and special handling agencies depending on the size of the airport and the operational philosophy adopted by the airport operating authority. For convenience of discussion, ground handling procedures can be classified as either *terminal* or *airside* operations. Such a division is however only a convention, in that the staff and activities involved are not necessarily restricted to these particular functional areas. Table 6.1 lists those airport activities normally classified under ground-handling operations. The remainder of this chapter deals with these activities, but for convenience the major areas of baggage handling, cargo, security, and load control have been assigned to other chapters to permit a more extensive discussion of these items.

TABLE 6.1 The Scope of Ground Handling Operations

Terminal
Baggage check
Baggage handling
Baggage claim
Ticketing and check-in
Passenger loading/unloading
Transit passenger handling
Elderly and disabled persons
Information systems
Government controls
Load control
Security
Cargo

Airside

Ramp services	Supervision
	Marshaling
	Start-up
	Moving/towing aircraft
	Safety measures
On-ramp aircraft servicing	Repair of faults
	Fueling
	Wheel and tire check
	Ground power supply
	Deicing
	Cooling/heating
	Toilet servicing
	Potable water
	Demineralized water
	Routine maintenance
	Non-routine maintenance
	Cleaning of cockpit windows, wings,
	nacelles and cabin windows
Onboard servicing	Cleaning
	Catering
	In-flight entertainment
	Minor servicing of cabin fittings
	Alteration of seat configuration
External ramp equipment	Passenger steps
	Catering loaders
	Cargo loaders
	Mail and equipment loading
	Crew steps on all freight aircraft

6.2 Passenger Handling

Passenger handling in the terminal is almost universally entirely an airline function. In most countries of the world, certainly at the major air transport hubs, the airlines are in mutual competition. Especially in the terminal area the airlines wish to project a corporate image and passenger contact is almost entirely with the airline, with the obvious exceptions of the governmental controls of health, customs, and immigration. Airline influence is perhaps seen at its extreme in the United States, where individual airlines on occasions construct facilities (e.g., the United and TWA terminals at New York JFK). In these circumstances, the airlines play a significant role in the planning and design of physical facilities that they will operate. Even where there is no direct ownership of facilities, industry practice involves the designation of various airport facilities that are leased to the individual airlines operating these areas. Long-term designation of particular areas to an individual airline results in a strong projection of airline corporate image, particularly in the ticketing and check-in areas and even in the individual gate lounges (Figure 6.1).

A more common arrangement worldwide is for airlines to lease designated areas in the terminal, but to have a large proportion of the ground handling in the ramp area carried out by the airport authority, a special handling agency, or another airline. At a number of international airports, the airline image is considerably reduced in the check-in area when *common user terminal equipment* (CUTE) is used to connect the check-in clerk to the airline computers. The use of the CUTE system can substantially reduce the requirements for numbers of check-in desks particularly where there is a large number of airlines, where some airlines have very light service schedules or where the airline presence is not necessary throughout the whole day. Desks are assigned by computers on a need basis. Check-in

Figure 6.1   Airline designated check-in desks.

Figure 6.2 Computer-assigned CUTE passenger check-in desks at Munich Airport.

areas are vacated by one airline and taken up by another based on departure demand. The airline's presence at check-in desks is displayed on overhead panels which are activated when an airline logs on to the CUTE system. The check-in desks at Munich shown in Figure 6.2 operate on the CUTE system. The airside transfer passenger steps (Figure 6.3) and loading bridges (Figure 6.4) might be operated by the airline on a long-term leasing arrangement or by the airport authority or handling agency at a defined hiring rate. Specialized passenger transfer equipment such as the overwing loading bridges at Schiphol (Figure 6.5) require experienced handling, but even these are normally operated by the airlines.

Apron passenger-transfer vehicles are usually of the conventional bus type. Both airline and airport ownership and operation are common, airline operation being economically feasible only where the carrier has a large number of movements. Figure 6.6 shows a typical airline operation. Where a more sophisticated transfer vehicle, such as the mobile lounges shown in Figure 6.7 are used, it is usual for the operation to be entirely in the hands of the airport authority[1].

---

[1]Although mobile lounges have been used for more than 30 years, they are still relatively rare. Many of the claimed advantages have not materialized and with the advent of high-capacity, wide-bodied aircraft, disembarkation time can be annoyingly slow for the passengers and expensive for the airline.

Figure 6.3 Airline passenger steps.

6.3 Ramp Handling

During the period that an aircraft is on the ground, either in transit or on turnaround, the apron is a center of considerable activity. Some overall supervision of activities is required to ensure that there is sufficient coordination of operations to avoid unnecessary ramp delays. This is normally

Figure 6.4 The elevating passenger bridge at Munich Airport. This is a mobile facility designed to accommodate a full range of commercial jet aircraft.

Figure 6.5 Overwing passenger loading bridge at Schiphol Amsterdam Airport.

Figure 6.6 Apron passenger transport bus.

carried out by a ramp coordinator or dispatcher who monitors departure control. Marshaling is provided to guide the pilot for the initial and final maneuvering of the aircraft in the vicinity of its parking stand position. In the delicate process of positioning the aircraft, the pilot is guided by internationally recognized hand signals from a signalman positioned on the apron (Figure 6.8). Where nose-in docking is used next to a building, self-docking guides such as the *Aircraft Parking and Information System* (APIS) using optical moiré technology or the *Docking Guidance System* (DGS) using sensor loops in the apron pavement enabling the pilot to bring the aircraft to a precise location to permit the use of loading bridges (Ashford and Wright 1992). Marshaling includes the positioning and removal of wheel chocks, landing gear locks, engine blanking covers, pitot covers, surface control locks, cockpit steps, and tail-steadies. Headsets are provided to permit ground to cockpit communication and all necessary electrical power for aircraft systems is provided from a ground power unit. Where the aircraft is

Figure 6.7 Mobile lounge for passenger transfer across apron.

Figure 6.8 Ground signalman marshaling aircraft. *(IATA)*

to spend an extended period on the ground the marshaling procedure includes arranging for remote parking or hangar space.

The ramp handling process also includes the provision, positioning, and removal of the appropriate equipment for engine starting purposes. Figure 6.9 shows an engine air-start power unit suitable for providing for a large passenger aircraft.

Safety measures on the apron include the provision of suitable fire-fighting equipment and other necessary protective equipment, the provision of security personnel where required, and notifying the carrier of all damage to the aircraft that is noticed during the period that the aircraft is on the apron.

Frequently there is a necessity for moving an aircraft, requiring the provision and operation of suitable towing equipment. Tow tractors might be needed simply for pushing out an aircraft parked in a nose-in position or for more extensive tows to remote stands or maintenance areas. Figure 6.10 shows a tractor suitable for moving a large passenger aircraft. It is normal aircraft-design practice to ensure that undercarriages are sufficiently strong to sustain towing forces without structural damage. Tow tractors must be capable of moving aircraft at a reasonable speed, (12 mph [20 kmh] approximately) over considerable taxiway distances. As airports grow larger and decentralized in layout, high-speed towing vehicles capable of operating in excess of 30 mph (48 kmh) have been developed although speeds of 20 mph (32 kmh) are more usual. Usually aircraft that are being towed have taxiway

Figure 6.9 Mobile apron engine air-start power unit.

Figure 6.10 Aircraft-tow tractor.

priority once towing has started. Therefore, reasonable tow speeds are necessary to avoid general taxiing delays.

6.4 Aircraft Ramp Servicing

Most arriving or departing aircraft require some ramp services, a number of which are the responsibility of the airline station engineer. When extensive servicing is required, many of the activities must be carried out simultaneously.

Fault servicing

Minor faults that have been reported in the technical log by the aircraft captain and that do not necessitate withdrawal of the aircraft from service are fixed under the supervision of the station engineer.

Fueling

The engineer, who is responsible for the availability and provision of adequate fuel supplies, supervises the fueling of the aircraft, ensuring that the correct quantity of uncontaminated fuel is supplied in a safe manner. Supply is either by mobile truck (Figure 6.11) or from the apron hydrant system (Figure 6.12). Many airports use both systems to ensure competi-

Figure 6.11 Mobile apron tanker.

Figure 6.12 Mobile dispenser to fuel aircraft from apron hydrant system.

tive pricing from suppliers and to give maximum flexibility of apron operation. Oils and other necessary equipment fluids are replenished during the
fueling process.

Wheels and tires

A visual physical check of the aircraft wheels and tires is made to ensure
that no damage has been incurred during the last takeoff/landing cycle and
that the tires are still serviceable.

Ground power supply

Although many aircraft have auxiliary power units (APU) that can provide
power while the aircraft is on the ground, there is a tendency for airlines to
prefer to use ground electrical supply to reduce fuel costs and to cut down
apron noise. At some airports the use of APUs is severely restricted on environmental grounds. Typically, ground power is supplied under the supervision of the station engineer by a mobile unit (Figure 6.13). Many airports
also can supply power from central power supplies that connect to the aircraft either by apron cable or by cable in the air-bridge structure.

Deicing and washing

Figure 6.14 shows a typical multiuse vehicle suitable for spraying the
fuselage and wings with deicing fluid and for washing the aircraft, especially the cockpit windows, wings, nacelles, and cabin windows. This self-
propelled tanker unit provides a stable lift platform for spraying or for

Figure 6.13 Mobile ground power unit.

Figure 6.14 Deicing washer vehicle.

various maintenance tasks on conventional and wide-bodied aircraft. At some airports, such as Munich and Luleî, aircraft are run through huge deicing gantries on specially designed deicing aprons sited close to the departure threshold. Such facilities permit the recapture and recycling of deicing fluid.

Cooling/heating

In many climates where an aircraft is on the apron for some time without operation of the APU, auxiliary mobile heating or cooling units are necessary to maintain a suitable internal temperature in the aircraft interior (Figure 6.15). The airline station engineer is responsible for ensuring the availability of such units.

With increasing fuel costs and environmental concern, much interest has been displayed in centralized compressed air units delivering air to the aircraft gate positions (usually called fixed air supply, or *preconditioned* air) and to mobile compressors at the gates (known simply as compressed air systems). Pneumatic systems can deliver high-pressure air for both heating and cooling and for air-starting the engines. Where fixed air systems are used, cockpit controls can ensure either internal heating or cooling on an individual aircraft basis, depending on the requirement. Studies indicate that the high cost of running aircraft APUs now mean that fixed air systems can completely recover capital costs from the savings of two years or normal operation (Krzac 1981).

Other servicing

Toilet holding tanks are serviced externally from the apron by special mobile pumping units. Demineralized water for the engines and potable water are also replenished during servicing.

Onboard servicing

While external aircraft servicing is being carried out, there are simultaneous onboard servicing activities, principally cleaning and catering. Very high levels of cabin cleanliness are achieved by:

- Exchange of blankets, pillow, and headrests
- Vacuuming and shampooing carpets
- Clearing ashtrays and removal of all litter
- Restocking seatback pockets
- Cleaning and restocking galleys and toilets
- Washing all smooth areas, including armrests

Catering

Personnel clear the galley areas immediately after the disembarkation of the incoming passengers. After the galley has been cleaned, it is restocked and a

Figure 6.15 Fixed ground cooling unit, attached to air bridge.

secondary cleaning takes care of spillage during restocking. Internationally agreed standards of hygiene must be met in the handling of food and drink from their point of origin to the passenger. Where route stations are unable to meet either quality or hygiene standards, catering supplies are often brought from the main base. Figure 6.16 shows the loading operation of a catering truck. These are usually constructed from a standard truck chassis with a closed van body that can be lifted up by a hydraulic scissor lift powered by the truck engine. Two different types of catering trucks are available: low lift vehicles suitable for servicing narrow-bodied aircraft up to 11.5 feet [3.5 m] doorsill height and high lift vehicles for loading wide-bodied jets.

6.5 Ramp Layout

During the design phase of a commercial air transport aircraft, considerable thought is given to the matter of ramp ground handling. Modern aircraft are very large, complicated, and expensive. Therefore, the apron servicing operation is also complicated and consequently time-consuming. Unless the ramp servicing procedure can be performed efficiently with many services being carried out concurrently, the aircraft will incur long apron turnaround times during which no productive revenue is earned. Inefficient ramp servicing can lead to low levels of aircraft and staff utilization and a generally low level of airline productivity. The complexity of the apron operation be-

Figure 6.16 Catering truck in loading position.

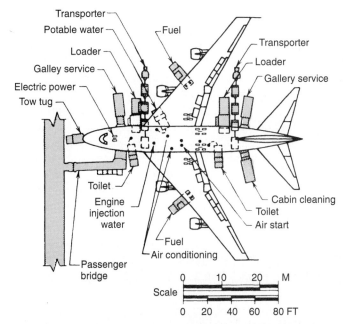

Transporter
Potable water
Loader
Galley service
Electric power
Tow tug
Fuel
Transporter
Loader
Gallery service

Toilet
Engine
injection
water
Cabin cleaning
Toilet
Air start
Fuel
Air conditioning
Passenger
bridge

Scale
0   10   20   M
0   20   40   60   80 FT

Figure 6.17 Ramp layout for servicing B747 SP. *Note:* Under normal conditions, external electrical power, air start, and air conditioning are not required when the auxiliary power unit is used. *(Boeing Airplane Company)*

comes obvious when Figure 6.17 is examined. This shows the apron positions typically designated for servicing and loading equipment for a Boeing 747. It can be seen that the aircraft door and servicing point layout has been arranged to permit simultaneous operations during the short period that the vehicle is on the ground during turnaround service. The ramp coordinator is required to ensure that suitable equipment and staff numbers are available for the period the aircraft is likely to be on the ground. Complicated as Figure 6.17 is, it hardly shows the true complexity of the problem. Because the ground equipment is necessarily mobile, the neat static position shown is made less easy by problems of maneuvering equipment into place. Positioning errors can seriously affect the required free movement of cargo trains, transporters, and baggage trains.

Particular attention must be paid to the compatibility of apron handling devices with the aircraft and other apron equipment. The sill height of the aircraft must be compatible with passenger and freight loading systems. In the case of freight, there is the additional directional compatibility requirement. Transporters must be able to load and unload at both the aircraft and the terminal onto beds and loading devices that are compatible with the vehicles' direction of handling. Many transporters can load or unload in the one direction only. The receiving devices must be oriented to accept this direction.

Most mobile equipment requires frequent maintenance. In addition to normal problems of wear, mobile apron equipment is subject to increased damage from minor collisions and misuse that do not occur in the same degree with static equipment. Successful apron handling might require a program of preventive maintenance on apron equipment and adequate backup in the inevitable case of equipment failure.

Safety in the ramp area is also a problem requiring constant attention. The ramps of the passenger and cargo terminal areas are high activity locations with much heavy moving equipment in a high noise environment. Audible safety cues, such as the noise of an approaching or backing vehicle, are frequently not available to the operating staff, who are likely to be wearing ear protection. Very careful training of the operating staff is required, and strict adherence to designated safety procedures is necessary to prevent serious accidents (IATA, 1995).

6.6 Departure Control

The financial effects of aircraft delay fall almost entirely upon the airline. The impact of delays in terms of added cost and lost revenue can be very high. Consequently, the functions of departure control, which monitors the conduct of ground handling operations on the ramp (not to be confused with Air Traffic Control departure), are almost always kept under the control of the airline or its agent. Where many of the individual ground handling functions are under the control of the airport authority, there will also be general apron supervision by the airport authority staff to ensure efficient use of authority equipment.

The complexity of a ramp turnaround of an aircraft is indicated by the critical path diagram shown in Figure 6.18. Even with the individual servicing functions shown in simple form, it is apparent that many activities occur simultaneously during the period the aircraft is on the ramp. This functional complication is a reflection of the physical complexity found on the ramp (Figure 6.17).

The ramp coordinator in charge of departure control must frequently make decisions that trade off payload and punctuality. Figure 6.19 shows the effect of intervention by the departure control in the case of cargo loading equipment breakdown. Figure 6.19a indicates satisfactory completion of task within a scheduled turnaround time of 45 minutes. Figure 6.19b shows a 10-minute delay due to equipment breakdown being reduced to a final ramp delay of only 5 minutes by the decision not to load nonrevenue airline company stores.

6.7 Division of Ground Handling Responsibilities

There is no hard-and-fast rule that can be applied to the division of liability for ground handling functions at airports. The responsibility varies not only

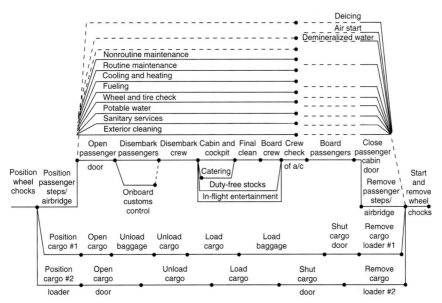

Figure 6.18 Critical path of turnaround ground handling for a passenger transport aircraft taking cargo.

from country to country, but also between airports in the same country. At some facilities (e.g., Frankfurt, Hong Kong, and Genoa), ground handling is largely an airport authority function. Authority staffs are entirely responsible for most ramp handling activities and for passenger baggage handling. A facility of this nature, dominated by airport authority activity, results in high employment levels for the airport with consequent personnel saving to airlines. In the United States, the converse situation is almost universal. The functions of ground handling are devolved from the airport authority. Handling is most usually carried out by the airlines. Some airlines, through interline agreements, carry out the handling activities for others. For example, USAir performs all ground handling at Los Angeles for British Airways, with the exception of ticketing/check-in, which is carried out by BA's own staff. At New York JFK, United Airlines handles a number of non-U.S. carriers. At some airports, ground handling companies have been set up to provide a service for airlines that would find it uneconomic to provide their own handling. For example, at Manchester International Airport, Servisair, a specialist company, handles all check-in except for the main carrier, British Airways. On the ramp, functions such as marshaling, steps, loading and unloading of baggage and cargo, and engine starts are carried out by the handling companies. At London Gatwick, two airlines set up a special handling company, Gatwick Handling, which performs all terminal and ramp handling functions for a number of other airlines. Handling com-

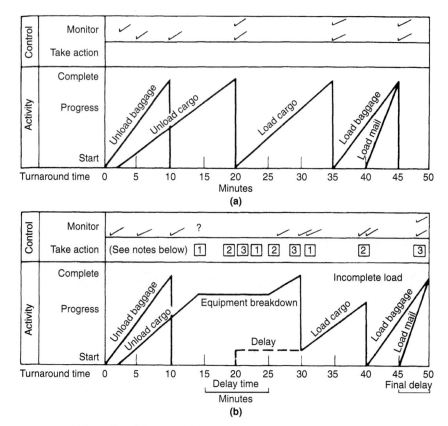

Figure 6.19 Effect of breakdown and delay on apron dispatch, *(a)* Activity normal, no control action; *(b)* delay through breakdown control required. Action 1 Assess the nature of the problem and how long the problem (breakdown of the cargo loader) will take to sort out. Action 2 Take corrective action immediately or call equipment base and ask engineer to come to the aircraft immediately or call up a replacement loader. Action 3 Advise all those other sections/activities that will be affected by the breakdown and give them instructions as necessary (e.g., notify movement control of a delay, tell passenger service to delay boarding, etc.)

panies also exist at U.S. airports. For example, Allied at New York JFK handles a number of nonbased foreign carriers. Table 6.2 shows how ground handling varies for a number of selected airports.

In theory, some economic advantages of scale could be expected from centralized ground handling operations. One body operating airportwide should be able to plan staff and equipment requirements with less relative peaking and probably with less duplication of facilities. Similar economies could be expected with standby equipment. Routine preventive maintenance should also be less expensive with a smaller proportion of the equipment out of service for repairs. However, the advantages that accrue from these areas are most likely to be countered by the disadvantages that accrue from too centralized an operation and lack of competition. Airports using only one ground handling organization are also vulnerable for severe industrial action from a relatively small group of workers. Gains in efficiency might be more than lost in unreasonable wage claims, and it might be difficult to introduce any level of competition once a monolithic agency has been set up. The Commission of the European Commission (EC) has introduced regulations that discourage or prevent monopoly positions of ground handling within the Community. The very scale of large airports to a degree negates the idea of being able to operate ground equipment from one pool. Physically, the total facility will probably have to be broken into a number of relatively self-sufficient and semiautonomous organizations based on the various parts of a single large terminal or on the individual unit terminals of a decentralized design.

In general, the ground handling function is not an area of considerable profit for an airport authority. Labor and equipment costs are high, and in general, revenues either barely cover attributable expenses or, as in many cases, are actually less than costs. These losses often are cross-subsidized using either revenues from other traffic areas, such as landing fees or non-traffic concession revenues.

6.8 Control of Ground Handling Efficiency

The extreme complexity of the ground handling operation requires skilled and dexterous management to ensure that staff and equipment resources are used at a reasonable level of efficiency. As in most management areas, this is achieved by establishing a system of control that feeds back into the operation when inefficiencies appear. The method of control used at any individual airport depends on whether the handling is carried out by the airline itself, by a handling agency such as another airline, or by the airport authority.

Four major reporting tools help to determine whether reasonable efficiency is being maintained and permit the manager to discern favorable and unfavorable operational changes.

Monthly complaint report

Each month, a report is prepared that shows any complaints attributable to ground handling problems. The report contains the complaint, the reason behind any operational failure, and the response to the complainant.

Monthly punctuality report

Each month the manager in charge of ground handling prepares a report of all delays attributable to the ground handling operation. In each case, the

TABLE 6.2 Distributions of Responsibilitie for Ground Handling Operations—Selected Airports

	Vancouver, Canada	Munich, Germany	Rio de Janeiro Int., Brazil	Manchester, U.K.	Bahrain, Bahrain	Birmingham, U.K.	Stockholm Arlanda, Sweden	London City, U.K.	Vienna, Austria	Copenhagen, Denmark	Shannon, Ireland	Boston, Logan, U.S.A.	Nice, France	Dusseldorf, Germany	Nairobi, Kenya	Bristol, U.K.	Changi, Singapore	Amsterdam, Netherlands	Dublin, Ireland	Toronto, Canada	Brussels, Belgium	Milan, Italy	Annaba, Algeria	Sydney, Australia	Atlanta Hartsfield, U.S.A.
Terminal																									
Baggage check-in	B	A	E	D	C	D	E	D	I	D	E	E	F	E	B	D	E	H	E	E	D	A	B	E	B
Baggage handling: Inbound/Outbound	B	A	E	J	C	A	E	A	A	D	E	E	D	A	E	A	E	D	E	E	D	A	B	E	B
Baggage claim	B	A	B	D	C	A	E	A	A	D	E	E	D	A	B	A	F	D	E	E	D	A	B	E	B
Passenger check-in	B	A	E	D	C	D	E	D	I	D	E	E	J	E	B	D	E	J	E	E	D	A	B	E	B
Transit passenger handling	B	B	E	D	C	D	E	D	A	D	E	D	F	E	B	D	E	E	E	E	D	A	I	E	B
Elderly and disabled passengers	B	B	D	L	C	D	J	D	A	D	E	F	D	G	E	I	E	K	E	E	D	A	A	E	B
Flight information systems	B	A	A	A	A	A	A	A	A	A	A	B	A	A	A	L	A	A	A	I	A	A	A	I	B
Ground transport information systems	A	A	A	A	C	A	A	A	A	A	A	A	A	A	A	A	E	A	A	A	D	A	A	A	A
Security if not police(P)	B	-	I	A	-	A	A	A	A	A	A	-	-	F	A	A	J	A	A	-	A	A	A	I	A
Airside																									
Ramp services																									
Supervision	B	A	I	D	C	F	J	A	A	B	E	A	A	F	A	A	E	J	E	E	D	A	B	E	E
Marshaling	B	A	B	A	A	A	A	A	A	A	M	B	-	A	A	A	D	A	C	E	A	A	B	E	G
Start up	B	B	B	D	C	E	E	A	A	D	E	B	B	-	B	L	E	E	E	E	A	A	B	E	-
Ramp safety control	A	A	A	A	A	A	A	A	A	I	C	A	D	A	A	A	F	A	I	A	D	A	B	F	G

Service	1	2	3	4	5	6	7	8	9	10	11	12	13	14	15	16	17	18	19	20	21	22	23	24
On-ramp aircraft servicing																								
Aircraft fault repair	B	E	D	B	B	B	E	E	E	D	D	B	D	B	E	B	E	E	E	M	E	D	B	B
Fueling	E	D	D	B	A	D	D	E	E	E	E	D	D	E	E	D	D	D	D	C	D	D	D	B
Wheel and tire check	B	B	A	B	A	B	D	E	E	E	E	B	B	E	E	B	E	B	B	M	B	D	D	B
Ground power supply	E	-	A	B	A	F	D	E	E	E	E	E	E	E	E	A	E	A	F	C	D	D	A	B
Deicing	H	-	D	-	-	-	D	E	E	E	E	-	-	E	E	D	D	E	-	-	-	D	A	B
Cooling/Heating	B	-	I	B		E	D	E	E	E	E	E	D	E	E	A	E	E	E	C	D	D	B	B
Toilet servicing	E	E	I	I	B	E	D	E	E	E	E	F	D	E	E	E	A	E	E	C	D	D	B	B
Potable water	B	E	I	I	A	F	D	E	E	E	E	B	B	E	E	E	A	E	E	C	D	D	A	B
Demineralized water	B	E	A	I	A	E	D	E	E	E	E	E	E	E	E	A	A	E	C	C	B	D	A	B
Routine aircraft maintenance	B	E	B	B	B	B	D	E	E	E	B	B	D	B	E	A	D	D	B	M	B	D	B	B
Non Routine aircraft maintenance	E	E	B	-	B	B	D	E	E	E	B	E	E	-	E	D	B	-	B	M	B	D	B	B
Exterior aircraft cleaning	E	E	B	B	B	E	D	E	E	E	E	E	E	B	E	D	I	D	D	M	D	D	B	B
On-board servicing																								
Cabin and cockpit cleaning	E	E	B	A	A	D	D	E	E	E	E	D	D	B	E	A	I	D	E	C	E	E	C	B
Catering	E	E	A	C	C	D	D	E	E	A	D	D	E	A	A	D	B	D	E	C	E	E	D	B
In-flight entertainment	E	E	B	B	B	E	D	E	E	B	B	B	B	B	E	B	B	B	B	B	B	B	D	B
Minor servicing of cabin fittings	E	E	B	B	A	B	D	E	E	B	E	E	E	B	E	B	B	B	B	M	B	E	B	B
Changing seat configuration	B	E	B	A	A	E	D	E	E	E	B	B	E	B	E	B	B	B	B	B	E	E	B	B
External ramp equipment provision and manning																								
Passenger steps	E	E	A	C	A	D	D	E	E	A	D	A	A	A	A	D	I	D	C	C	C	D	A	B
Catering loaders	E	E	A	C	C	D	D	E	E	A	D	A	D	D	A	D	D	D	D	C	D	E	D	B

TABLE 6.2 Distributions of Responsibilities for Ground Handling Operations—Selected Airports (Continued)

	Atlanta Hartsfield, U.S.A.	Sydney, Australia	Annaba, Algeria	Milan, Italy	Brussels, Belgium	Toronto, Canada	Dublin, Ireland	Amsterdam, Netherlands	Changi, Singapore	Bristol, U.K.	Nairobi, Kenya	Dusseldorf, Germany	Nice, France	Boston, Logan, U.S.A.	Shannon, Ireland	Copenhagen, Denmark	Vienna, Austria	London City, U.K.	Stockholm Arlanda, Sweden	Birmingham, U.K.	Bahrain, Bahrain	Manchester, U.K.	Rio de Janeiro Int'l., Brazil	Munich, Germany	Vancouver, Canada
Mail and equipment loaders	C	E	B	A	D	E	E	D	D	D	E	A	D	E	E	D	A	A	D	-	C	C	E	A	B
Crew steps	E	E	B	A	D	E	E	D	D	F	E	I	D	E	E	D	A	A	D	E	C	C	D	B	B
Apron passenger busses and mobile lounges	C	F	A	A	D	E	E	A	D	A	A	A	A	L	D	A	A	A	A	A	C	A	C	A	B

KEY:

A ... Airport
B ... Airlines
C ... Handling company for airport
D ... Handling company for airline
E ... Combination of B & D
F ... Combination of A, B & D
G ... Combination of A,B, C & D
H ... Combination of B, C & D
I ... Combination of A & B
J ... Combination of C & D
K ... Combination of A, B & C
L ... Combination of A & D
M ... Combination of B & C

particular flight is identified, with its scheduled and actual time of departure. The reason for each delay is detailed. The monthly summary should indicate measures taken to preclude or reduce future similar delays. Typical aircraft servicing standards are 3 to 60 minutes for a transit operation and 90 minutes for a turnaround.

Cost analysis

The actual handling organization will, at least on a quarterly basis, analyze handling costs. These costs should include capital and operating costs. For large airlines, airports, and handling agencies, this can be achieved fairly easily on a monthly basis by a computerized management reporting system that allocates expenditures and depreciation to management cost centers. For smaller organizations, computerized financial systems might not yet be available. Manual checks of handling costs should be computed at least on a quarterly basis. It is normal practice to have budgeted expenditures in a number of categories. Variances between budget and actual expenditure require explanation.

General operational standards

To ensure an overall level of operational acceptability, periodic inspections of operations and facilities must be made. This is especially important for airlines carrying out their own handling away from their main base or at airports where they are handled by other organizations. For the airport operation, it is equally important. Whether or not the handling is carried out by the airport, the general standards reflect on airport image. Inspections ensure that agreed standards are maintained and highlight areas where standards are less than desirable. Table 6.3 shows the form of checklist used by an international airline to ensure that handling at outstations conforms with company standards. The airport operator should maintain a similar checklist for all major airlines operating through the airport that do their own handling, omitting the areas related to administration and accounting. In all areas possible, the evaluation should be carried out using quantitative measures. Subjective measures should be avoided since they are not constant between evaluators and might not be constant over time even with a single evaluator.

6.9   General

Ground handling of a large passenger aircraft requires much specialized handling equipment and the total handling task involves considerable staff and labor inputs. Good operational performance implies a high standard of equipment serviceability. In northern climates, it is usual to assume that

TABLE 6.3 Checklist for Monitoring the Adequacy of Ground Handling

Passenger Services: Check-in
  Operational adequacy of general check-in desks
  First-class check-in service
  Waiting time at check-in
  Seat selection procedure
  Information display
  Courtesy and ability of check-in staff
  Passenger acceptance control
  Standby control, late passengers, overbooking, rebates
  Acceptance of excess, special, and oversized baggage
  Baggage tagging, including transfer, first class
  Security of boarding passes/ticketing/cash and credit vouchers
  Minimum and average check-in times
  Preparation of passenger lists
  Control of catering orders
  Ticket issues and reservations
Passenger Services: Security
  Personal search or scan efficiency
  Hand baggage search efficiency
  Inconvenience level and waiting times
Passenger Services: Escort and Boarding
  Effectiveness of directions and announcements
  Staff availability for inquiries at waiting and boarding points
  Assistance at governmental control points
  Control of boarding procedure
  Liaison level between check-in and cabin staff
  Service levels of special waiting lounges for premium ticket holders
  Special handling: minors, handicapped
Passenger Services: Arrivals
  Staff to meet flight
  Information for terminating and transfer passengers
  Transfer procedures
  Assistance through government control points
  Special passenger handling: minors, handicapped
  Baggage delivery standards
  Assistance at baggage delivery
Passenger Services: Delayed/Diverted/Canceled Flights
  Procedures for information to passengers
  Procedures for greeters
  Messages including information to destination and en route points
  Procedures for rerouting and surface transfers
  Meals, refreshments and hotel accommodation
Passenger Services: Baggage Facilities
  Compilation of lost or damaged reports
  Baggage tracing procedures
  Claims and complaints procedures
Passenger Services: Equipment
  Check security and condition of all equipment: scales, reservations printer, seat plan stand,
    ticket printer, credit card imprinter, calculators, etc.
  Condition and serviceability of ramp vehicles
  Serviceability and appearance of ramp equipment
  Maintenance of ramp equipment and vehicles
  Control of ramp equipment and vehicles

TABLE 6.3 Checklist for Monitoring the Adequacy of Ground Handling (Continued)

Driving standards and safety procedures
Communications: telephones, ground-air radio, ground-ground radio
Ramp Handling: Aircraft Loading/Unloading
  Care of aircraft exteriors, interiors and Unit Load Devices
  Adequacy of loading instructions and training
  Ramp equipment planning and availability
  Positioning of equipment to aircraft
  Loading and unloading supervision
  Securing, restraining and spreading loads
  Operation of load equipment
  Operation of aircraft onboard systems
  Securing partial loads
  Ramp security
  Ramp safety
  Pilferage and theft
Ramp Handling: Cleaning/Catering
  Standard of cockpit and cabin cleaning/dressing
  Toilet/potable water servicing
  Catering loading/unloading
  Availability of ground air
  Air jetty operations
Ramp Handling:Load Control (for airline only)
  Load sheet accuracy and adequacy of presentation
  Load planning
  Advance zero fuel calculation and flight preparation
Ramp Handling: Aircraft Dispatch
  Punctuality record
  Turnaround/transit supervision
  Passenger release from aircraft
  Passenger waiting time at boarding point
  Logs and message files
  Accuracy of records of actual departure times
  Flight plan, dispatch meteorological information
Ramp Handling: Postdeparture
  Accuracy and time of dispatch of postdeparture records and messages
Cargo Handling: Export
  Acceptance procedures
  Documentation: procedures and accuracy
  Reservations: procedures and performance
  Storage: procedures and performance
  Makeup of loads: procedures and performance
  Check weighing
  Palletization and containerization: procedures and performance
Cargo Handling: Import
  Breakdown of pallets/containers: procedures and performance
  Customs clearance of documents
  Notification of consignees
  Dwell time of cargo
  Lost/damaged cargo procedures
  Proof of delivery procedures
  Handling of dangerous goods procedures
  Handling of restricted goods procedures
  Handling of valuable consignments procedures

TABLE 6.3 Checklist for Monitoring the Adequacy of Ground Handling (Continued)

Handling of live animals procedures
Handling of mail
Administration of Ground Handling
   Office appearance
   Furniture and equipment condition
   Inventory records: ramp equipment/vehicles/office equipment/furniture
   Budgeting: preparation and monitoring
   Control of cash/invoices/tickets/accounting/sales returns/strongboxes/keys/airport records/
     stationery
   Complaints register
   Staff appearance
   Condition of manuals/local instructions/emergency procedures/standing orders/general
     office files

equipment will be serviceable for 80 percent of the time during the winter and 85 percent during the summer. Backup equipment and maintenance staff must be planned for the periods of unserviceability. Most airlines operate failure maintenance procedures rather than extensive preventive maintenance programs. Availability of equipment and staff becomes a problem where airports do not designate individual gates to specific airlines. Some airlines operating with low frequencies into airports, such as Amsterdam, Brussels, and Paris, find that there is no policy of preferred gates for many airlines. This can mean considerable movement of airline equipment and staff around the airport. In the United States, a different handling problem can occur. U.S. ground handling, passenger handling, load control, and baggage-handling crews have a system of bidding for shift choice based on seniority. Consequently a non-U.S. airline being handled by a U.S. carrier might well find that there is a seemingly continuous training problem as ground handling crews change. It is even possible on a long stopover to have two different crews that each need instruction on equipment operation.

References

Ashford, N. J., and P. H. Wright. 1992. *Airport Engineering, 3rd edition.* New York: Wiley-Interscience.
British Airport Equipment Catalogue. 1995. Farnborough: Combined Service Publications.
International Air Transport Association (IATA). 1995. *Principles of Airport Handling.* Montreal.
Krzac, David W. 1981. Fixed air supply offers quick returns but what system is best? *Airports International.*

7

Baggage Handling

7.1 Introduction

Within the total airport operations system, an essential element is the handling of passengers' luggage. If there are any difficulties with the processing of baggage, either on departure or arrival, it can have repercussions across a wide range of airport operations. If, for instance, baggage for departing flights is delayed then aircraft are kept at the gate longer than planned, and extended parking on the ramps inevitably leads to congestion and a general slowing down of airside operations and with this, possible delays also to the parking of arriving aircraft. It has also been established by past studies in the U.S. that terminal and roadway congestion can result from delays in processing arriving baggage.

Furthermore, baggage handling is a particularly sensitive issue from the passengers' point of view, as indicated by numerous surveys, which place the subject very high, if not at the top, of the passengers' priority list. As a consequence, the subject figures predominantly in correspondence between passengers and airport/airline management. Even though the handling of baggage is more often than not performed by non-airport personnel—airline or handling company—it is still all too often perceived by passengers as an airport operational responsibility.

For the airlines, the cost of irregularities can be substantial. Typical costs of (US)$150 upwards can result from the temporary loss of a bag and the necessity subsequently to deliver it to the passenger's home. Station costs can quickly get out of hand if this is a frequent occurrence, to say nothing of costs if the baggage is irretrievably lost. One major North Atlantic carrier has estimated that it needs to earn an additional (US)$30 to $40 million in revenue to cover costs under this heading.

The nature and amount of passengers' baggage changed dramatically during the early 1970s as a result of three influences:

1. Aircraft sizes increased.

2. Baggage allowance criteria were changed.

3. Low fares were introduced.

Until the advent of the wide bodies, baggage seemed to be a manageable element of passenger service. It was subject to specific controls, and these were generally accepted by the traveling public. Manifestly there was a limit to what could be carried on a narrow-bodied jet, especially on a long-distance flight. Excess baggage charges were imposed for baggage above a specific weight (Economy, 44 pounds [20 kg]; First Class, 66 pounds [30 kg]), whether it was contained in one or more bags. The greatly increased capacity available on the wide bodies, however, brought about considerable relaxation of baggage constraints in the interest of simplified procedures. On many long distance routes, the limitations now related only to the number of bags checked, a maximum of two was allowed. There was no longer the weight restriction in regard to these checked bags. In addition, low fares brought with them large numbers of low-budget travelers, including vastly increased numbers of young people carrying with them all they required to be self-sufficient, including such items as bedrolls and rucksacks/backpacks. It became a problem not only of handling greatly increased volumes of baggage, but also of dealing with baggage of every size, shape, and description. It was necessary to make provision for this oversized, or in the European terminology *out of gauge* baggage. There was initially a failure on the part of the industry in the early 1970s, to appreciate that the flood of passengers would bring with them an even greater volume of baggage. The typical average on all but the shortest range business flight is in the region of 1.3 checked bags per passenger. There are certain routes throughout the world where checked baggage, regardless of excess charges, far exceeds this average. Typically, this phenomenon is associated with routes to and from third-world countries where, for one reason or another, consumer goods are not readily available. The overall effect over recent years has been a truly massive increase in the amount of baggage presented by passengers to accompany them on their journey, with a significant impact on all aspects of baggage handling. Foremost among these has been the introduction of containers for the carriage of baggage.

With the gradual increase in the numbers of wide-bodied aircraft, containerized baggage is now becoming the industry norm. It is interesting to note that there is still no universal standard size for wide-bodied baggage containers, although the LD3 is the most commonly used size, designed that two, positioned side by side, fill the underfloor cross section of most wide-bodied aircraft.

7.2 Baggage Operations

A certain number of tasks have to be carried out at every airport, and they are essentially similar whether the airport is small or large. The differences will emerge in the means employed and the procedures adopted. Baggage operations may be conveniently divided into two broad areas, departures and arrival:

Departure

- Carriage of baggage to check-in
- Check-in procedures including tagging and on occasions weighing
- Conveyance of baggage to airside
- Sortation and makeup into aircraft loads
- Transport of baggage to planeside
- Loading onto aircraft

Arrival

- Unloading from aircraft
- Transport to terminal airside
- Sortation—loading onto claim devices
- Conveyance to reclaim area
- Presentation of baggage to passengers for reclaim
- Carriage from reclaim area

These operations are indicated in Figure 7.1, which shows the baggage loading and unloading sequence.

Carriage of baggage to check-in

The majority of passengers need to check in one or more bags at the commencement of their air journey and the aim of most baggage handling systems is to enable this to be done as soon as possible after the passenger enters the airport in order to ensure there is the shortest possible distance for the passenger to carry heavy baggage. This does not, of course, take into account the fact that there might be off-airport check-in facilities.

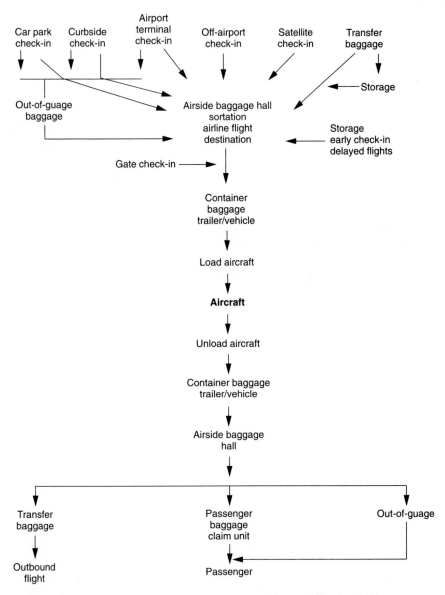

Figure 7.1 Baggage loading and unloading sequence. *(Ashford and Wright 1979)*

The concept of off-airport check-in facilities has entered a new phase changing from the older concept of dedicated downtown terminals to a much more varied system. In some cases in Europe hotels have facilities to check-in baggage for their guests. Where there are direct rail links to airport terminals, baggage can often be checked-in at the passengers' origination rail station; (e.g., Switzerland) and at some major city rail stations in England and Germany. In addition, several major European airlines now allow passengers with only hand luggage to check-in by telephone, or in one case by fax (Air France).

Within individual airports there will be considerable variation in the distances over which a passenger has to carry baggage to the check-in desks. Some might have to carry baggage from nearby public transportation (e.g., at Washington National Airport, this distance involves several hundred yards and a change of level). In other instances, an airport or airline will provide curbside check-in or, at very least, porter/skycap service to transport the baggage to the check-in desk for the passenger. Curbside check-in usually provides for the bag to be conveyed directly to the baggage sort/make-up area adjacent to the ramp.

Additionally, the airports might provide self-help baggage trolleys—usually called baggage carts in the U.S.—for passengers. The practice is for them to be provided free of charge at European airports, although at many U.S. airports a small charge—usually (US)$1.00 is made. Again in Europe, the type of trolley is generally longer/heavier than the U.S. counterpart. In the case of one airport—Frankfurt, Germany—the trolleys are specially designed to allow their use with safety on escalators. While the cost of such trolleys is relatively high, more than (US)$600 each, they are very effective at bringing about a smooth passenger flow in multilevel terminals. Many airports consider the cost of providing several thousand trolleys, although substantial, to be money well spent in the interests of passenger service. When large numbers of trolleys are in use, however, the collection after use has to be well managed and ongoing, as also their placement again ready for re-use at strategic points. Such points might be near the curbside drop-off, when passengers have some distance to walk to check-in. In the case of arriving passengers, where again walking distances may be long, such points might be in the terminals where passengers emerge from their aircraft, especially in view of the current trend for passengers to carry increasingly heavy hand baggage; these points would be in addition to the large number of trolleys required in the baggage reclaim areas. Increasing numbers of passengers are going one step further and using their own small collapsible trolleys or baggage cases with built-in wheels and handles.

Appropriate signing can also help reduce unnecessary walking distances for passengers carrying heavy baggage. Where multilevel terminals are involved, it is especially important to indicate clearly the location/level of the actual departure facilities to avoid confusion and congestion of vehicles on the approach road and at the curbside. There will then be the need to provide further detailed signing indicating the position or nearest entrance to various individual check-in desks. Overall, the effort will need to be directed toward achieving as smooth a flow as possible of passengers with their baggage from drop-off points within the airport to check-in desks. At smaller airports, particularly with single-level terminals, special care will need to be taken to separate the check-in flow of passengers from the flow of passengers leaving the airport with their baggage. In one instance, San Diego, the reclaim building has been positioned on the opposite side of the access road from the main departure terminal.

Check-in procedures

One of the several tasks involved in the check-in procedure is to ensure control of the numbers and weight of passengers' baggage. The number of bags carried by the individual passenger is recorded on that person's ticket, together with the weight when this is required. It is at this point that the airline (or handling agent acting on behalf of the airline) takes charge of the baggage and assumes responsibility for it, issuing reclaim tags to the passenger as appropriate. Prior to accepting the baggage, the airline should take the necessary steps to warn a passenger against including any dangerous or hazardous articles in the checked baggage.

These procedures will invariably lead to waiting lines and, for wide-bodied aircraft in particular, waiting lines might be very long. Typically, individual check-in time per passenger is somewhere between 45 seconds and 3 minutes. This can, however, be seriously disrupted if any query or problem arises, and because of this, most airlines/handling agents will have a procedure whereby the passenger with a query is removed from the check-in line and dealt with separately at another desk. Every effort must be made by airline and airport authorities to exercise control over the check-in lines, and this might be accomplished by stationing additional staff in front of the desks to direct passengers or, alternatively, by providing light barriers of one kind or another (Figure 7.2).

Many passengers make use of the widespread airline practice of allowing those carrying only hand baggage to avoid the ticket desk queue and proceed directly to the gate for their flight. There is also the added advantage that passengers then have no need to wait for checked baggage on arrival at their destination. However, as a result, there has been a growing tendency for carry-on items to become larger and heavier, leading to difficulties with storage in the overhead bins in the aircraft cabin. This, in turn, can lead to congestion and delays in the boarding area. Airlines are aware of the problem and have attempted to control it but with limited success, largely due to the highly competitive nature of the airline business and their desire to retain the goodwill of their passengers.

7.3 Operating Characteristics of Baggage-handling Systems

The manner of operation of baggage-handling systems can differ quite substantially from airport to airport, and certainly from country to country.

Figure 7.2 Check-in counter for American Airlines.

Costs of baggage handling depend on the following factors:

1. *Airport design*. Single or unit terminals; access modes, space extensive of intensive design; airport design volumes.

2. *Terminal design*. Centralized or decentralized check-in; gate arrival; single level or multiple level; centralized, decentralized, or remote outbound bag rooms; mechanization; terminal design volume.

3. *Traffic makeup*. Scheduled or charter; standby arrangements; transfers as a percentage of originating or destined passengers; short or long haul; domestic or international; nature of baggage to be handled.

4. *Baggage System Design. Outbound*—centralized, decentralized, or remote operations; manual, semiautomated, or automated; containerization. *Inbound*—type of claim device; design of bag claim feed areas.

5. *Employee Motivation*. Handling by airline, handling agent, or airport.

6. *Airport service standards*. Competition; internally set standards; airport policies.

7. *Flight schedules*. Peaking through the day and throughout the peak hour.

8. *Type of aircraft*. Wide or conventional body; aircraft size.

Recognizing that there are very wide variations between the operational standards at different airports and different airlines, it is difficult to draw many general conclusions that can be applied directly to an individual case.

Overall baggage handling costs reflect the high labor input that is the norm in the industry at the moment; almost three-quarters of handling costs are directly attributable to labor, and almost one-quarter of the costs are due to irregularities that are largely caused by human error. Procedures of the check-in, sortation, airport transportation and the loading/unloading of aircraft make the most demands on labor. In the last two years the process of sortation has been expanded to include a security element—passenger/baggage reconciliation—and this is dealt with in Chapter 9 in the discussion on security.

Overall costs could be significantly reduced if a way could be found to abandon the present system of separating passengers from their bags and mixing all bags together prior to sortation and transportation to the aircraft. The air passenger is in a unique position vis-à-vis other travelers, in being able to transfer responsibility for the personal baggage to the carrier for most of the journey. This not only requires a higher labor input at the airport but also opens up the possibility of baggage irregularities, either by mishandling or by loss and pilferage, which can be quite high at some airports. In attempting to lower operating costs, new baggage systems and procedures are constantly being considered. Among these is the possibility of taking the baggage from the passenger at the last possible moment for direct loading onto the aircraft. Such a procedure is, in fact, adopted at some airports for "last minute" passengers who, on arriving late for check-in, are advised to carry their baggage to the aircraft boarding gate. This presupposes the provision of the slides beside each boarding gate and, on wide-bodied aircraft, the capability to "loose-load" baggage, i.e., without using a baggage container. It also presupposes the existence of an intervening security check.

7.4   The Inbound Baggage System

The objective of the inbound baggage system is to provide, in an economical and efficient manner, a fast method of unloading and delivery of baggage to the terminal and displaying it in the reclaim hall so that the passenger can easily retrieve it. Clearly, as aircraft have become larger, the amount of baggage to be displayed has grown greater, and thus the display area has grown at least proportionately. To avoid excessive confusion of many passengers individually moving among and examining many pieces of baggage, mechanical claim devices are usually used to carry the bags in a continuous display before the stationary passenger.

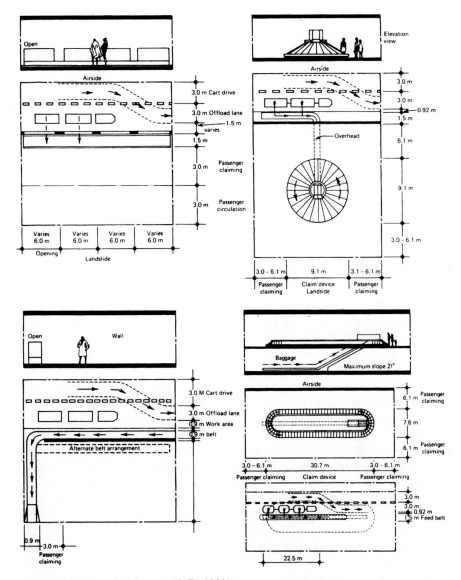

Figure 7.3 Baggage claim layouts. *(IATA 1989)*

Figure 7.3 shows four typical layouts of claim devices. They are:

- Linear counters
- Linear conveyors
- Carousels
- Racetracks

Linear counters are nonmechanized displays that are suitable for small aircraft only and at most airports where passenger flows are very low. The *linear conveyor* is slightly more sophisticated in that it moves the baggage past the passenger rather than making the passenger hunt for baggage up and down a counter. Normally there is a roller storage at the end of the conveyor for bags that remain unclaimed for a time. *Carousels* and *racetracks* are continuous devices that continue to circulate unclaimed baggage.

One of the decisions facing management in regard to baggage reclaim operations is the numbers of claim devices to be provided. This will obviously relate to the anticipated volume of baggage, bearing in mind that a typical wide-bodied load can be some 800 bags. Reclaim devices are of necessity tailored to the space available in the reclaim area, and it is usual for suppliers to manufacture devices to fit these dimensions. Bag capacity or storage (i.e., the maximum number of bags that can be displayed) on any device will depend on its width and length.

There are two basic types of mechanical claim devices to be considered. The simplest is the flatbed device by means of which a moving belt, usually between 2½ and 3¾ feet (0.8 and 1 m) wide and standing about 1½ feet (0.5 m) high, circulates bags in a chosen pattern around the reclaim area. It is generally considered that the speed of the belt should be no faster than 1½ feet (0.45 m) per second. The maximum number of bags that can be displayed on a claim device is based on the assumption that none are taken off by passengers and that the belt is loaded until it is covered. On this basis an FAA report gives two important operational factors: the length of the belt or frontage available for passengers to view the whole belt and the total space available on the belt for baggage. Some typical figures for both values that have been published by the FAA are given in Table 7.1. Somewhat lower capacities are used by one larger European airport authority, which calculated 1½ bags/m for flatbed devices and 2½ bags/m for sloping bed devices. The length of belt in public view in front of which passengers are allowed to stand to identify and retrieve their baggage is the *claim frontage* and the total number of bags that the claim device can display is the *bag storage* capacity. Once this capacity is reached, no more bags can be delivered until some have been taken off the claim device. In normal operation, it is generally assumed that bags can be unloaded onto a claim device at the rate of approximately 20 per minute. However, problems can still arise when bags are off loaded unevenly onto the airside delivery belt and arrive in heaps on the carousel, making it difficult for passengers to extricate their bags. Some arrival baggage systems are designed to ensure an even distribution of bags on the carousel by the use of photo electric sensors.

Within the baggage reclaim hall of any medium to larger airport, there will be several baggage conveyors or carousels. Figure 7.4 shows a typical layout. It is therefore necessary to have some indication of flight allocations to individual carousels. Often this is in the form of a TV monitor or

TABLE 7.1   Bag Storage Capacities Related to Various Reclaim Configurations

Shape	L&W (ft)	Claim Frontage (ft)	Bag Storage[a]
		Flatbed - Direct Feed	
Oval	65 × 5	65	78
`	85 × 45	180	216
`	85 × 65	22	264
⌴	50 × 45	190	228

Diameter (ft)	Claim Frontage (ft)	Bag Storage[a]
	Circular Remote Feed Sloping Bed	
20	63	94
25	78	132
30	94	169

L&W (ft)	Claim Frontage (ft)	Bag Storage[a]
	Oval Remote Feed Sloping Bed	
36 × 20	95	170
52 × 20	128	247
68 × 18	156	316

[a] Theoretical bag storage - practical bag storage capability is one third less.

split flap type indicator above each carousel. Airports might also have indicators at some point prior to the passengers' entry into the reclaim hall. This is particularly important at international airports, where such prior notice can save a great deal of possible congestion, which would otherwise occur as passengers circulate in the reclaim hall looking for the appropriate carousel for their baggage. Some airlines go one step further and have their flight crew make an announcement shortly before landing, reminding passengers of their flight number and, where possible, the carousel on which they can expect to find their baggage. It would clearly be an advantage for airports and airlines if this information could be passed to airlines prior to their flight arrivals, so that such announcements could be made in the aircraft.

Where only domestic operations are concerned, the operation of the baggage reclaim hall can be seriously compromised by the presence of greeters who are given free access to the hall. These visitors might cause severe overcrowding in the hall. This can hamper the rapid identification and retrieval of baggage and further add to the congestion. From the viewpoint of both efficiency and security, many airport authorities ban all visitors from the baggage claim area unless they are aiding an elderly or handicapped

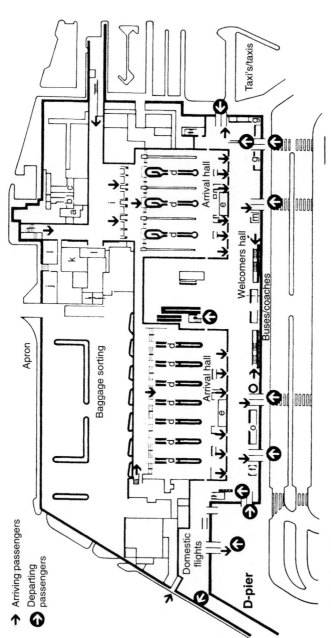

Figure 7.4 A typical baggage handling layout.

traveler. Airport authorities have found that such bans materially improve the efficiency of the system and also cut down on baggage theft.

To some extent, the customs examination and controls for international passengers removes this problem, certainly from the immediate vicinity of the reclaim area. However, serious congestion might still occur in adjacent areas where there is a solid wall between these areas and baggage reclaim. Random groups collect and usually seriously impede the progress of anyone who wishes to pass through that part of the terminal. Where part of the wall is glass, for example at the Rio de Janeiro, Frankfurt, and Amsterdam airports, waiting groups tend to line up against the glass screens, thus leaving space for circulation behind them.

7.5 The Outbound Baggage System

The outbound baggage system consists of *check-in, carriage to the outbound bag room, sortation*, and *carriage to the aircraft.*

Check-in

Three different check-in counter configurations are shown in Figure 7.5. The *linear counter* arrangement, with a backing baggage conveyor, is perhaps the most traditional. Although it has the advantages of good visual presentation to the passenger, the waiting lines can make for an inefficient use of space, and passengers once served must backtrack and cut through lines of those still waiting. In narrow-gate arrival terminals, the lineups of waiting passengers tends to interfere with free movement of passengers through the terminal, and special single-line multichannel lineup arrangements have become necessary to avoid unnecessary congestion, as shown in Figure 7.2. The *island* check-in arrangement makes more efficient use of the baggage conveyor, which is loaded from both sides; there is a consequent saving in space. However, the flow patterns of waiting passengers and those seeking to leave the area after service offer many points of conflict. With island configurations, however, there is little interference between waiting passengers and general flows along the longitudinal terminal axis. To overcome these problems, the *flow-through* arrangement of desks provides flow patterns in which passengers move in one direction without backtracking. Flow-through systems are not necessarily more space extensive than the linear counter system, but because they require deep check-in halls, they are usually feasible only if considered in the design stage of a facility. They have the added advantage that if designed as shown, there is no need for the check-in attendant to lift baggage.

Baggage is normally conveyed directly by belt from the check-in area to the outbound bag room or baggage make-up area, where it is sorted into the appropriate bag cart or container to be conveyed to the aircraft. Currently,

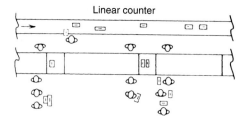

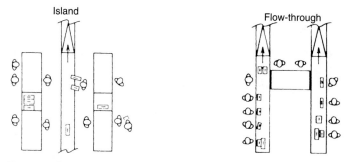

Figure 7.5 Check-in counter configuration. *(FAA 1988)*

three principles of sortation are in use: *manual, semiautomated,* and *fully automated.* Figure 7.6 indicates schematically how these make-up systems operate. The manual system is the simplest, and it accounts for more than 98 percent of all operating systems. Bags arriving on a belt are manually sorted into the appropriate bag cart by operators, who read the baggage tags. At large installations, the incoming baggage might feed onto a race-track, from which the sorting is carried out. *Semiautomated* systems, such as the one shown in Figure 7.6, also rely on an operator reading the baggage tag. The tag code is punched into a computer that causes a pulley arm to di-vert the bag at the appropriate point along the belt. Bags are then manually loaded into the bag cart. Other semiautomated systems exist that use either a tilting tray or a slat conveyor to effect the sort. Both are controlled by the computer, which must be controlled by a worker.

Automated baggage systems are more commonly used for the handling of departure baggage. They offer two advantages: reduction in labor costs and speed of conveyance. The elements of the handling operation for departure baggage that come between check-in and make-up into flight loads were the last ones to be automated. They involve "recognition" of the bags' even-tual destination and conveyance to the appropriate loading point for that particular flight. If the bag is conveyed on a belt, a recognition code can be incorporate in the baggage tag. One of the systems developed in the United States in recent years utilizes the commodity bar code, which is read by laser beam(s) as in many supermarket checkout points. The original format used for the bar code was in the shape of a bull's-eye, but this occasionally gave rise to difficulties if the bag was upside down or in some unusual position that caused the laser beam to miss the label. This has now been resolved by increasing the number of laser beams at the read position and using a new strip label and also, for security purposes, adding check-in information.

If, as in some systems, it is conveyed on a self-propelled cart or bin, then the code is electronically imprinted on the bin by check-in staff. This code can

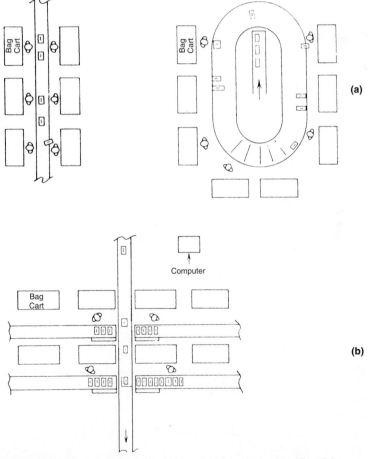

Figure 7.6 *(a)*Manual systems; *(b)* semiautomated systems sort puller; *(c)* Automated system—Frankfurt Airport.

(c)

Figure 7.6 *Continued.*

then be read out by a variety of systems and the bag or bag container directed along an appropriate route to a sorting area or to the departure gate along the automated conveyor lines. If these lines, in effect miniature rail systems, are routed to individual gates, then the bag can be processed automatically from the check-in desk to the departure gates. Such a system is used in Frankfurt Airport, Germany[1] and it involves 30 miles (50 km) of track. A recent development has been the requirement to ensure that no unaccompanied bags are loaded onto an aircraft, which is achieved by reconciling bags and individual owners. An automated system, introduced at Frankfurt Airport in 1994, enables passenger/baggage reconciliation to be carried out at the final makeup state of baggage loading. A hand-held laser gun is used to scan the bar code label that contains confirmation of check-in, as well as destination information. This system is described more fully in Chapter 9, "Security".

Provision is made for connecting/interline baggage to be fed into the system and the running connect time for Frankfurt Airport interline baggage is 40 minutes, an important factor for an airport where 40 to 45 percent of its passengers transfer to other flights.

With manual systems where there is only a single sortation belt serving a large number of check-in desks, serious delay problems can arise if the belt breaks down.

If there are two or more sortation recirculating belts, some check-in desks can still operate even if one belt breaks down. Especially in larger ter-

---

[1]In 1994 Frankfurt had 365,000 air transport movements and handled 35 million passengers.

minal buildings, it is advisable to safeguard the continuation of the check-in process by making suitable provision for sortation in the event of a partial breakdown.

The design of the passenger terminal complex itself can radically affect the outbound baggage system. Conventional centralized pier finger airports, such as Chicago O'Hare, Schiphol, Amsterdam, and Manchester International, operate on one or more *central* bag rooms in the main terminal area. These require elaborate sorting systems, but can be efficient in the use of personnel that is released when not necessary in off-peak periods. Decentralized facilities, such as Kansas City and Dallas-Fort Worth, have a number of *decentralized* bag rooms that are closely associated with a few gates. The sortation requirements of these makeup areas is minimal, but it is more difficult to use staff efficiently in the decentralized situation where there are substantial variations in workload between peak and off-peak periods. A third concept of baggage makeup area is the remote bag room. In an airport like Atlanta, where three-quarters of the traffic is transfer, there is considerable cross-apron activity. Remote bag rooms provide for the complex sortation necessary without transporting all baggage back to the main terminal. Schematics of the three systems are shown in Figure 7.7.

The effect of a large volume of traffic can be seen in Figure 7.8. Passengers who are transferring from other flights tend to arrive at the transfer airport much later than originating passengers[2]. The rightward shift of the cumulative curve from the 30-percent transfer situation, which is a typical international hub situation, to the 75-percent transfer shown for the Delta operation in Atlanta, puts considerable pressure on sorting transfer baggage because of the considerable time constraint. In fact transfer baggage represents a critical performance area for airlines or handling companies, as IATA notes in its *Airport Handling Manual:* "Transfer baggage accounts for most of the airlines baggage mishandling." It therefore calls for particular diligence on the part of staffs dealing with transfer baggage. There can be time pressures, when arriving and departing times are close, or alternatively problems of timely identification and retrieval from storage where there is a long interval between connecting flights.

Because of the acknowledged importance of this subject, IATA provides extensive guidance (IATA 1995) to its members including procedures for setting up a local baggage committee at airports served by its own and ATC (Air Traffic Conference of America) members. Among other activities, such committees are to review problems that cause baggage misconnections and generally monitor the efficiency of procedures for the transfer of baggage in-

---

[2]The reader is advised to compare the cumulative graph of Figure 7.8, with those of Figures 13.6 and 13.7. The former includes transfer passengers, whereas they are excluded in the last two.

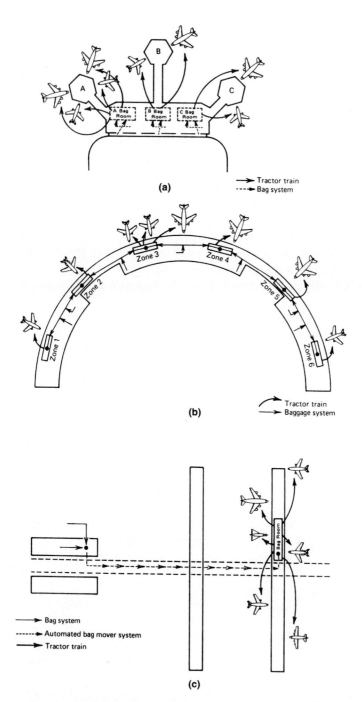

Figure 7.7 Outbound bag room arrangements: *(a)* central bag room; *(b)* decentralized bag rooms; *(c)* remote bag rooms. *(FAA 1988)*

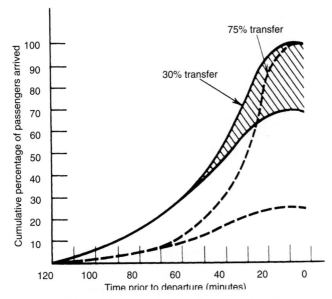

Figure 7.8 U.S. arrival patterns for departing passengers at airports with 30- and 75-percent transfers.

cluding performance in relation to established Minimum Connecting Times (MCTs) (the time taken to off-load, transfer and load transfer baggage).

7.6 Operating Performance

The operating performance of a baggage system should be judged against an appropriate set of criteria that an individual airline or airport chooses relative to its performance objectives. Such criteria are likely to be ideal standards set in terms of dealing with individual flights arriving in an orderly fashion. However, conditions are rarely ideal; delayed flights, diversions, and the late arrival of passengers all introduce complications. Although such considerations point to some flexibility in the manner of evaluation, they do not obviate the need for performance criteria that can be developed for each of the following stages of the baggage handling procedure: *check-in, processing, carriage*, and *arrivals*.

Check-in

Because the departure formalities of air transport are the most complex of any mode of public transport, the pressure on check-in procedures is severe. Although it is important to have close control over what is carried in terms of checked and cabin baggage, there is a need to provide a reasonably rapid service. Performance standards in the check-in phase should be set for *lengths of waiting lines* (which ideally should be no more than five passengers) and for *average waiting and processing time*. Standards for the latter vary according to the type of journey. In shuttle operations, processing time can well be set at 30 to 40 seconds, whereas for an international journey a processing time of 2 minutes is acceptable. The figure set for total waiting and processing time should conform with local standards.

It is interesting to note that the use of computer terminals for check-in does not always lower processing times. A manual check-in for a short distance domestic passenger is likely to be faster than a computerized operation.

The new procedure of *ticketless check-in* introduced in early 1995 by United Shuttle and Southwest on the U.S. West Coast should produce even faster times, especially for short-range domestic flights where only hand baggage is involved.

Processing

For the processing functions, criteria should be developed that relate to:

- Correct destination tagging
- Correct use of transit and interline
- Labeling of packages with fragile contents
- Suitable marking of bags received in a damaged condition
- Labeling heavy bags that might injure or strain staff when lifting
- Ensuring that bags have adequate external identification of owner

Carriage

This includes not only the air portion of the journey, but more importantly, from the control viewpoint, the ground transport of checked baggage. With containerized baggage this part of the process—the ground transport of baggage to the planeside—is usually straightforward. However, problems can arise when baggage is conveyed to the aircraft on open bag carts. There is always the risk of bags falling off and being damaged if the load is not properly secured, although the observance of speed limits by baggage-train drivers should minimize this risk. When weather is inclement extra care is needed to cover bags on open carts to avoid damage due to the effects of weather. Normally a record is kept of the number of mishandled, damaged or lost bags per 1000 passengers handled; this allows a common measure to be applied regardless of whether it refers to a high or low activity airport. Control charts can be used as continuous performance monitors. An airline might have an oversight of several stations' performance in relation to one aspect "Damaged Bags" (Figure 7.9). At the same time there would be detailed analysis of several aspects for each individual station (Figure 7.10).

Arrival

The handling of baggage on arrival is of critical importance because the service is provided at the conclusion of a journey and inevitably forms a lasting impression in the passenger's mind of the overall quality of a trip. Because the passenger is not likely to be aware of the division of the responsibility of handling arrival baggage, blame for poor performance might accrue to the wrong party. The appearance of good performance depends on being able to present arriving baggage at the reclaim units soon after the passenger's arrival in the area. Among the factors having a bearing on the performance are: *airport layout, density of airside ground traffic, baggage handling equipment, availability and allocation of personnel.* A passenger at a "gate arrival" airport, such as Kansas City, is likely to have only a short walking distance from the aircraft to baggage reclaim. Even with the fastest aircraft unloading procedures, a passenger might still have an appreciable waiting time in the reclaim area, although delivery time might have been under 10 minutes. On the other hand, an arrival at Paris Orly or Atlanta will have to travel very much farther from the aircraft to baggage reclaim. In the International Terminal 3 at Heathrow it would be normal for even the first passenger in the reclaim unit to have taken 15 to 20 minutes to reach the reclaim area and could well find his or her bags waiting. To the passenger, this would seem a better performance than the earlier example. Performance criteria based on complaint levels might therefore be inadequate. For example, U.S. immigration procedures are so slow that they would mask appalling arrival delivery standards for international arrivals. Therefore, the following two objectives criteria are often set: time taken to unload the first and last bags and percentage of bags delivered to reclaim area in a given time.

Number of damaged bags – Departing station

Station	May	Jun	Jul	Aug	Sep	Oct	Total	Average per 1000 PAX
A	72	60	67	96	74	41	409	0.70
B	63	87	50	71	76	39	386	0.86
C	23	25	32	28	37	23	168	0.84
D	39	42	29	40	31	21	202	1.10
E	10	4	6	5	7	6	38	0.48
F	4	2	8	4	5	7	30	0.97
							1233	0.81

Figure 7.9 Baggage mishandled (damaged) control chart—multi station.

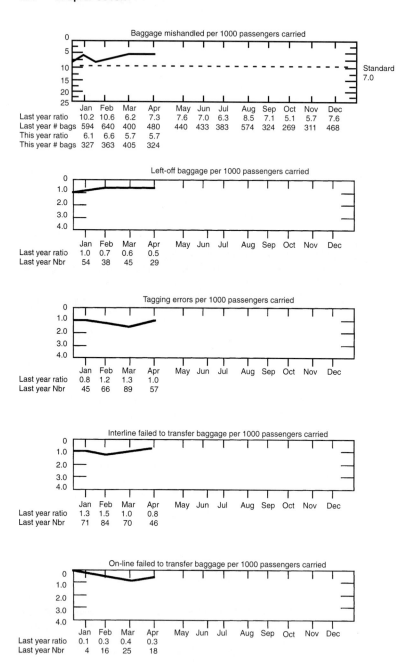

Figure 7.10 Baggage mishandled control charts—single station.

TABLE 7.2   **Minimum and Maximum Observed Aircraft Unloading Times**

	DC9	B737	B727	B707	B747	L-1011
Minimum	5	5	6	5	12	10
Maximum	12	10	22	15	36	40

SOURCE: Bagging handling survey, Air Transport Training board.

Time to unload bags

As a measure of performance, time to unload the last bag is obviously more reliable than the time for the first bag. Table 7.2, which shows the results of a British survey, indicates that considerable variation in unloading time can be encountered even with identical aircraft with similar loads, and that the size of aircraft is not necessarily a good indication of speed of unloading. Typically a performance standard is set, such as percentage of flights failing to meet 20 minutes delivery time for first bag and 35 minutes for last bag.

Percentage of bags delivered in given time

Alternatively, the criteria can be set to specify a percentage of bags within a certain time limit, as shown in Figure 7.11.

7.7   Organizing for the Task

Growth in the volume of baggage handled, coupled with the constant search for economies by airports and airlines, has led to gradual changes in the organization for this task. There has been a growing tendency for airlines and airports who have previously carried out the task of baggage handling to transfer it to handling agents, whether to another airline or an independent company such as Allied (United States) or Ringway Handling (Manchester Airport, UK). The tendency for airports in Europe to enjoy monopoly handling rights is being challenged by the European Commission. There is increasing pressure for the establishment of competing companies to carry out ground handling, including baggage handling, based on the argument that such competition would result in lower costs to airlines together with improved efficiency. Where an airline is a major operator at a particular airport, however, it is more usual for it to use its own personnel for baggage handling (e.g., British Airways at London Heathrow, USAir at San Francisco, KLM at Amsterdam).

Staffing

As with all other aspects of air transport operation, the peaks and troughs of traffic so typical of the industry, present problems to management when

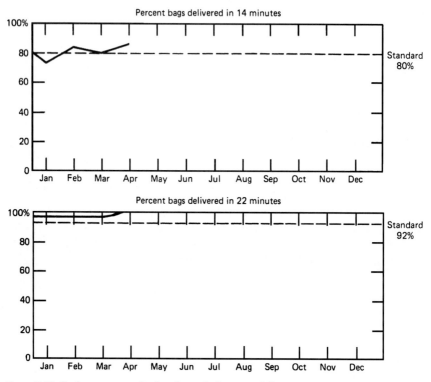

Figure 7.11 Performance monitoring charts for baggage delivery.

attempting to determine the level of staff needed for any operation. There are obvious constraints in terms of costs, and as a result there can be only limited response to the possibility of diversions or bunching of arriving flights. Where premium service is demanded and paid for, then special effort can be made, as it has been with the Concorde service, where a high level of staff is assigned. Normally, however, there will be a compromise and a tacit acknowledgment that there probably will be a few occasions when staffing levels will be inadequate when abnormal demand arises.

The largest group of personnel engaged in handling baggage are those who deal with it on the ramp, transporting baggage to and from the aircraft and loading and unloading the hold (Figure 7.12).

Ramp personnel must be allocated by some system to individual flights and this necessitates an oversight of ramp activity. The practice at many airports is to have a central observation point, usually on the top of a terminal building, or pier or on the face of a terminal building overlooking the ramp, as at Copenhagen Airport. In this way a good view is obtained of aircraft arrivals and departures and a rapid response can be made to changing circumstances. Alternatively, TV cameras could be positioned at strategic sites round the ramp with an input to video displays in a central control room, which might be within the terminal building. The control room would also be equipped with displays of flight schedules and relevant airport (gate) information and the necessary communications, as well as having listings of the personnel available from which to allocate required teams.

The basic method of allocating staff to flights is tackled in a variety of ways. At low-activity stations this is not a complicated procedure and merely requires the lead hand (head loader) personally to allocate staff based on personal experience. At higher-activity stations, where handling staff might number several hundred, it is usual to find specialist staff employed as allocators. Their task is not only to assure the necessary number of staff for a particular flight, but also to ensure a reasonably fair distribution of the workload. In order to satisfy these requirements it is essential for staff allocators to have available up to the minute details of flight arrivals

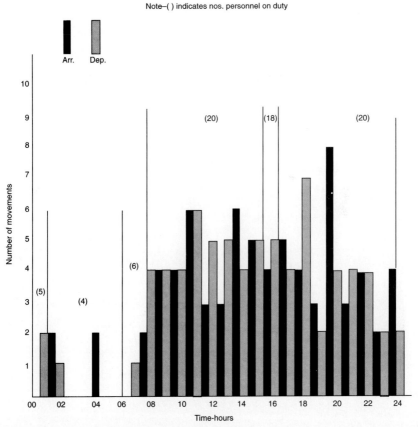

Figure 7.12 Ramp baggage handling staff compared with aircraft operations. *(Baggage Handling Survey, Air Transport Training Board)*

and departures, as well as prior notice of the load on board an arrival or the load planned for a departure. There is less of a problem in this respect if an airline is doing its own handling but information can easily be delayed or forgotten when it has to be passed to another organization. All too often this is manifested by the unannounced flight. The establishment of a direct link between staff allocators and ATC, where possible, should ensure accurate up to the minute times are available.

Computer technology is increasingly proving a useful tool with the availability of databases such as airline schedules, personnel and equipment data, although still dependent on timely and accurate input. The availability of mobile radios and telephones has also greatly assisted with on the spot last-minute changes, or problems, being encountered at the planeside. The whole system of allocating staff for baggage handling plays a vital part in achieving the effectiveness of the overall operation of aircraft turnaround.

References

Ashford, N. J., and P. H. Wright. 1992. *Airport Engineering, 3rd edition.* New York: Wiley-Interscience.

Federal Aviation Administration (FAA). 1988. *Planning and Design Guidelines for Airport Terminal Facilities, AC150/5360-13.* Washington, DC.: Department of Transportation.

International Air Transport Association (IATA). 1995. *Principles of Airport Handling.* Montreal.

International Air Transport Association (IATA) 1989. *Airport Terminals Reference Manual* 7th ed. Montreal.

Passenger Terminal Operations

8.1 Functions of the Passenger Terminal

Analysis of the operation of an airport passenger terminal leads to the conclusion that three principal transportation functions are carried out within the terminal area (Ashford and Wright 1992):

1. *The processing of passengers and baggage.* This includes ticketing, check-in and baggage drop, baggage retrieval, governmental checks, and security arrangements.

2. *Provision for the requirement of a change of movement type.* Facilities are necessarily designed to accept departing passengers, who have random arrival patterns from various modes of transportation and from various points within the airport's catchment area at varying times, and aggregate them into planeloads. On the aircraft arrivals side, the process is reversed. This function necessitates a holding function, which is much more significant than for all other transport modes.

3. *Facilitating a change of mode.* This basic function of the terminal requires the adequate design and smooth operation of terminal facilities of two mode types. On the airside, the aircraft must be accommodated, and the interface must be operated in a manner that relates to the requirements of the air vehicle. Equally important is the need to accommodate the passenger requirements for the landside mode, which is used to access the airport.

An intimation of the complexity of the problem can be grasped from an examination of Figure 8.1, which is admittedly a simplification of the flow process for passengers and baggage through a typical domestic-international airport passenger terminal. When examining a chart of this nature, it must be remembered that the representation can only be in generalized terms and that the complexities of operation are introduced by the fact that flows on the airside are discrete and those on the landside are continuous. The substantial growth rate of air transportation since World War II has meant that many airports around the world are now large operations. Unlike the pre-1940 period when air transportation was a fringe activity on the economy, the air mode is now a well-established economic entity. The result on passenger terminals has been dramatic. A scorel of large international terminals are handling more than 30 million passengers per year or

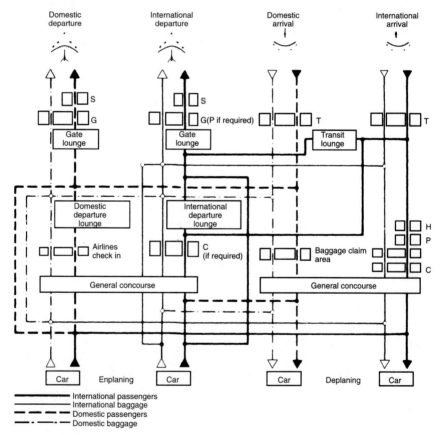

Figure 8.1 Schematic of the passenger baggage flow system. G = gate control and airline check-in (if required); P = passport control; C = customs control; H = health control (if required); T = transfer check-in; S = security control. *(Adapted from IATA 1976)*

will be doing so in the near future. Operations of this scale are necessarily complex.

The relatively recent development of large air passenger volumes has required the provision of increasingly large facilities to accommodate the large peak flows that are routinely observed (see Section 2.2). Single terminals designed for capacities in the region of 10 million passengers per annum often have internal walking distances of 3500 feet (1100 m) between extreme gates. Where capacities in excess of 30 million annual passengers are involved, largely single-terminal complexes, such as Chicago O'Hare and Amsterdam, are likely to have internal gate-to-gate distances in the region of 5000 feet (1500 m). To overcome problems such as this, and to meet IATA recommendations on passenger walking distances, several "decentralized" designs were evolved, such as those that are now in operation at Kansas City, Dallas-Fort Worth, and Paris Charles de Gaulle II. Decentralization is achieved by:

1. Breaking the total passenger terminal operation into a number of unit terminals that have different functional roles (differentiation can be by international/domestic split, by airline unit terminals, by long-haul/short-haul divisions, etc.)

2. Devolving to the gates themselves a number of handling operations that previously were centralized in the departure ticket lobby (ticketing, passenger and baggage check-in, seat allocation, etc.)

Coupling a decentralized operational strategy with a suitable physical design of the terminal can result in very low passenger walking distances, especially for the routine domestic passenger. Where considerable interlining takes place, or where the passenger's outbound and inbound airlines are likely to differ, decentralization is likely to be less convenient to the traveler. For example, one of the earlier decentralized designs, Dallas-Fort Worth (Figure 8.2a) can be more convenient for an interlining passenger than the newer Atlanta design (Figure 8.2b). International operations significantly affect the design of terminal facilities and the procedures used. From this viewpoint, the airport planner and operator must be extremely careful in extrapolating U.S. experience, which although well documented might be based mainly on domestic operations. The infusion of the governmental requirements necessarily associated with international operations (customs, immigration, health, agricultural controls, and especially security) can add considerable complications to the layout and operation of a terminal. The new Eurohub terminal at Birmingham has a most complex arrangement of the interlocking doors to allow for flows among international, domestic and "common travel" passengers, who must be segregated. The complicated door system is centrally operated by a computerized control room with extensive closed circuit TV monitoring. Figures 8.3a and 8.3b show conceptualized pro-

Figure 8.2a Decentralized terminals of Dallas-Fort Worth International Airport. *(Dallas-Fort Worth International Airport)*

Figure 8.2b This aerial view shows the connected North Terminal (on the left) and South Terminal (right), International Terminal and concourse (adjacent to North Terminal) and four domestic concourses—the largest passenger terminal complex in the world. *(Hartsfield Atlanta International Airport)*

cessing outbound flow patterns for centralized and decentralized facilities. In almost all countries, it is not possible for outbound passengers to pass back through the governmental controls, and universally, airport visitors are precluded from the international departure lounges. As a result, many passenger- and visitor-related facilities must be duplicated, as will be discussed in Sections 8.4 and 8.5. In many countries, there is also a governmental requirement for security purposes to separate international arriving and departing passengers. In terms of space, this can be very expensive, leading to considerable duplication of facilities and staff. Mixed arrival/departure areas are now widely accepted at European airports, such as London, Paris, and Amsterdam in the international terminals, but in Britain, for example, there are requirements for older terminals to introduce separation as rapidly as this becomes feasible. As a general rule, however, the inclusion of international operations must be seen as a complication of terminal processing activities that will cut down on the use of multiple purpose space, require duplication of facilities, necessitate additional processing space and, in most cases, increase the number of languages involved in the operation.

8.2 Terminal Functions

Transportation planners use the term *high activity centers* to describe facilities such as airport terminals that have a high throughput of users. In the peak hour, the largest passenger airports process well in excess of 10,000 passengers. Departing international passengers are likely to spend on average more than 1 hour in the terminal facility, and arriving international passengers at least 30 minutes. During the period that they spend in the terminal, passengers are necessarily engaged in a number of processing activities and are likely to use a number of subsidiary facilities put in the airport for their comfort and convenience as well as for the airport's profit. Before discussing in some detail these individual activities, it is worth classifying the terminal activities into five principal component groups:

- Direct passenger services
- Airline-related passenger services
- Governmental activities
- Nonpassenger related airport authority functions
- Airline functions

Either directly or indirectly, these functions, where conducted in the passenger terminal area, will involve some responsibility on the part of the terminal manager. Figure 8.4 shows the organization of these responsibilities for the terminal operation of a major international airport. The individual terminal functions are discussed in more detail in Sections 8.4 to 8.12.

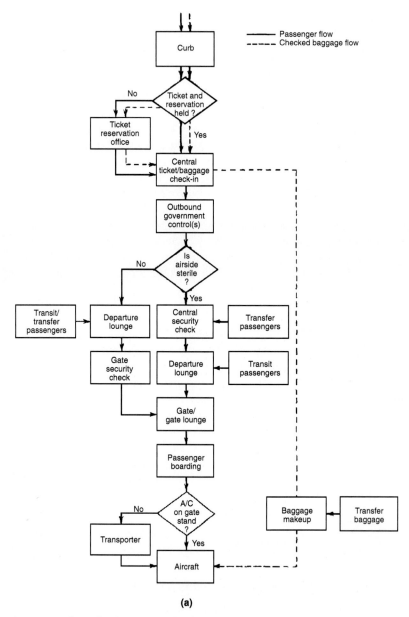

(a)

Figure 8.3a Centralized processing (outbound). *(IATA 1989)*

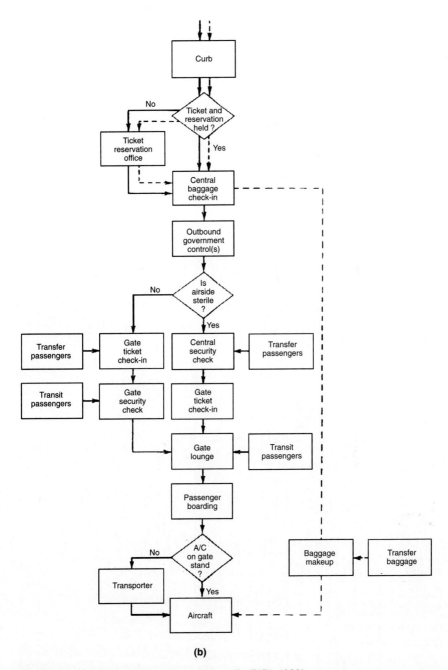

(b)

Figure 8.3b Decentralized processing (outbound). *(IATA 1989)*

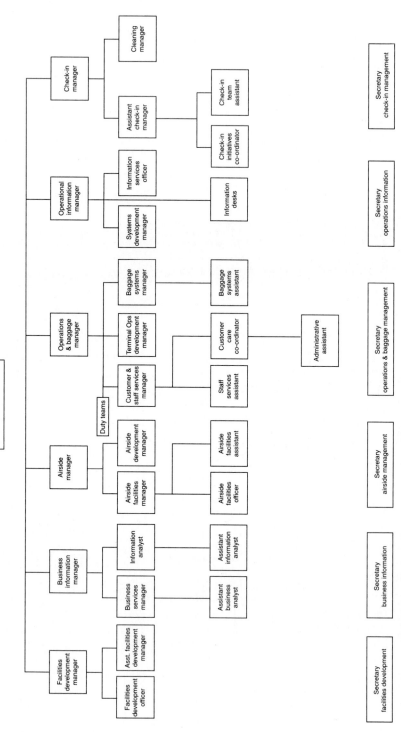

Figure 8.4 Organizational structure of terminal management—London Gatwick.

8.3 Philosophies of Terminal Management

Although the basic operational procedures of airports as they relate to safety are generally rather similar throughout the world, the manner in which those procedures are operated and the organization used to effect them can differ quite radically. Perhaps nowhere in the airport do the operational philosophies differ as much as in the terminal area. The two extreme positions may be designated as:

- Airport dominant
- Airline dominant

Where terminal operations are airport dominant, the airport authority itself provides the staff to run terminal services. Apron, baggage, and passenger handling are either entirely or largely carried out by airport authority staff. Services and concessions within the terminal are also mainly authority operated. Airport-dominant operations are sometimes called the European model, although similar arrangements are found throughout the world. Frankfurt is perhaps the best example of this form of operation, which involves high airport-authority staffing levels and high authority equipment costs with concomitant savings to airlines.

At the other end of the scale is the airline-dominant operation, sometimes called the U.S. model, where the airport authority acts almost as a broker, providing only the most basic facilities in the terminal. Much of the internal furnishing and all necessary operational equipment and staff are provided by airlines or concessionaires. At some U.S. airports, the airlines are integrally involved in financing and building the terminals themselves, and are legally part of the decision process that determines the policy of the airport. In contrast to the authority-dominant model, the individual airline operations are closely associated and identified with individual company images. Where an airline-dominant facility is operated, the need for airport-authority staff is significantly reduced.

Most major airports around the world work on a mixed model, where the airport authority takes care of some terminal operations and airlines and concessionaires operate other facilities. In some airports, competitive facility operation is encouraged to maintain the high service standards usually generated by competition. In the EC, Commission directives are forcing airports to introduce competition where operation has previously been by a single organization. Competitive handling operations are also less vulnerable to a complete shutdown by industrial action. The final choice of operational procedure will depend on a number of factors, including:

- Philosophy of the airport authority and its governing body
- Local industrial relations

- Financial constraints
- Availability of local labor and skills

8.4 Direct Passenger Services

Those terminal operations that are provided for the convenience of the air traveler and are not directly related to the operations of the airline are normally designated *direct passenger services.* It is convenient to further divide this category into *commercial* and *noncommercial* services. There is no hard-and-fast division between these two subcategories, but noncommercial activities are usually seen as being entirely necessary services that are provided either free of charge or at some nominal cost. Commercial activities, on the other hand, are potentially profitable operations that are either peripheral to the transportation function of the airport (e.g., duty-free shops), or avoidable, and subject to the traveler's choice (e.g., car parking and car rental).

Typically, at a large passenger terminal, the following noncommercial activities will be provided, usually by the airport authority:

- Portering
- Flight and general airport information
- Baggage trolleys[1]
- Left luggage lockers and left luggage rooms[2]
- Directional signs
- Seating
- Toilets, nurseries, and changing rooms[1]
- Rest rooms[1]
- Post office and telephone areas
- Services for disabled and special passengers

Depending on the operating philosophy of the airport, commercial facilities will either be operated directly by the authority itself or leased on a concessionary basis to specialist operators. Typically, at a large airport, the following commercial activities can be expected to play and important part in the operation of the passenger terminal:

- Car parking
- Duty-free shops

---

[1]Some airports make nominal charges for some of these facilities.

[2]Airports usually make commercial charges on these facilities.

- Other shops (book shops, tourist shops, boutiques, etc.)
- Car rental
- Insurance
- Banks
- Hairdressers, dry cleaners, valet services
- Hotel reservations
- Amusement machines
- Advertising
- Business center facilities

Figures 8.5 and 8.6 show examples of commercial duty-free shops and advertising at airports having an aggressive commercial policy. The degree of commercialization of airports varies substantially. Airports that have adopted policies promoting such activities, such as Frankfurt, Miami, Orlando, Amsterdam, London, and Paris, have commercial revenues that account for up to 60 percent of total revenues. Other airports that have no strong commercial development, either as a policy decision or lack of opportunity, would typically expect only 10 percent of their income to come from commercial sources.

There has been much discussion on the wisdom of commercial operations at airports. Opponents claim that commercial facilities produce unneces-

Figure 8.5 Duty-free shops—Bahrain.

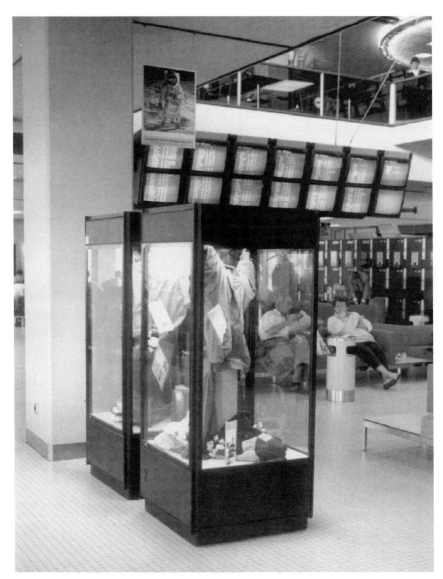

Figure 8.6 Advertising display case—Amsterdam.

sary and undesirable obstructions to the principal passenger-processing function of terminals; opponents also claim that airports, as instruments of governmental transport policy, should not engage in unnecessary commerce and that commercial operations therefore should be restricted to the absolute minimum. The proponents of commercial facilities argue that there is a demand for such facilities generated by high volumes of passengers, who spend on the average 1 hour in a terminal and of this time only 40 percent is required for processing. The high volumes of passengers, meeters, senders, and visitors constitute a strong potential sales market that invariably can be developed if desired. Furthermore, the revenues generated by commercial operations can eliminate or reduce governmental subsidy and can cross-subsidize airside operations. The argument against commercialization of the airport operation seems to have been lost. With the trend toward privatizing airports, most airports that have remained in the public domain have been required to adopt a more commercial approach to their operation.

The debate will not be engaged here. The policy adopted toward commercialization and privatization is likely to depend much more on political views than economic issues, and the question is therefore considered beyond the scope of this book.

If however, commercial exploitation of the airport is decided upon, a number of operational policy decisions must be made. First, a decision must be made on the mode of operation. Five different modes are common; these are operation by:

- A department of the airport authority directly

- A specially formed fully owned commercial subsidiary of the airport authority

- A commercial subsidiary formed by the airport authority and the airlines

- A commercial subsidiary formed by the airport authority and a specialist commercial company

- An independent commercial enterprise

Some publicly owned airports choose to retain direct control of commercial operations. This option, however, is unusual. Most airports that run highly successful commercial operations, such as Dubai, Heathrow, and Frankfurt, prefer to use an approach of granting controlled concessions to independent enterprises with commercial experience in the particular area. However, Aer Rianta, the Irish Airports Authority, operates many of its own concessions through its highly successful commercial division, which also acts as a concessionary management organization to other airports. The contractual arrangements between the concessionaires and the authority ensure certain standards of service to the consumer and guaranteed profit levels to the authority: beyond these guarantees, the concessionaire is free to use his enterprise to maximize commercial opportunities and therefore profit. Hybrid arrangements in which the authority collaborates either with the commercial departments of the airline or directly with specialized enterprises have been equally successful. Table 8.1 shows how various concessionary arrangements have been handled in a number of international airports.

TABLE 8.1 Operational Mode of Concessions at Selected Airports

Facility	Atlanta Hartsfield, U.S.A.	Sydney, Australia	Rome, Italy	Linate Milan, Italy	Malpensa Milan, Italy	Oslo, Norway	Bristol, U.K.	Toronto, Canada, T3	Vancouver, Canada	Chicago O'Hare, U.S.A.	Gatwick London, U.K.	Boston Logan, U.S.A.	Lagos, Nigeria	Pittsburgh, U.S.A.	Annaba, Algeria	Toronto, Canada T1	Changi, Singapore	Dublin, Ireland	Dublin, Ireland	Nice, France	Dusseldorf, Germany	Shannon, Ireland	Oakland, U.S.A.	London City, U.K.	Manchester, U.K.	Amsterdam, Netherlands	Bahrain, Bahrain	Birmingham, U.K.	Vienna, Austria	Marseilles, France	Toronto, Canada T2	Madrid, Spain	Rio de Janeiro, Brazil	Munich, Germany	London Heathrow, U.K.
Duty and Tax Free																																			
Duty free liquor & tobacco	A	A	B	A	A	C	A	A	A	A	A	A	A	A/C	A	A	A	A	B	A	B	A	X	A	A	A	A	C	A	A	A	A	A	C	A
Tax free shop (other goods)	A	A	A	A	A	C	A	A	A	A	A		A		A	A	A	A	B	A	B	A	X	A	A	A	A	A	A	A	A	A	A	X	A
Specialized Shops and Facilities																																			
Bookstalls	A	A	A	A	A	C	A	A	A	A	A	A	A	A	A	A	A	A	A	B	A	A	A	A	A	A	A	A	A	A	A	A	C	A	A
Gift shop	A	A	A	A	A	C/D	A	A	A	A	A	A	A	A	A	A	A	A/B	A	A	B	A	X	A	A	A	A	A	A	A	X	A	A	C	A
Jewellry & gems	X	A	A	A	A	D	X	A	X	A	A	A	A	A	X	A	B	A	X	A	A	A	A	A	A	A	A	A	A	X	A	A	X	A	A
Clothing	X	A	B	A	A	C/D	X	A	A	A	A	A	A	A	X	X	A	B	A	A	B	A	X	A	A	A	A	A	X	X	X	A	A	C	A
Confectionery, etc.	A	A	A	A	A	C/D	X	A	A	A	A	A	A	A	A	A	A	A/B	A	A	B	A	X	A	A	A	A	A	A	A	X	A	A	X	A
Pharmacy	X	A	A	X	X	C	X	X	A	X	A	X	A	X	X	X	A	A	A	A	X	X	X	A	A	A	A	A	A	A	A	A	A	X	
Perfumes	A	A	B	A	A	C	X	X	A	X	A	A	X	A	X	X	A	B	A	A	B	X	X	A	A	A	A	A	X	X	A	A	A	X	A

Airport	Food	Flowers	Photographic equipment	Electrical goods	Hairdressing	Dry cleaners	Other	Car hire	Hotel reservations	TV	Cinema
Atlanta Hartsfield, U.S.A.	A	X	X	X	A	X		A	X	A	X
Sydney, Australia	A/C	A	A	A	A	X		A	A	A	X
Rome, Italy	A	X	A	X	A	X		A	A	X	X
Linate Milan, Italy	A	X	X	X	A	X		A	B	X	X
Malpensa Milan, Italy	A	X	X	X	X	X		A	B	X	X
Oslo, Norway	C/D	C	C/D	X	C	X		A	C	X	X
Bristol, U.K.	A	X	X	X	X	X		A	A	X	X
Toronto, Canada, T3	A	A	A	A	A	A		A	A	X	X
Vancouver, Canada	A	A	A	A	A	A		A	A	X	X
Chicago O'Hare, U.S.A.	A	X	X	X	X	X		A	X	X	X
Gatwick London, U.K.	A	X	A	X	X	X		A	A	B	X
Boston Logan, U.S.A.	A	A	A	A	A	A		A	A	D	X
Lagos, Nigeria	A	X	A	A	A	A		A	X	B	X
Pittsburgh, U.S.A.	A	A	A	X	A	X		A	A	X	X
Annaba, Algeria	B	X	X	X	X	X		A	X	X	X
Toronto, Canada T1	A	X	X	X	A	X		A	A	X	X
Changi, Singapore	A	A	A	A	A	A		A	A	A	X
Dublin, Ireland	A	A	B	B	A	X		A	A	X	X
Nice, France	A	A	A	A	X	X		A	A	X	X
Dusseldorf, Germany	A	A	A	A	A	A		A	A	X	X
Shannon, Ireland	B	A	B	B	B	X			C	B	X
Oakland, U.S.A.	A	A	X	X	A	A		A	X	D	X
London City, U.K.	A	A	X	X	X	X		A	A	X	X
Manchester, U.K.	X	A	X	X	A	A		A	A	X	X
Amsterdam, Netherlands	A	A	A	A	A	X		A	A	A	X
Bahrain, Bahrain	A	A	A	A	X	X		A	B	B	B
Birmingham, U.K.	X	A	A	A	X	X		A	B	X	X
Vienna, Austria	A	A	A	A	A	X		A	A	X	X
Marseilles, France	A	A	X	X	X	X		A	B	X	X
Toronto, Canada T2	A	A	X	X	A	A		A	B	X	X
Madrid, Spain	A	X	X	A	A	X		A	A	X	X
Rio de Janeiro, Brazil	A	A	X	A	X	X		A	A	X	X
Frankfurt, Germany	A	A	A	A	A	A		A	A/B	X	A
Munich, Germany	C	A	C	C	C	X		A	A	X	X
London, Heathrow, U.K.	A	X	A	A	X	X		A	A	X	X

Key:

A. Concessionery operation. B. Operation by direct airport authority. C. Operation by subsidiary company of the airport. D. Operation by airline. X. Not available at this airport.

It is also interesting to compare the way in which concessionaires are selected. Some airport authorities are required by law to accept the highest bid for a concession. Schiphol Airport in Amsterdam has developed a successful commercial policy based rather on maximizing the level of airport control on operating standards and pricing. In this way, the airport feels it is more able to attain its own commercial ends while still using the expertise of the individual concessionary enterprises. Concessions at airports may be leased in a number of ways:

- Open tender
- Closed tender
- Private treaty

Of these three, it is most likely that the second option, Closed tender, will meet the airport's requirements. Private treaty is likely to be seen to be a too restrictive manner of handling public funds, leading to charges of preferential treatment. Open tender, on the other hand, while giving a free hand to competition, may well lead to bidding by organizations that will prove to be incompetent in reaching necessary performance standards. In some countries, however, open tenders are legally required where public funds are involved. Under these conditions, it is sometimes permissible to have a prequalification arrangement to ensure that only competent and financially stable enterprises enter the bidding procedure.

Other methods of control that have been used successfully are:

1. *Length of lease.* Medium term leases of 5 to 10 years have several advantages. They permit the concessionaire to run an established operation with medium-term profits. Successful operators are usually able to renegotiate for renewed concessionary rights. Unsuccessful operators can be removed before long-term financial damage accrues to the airport.

2. *Exclusive rights.* In return for exclusive rights on the airport, the authority can demand contractual arrangements that protect the airport's financial and performance interests. There is a significant recent move away from granting exclusive rights in shopping concessions in order to encourage competitive pricing.

3. *Quality of service.* Many airports, require contracts that restrict the concessionaire's methods of operation. These constraints include: authority control over the range of goods to be stocked, profit margins and prices, and staffing levels as well as detailed operational controls on such items as advertising, decor, and display methods.

Advertising is an area of financial return that has not been fully explored by many airports. The advertising panel shown in Figure 8.6 is an example of a very satisfactory modern display that adds to the decor of the terminal without clutter, while paying a handsome return to the authority from a little financial outlay. Care must be taken in selecting advertising so that the displays do not interfere with passenger flow or obstruct necessary informational signs. Significantly, there are airports that ban internal advertising on aesthetic grounds, but these are growing fewer in number.

8.5 Airline Related Passenger Services

Within the airport passenger terminal, many operations are usually handled entirely by airlines or their agents, including:

- Airline information services
- Reservations and ticket purchase
- Baggage check-in and storage
- Loading and unloading baggage at aircraft
- Baggage delivery and reclaim (reclaim is often under authority control)
- Airline passenger "club" areas, sometimes called CIP, commercially important persons, facilities

These areas are part of the service offered to the traveler by the airline, and as such the airline has an interest in retaining a strong measure of control in the service given. Such control is most easily obtained by carrying out this particular area of the operation. It is important to remember that the basic contract to travel is between the airline and the passenger. The airport is a third party to this contract, and as such should not intrude into the relationship more than is necessary. Where airports remove the general handling responsibility from the airlines, there might be a lowering of passenger service, since there is no overt contract between the passenger and the airport. Service levels are more likely to be maintained where the direct customer relationship has some influence on services performed. The relationship becomes complicated when the airport is privatized and has extensive terminal commercial operations. The passenger in this case also becomes the airport's client in a very real sense. Figures 8.7 and 8.8 show check-in and baggage delivery areas where the design of the facility emphasizes that the passenger is still under the care of the airline. A more common arrangement for baggage claim areas outside the U.S. is that the claim area is operated by the authority, whereas the airlines have the responsibility of delivering bags to the claim area. This more common arrangement often results in authority staff receiving passenger abuse for delayed, lost, or damage baggage when, in fact, the receiving airport has had no involvement in its handling and thus bears no responsibility for the default.

Figure 8.7 Check-in showing area under care of airline.

Figure 8.8 Designated baggage area.

In recent years, airports have attempted to obtain better usage of the check-in desks by the adoption of the common user terminal equipment (CUTE) in the check-in area. Use of CUTE technology permits the switching of desks among airlines according to their real demand for desks, which is likely to vary both seasonally and over the day. Many airlines have resisted the introduction of CUTE because it prevents an airline having a permanent presence in the terminal, whether or not it has operations at a particular time. Most new terminals are being designed with CUTE systems where there are shared facilities.

8.6 Airline related Operational Functions

Flight dispatch

A major preoccupation for airline management in relation to airport terminal operations is the achievement of on-time departures. Many of the activities associated with this, such as the refueling and cleaning aircraft together with the loading of food supplies, are carried out on the ramp and are familiar to most airport staff. There is, however, a less familiar procedure that covers all the necessary technical planning without which the flight could not depart. The main activities associated with this procedure of flight dispatch are:

- Flight planning
- Aircraft weight and balance
- Flight crew briefing
- Flight watch

In the United States, this is a long-established procedure and the work is carried out by aircraft dispatchers who work in close cooperation with the aircraft captain. Although aircraft dispatchers are used by many international airlines there is also the designation of flight operations officer for staff who carry out this work.

The airline departments at airports concerned with flight dispatch will need access to airport operations departments, air traffic services, meteorological services, and communications facilities, including teleprinter, telephone, and radio. Depending on the extent of their activities, many airline operations' offices will also use a variety of computer facilities, though these latter might not necessarily be in-house systems.

Flight planning

The primary purpose of flight planning is to determine how long an individual flight will take and how much fuel will be required. For long-range flights there will be a variety of options in terms of altitudes, tracks, and aircraft power settings and speeds. Variations in weather, wind, and temperature also will have to be taken into account. Because of these various factors, it is usual for an initial evaluation or preflight analysis to be carried out. This examines all feasible options so that a decision can be taken as to the most appropriate of the several alternatives. The evaluation might include an indication of comparative costs: a slower flight might prove desirable from a cost point of view. The analysis would include several altitude options. This often proves useful if, due to density of traffic, ATC has to impose a last minute altitude change. An example of a typical altitude comparison is given in Figure 8.9 for a flight from Los Angeles to Chicago. Once it has been decided which alternative seems most appropriate, an expanded format is used for a flight plan that shows individual segments of the flight between the various en route reporting points. An example of a full flight plan is given in Figure 8.10.

For short-range flights there are generally very few options, and in areas of very dense traffic, routings for all practical purposes are predetermined by the structure of the airways. In such cases, as for example in Europe, the flight plans will usually be standardized to the extent that relevant extracts

A	B	C	D	E	F	
37	3818	03:12	508	000	M83	➤ Base flight plan
33	3836	03:10	526	900	M83	
29	3879	03:09	569	2033	M83	
27	3909	03:08	599	2275	M83	
25	3941	03:07	631	2460	M83	

A – Highest flight level (thousands of feet)

B – Planned takeoff gross weight (hundreds of pounds)

C – Total flight time enroute

D – Total fuel burnoff (hundreds of pounds)

E – Amount of fuel burnoff (lbs) for each minute gained on Base Flight Plan.*

F – Mach number used to compute analysis at various altitudes

*Base Flight Plan – Standard used for comparison (analysis)

Figure 8.9 Flight analysis—altitude comparison. *(United Airlines)*

Flight plan

DC10 Los Angeles (LAX) to Chicago–O'Hare (ORD) –1580 n.m.
Route–LAX.. DAG.J146.GLD.J192.PWE.J64.BDF.V10.VAINS..ORD

⎡ Daggett vio Jet Airway 146 to Goodland vio Jet Airway ⎤
⎢ 192 to Pawnee City via Jet Airway 64 to Bradford via ⎥
⎣ Victor Airway 10 to Vains and Chicago–O'Hare.   ⎦

A	B	C	D	E	F	G	H	I	J	K	L	M
RCA	256	37	828	09	51	486	26045	044	530	36	148	588
DVC	308	37	827	07	50	483	29071	056	539	35	91	497
GUC	98	37	826	07	49	482	29086	056	538	11	28	469
GLD	257	37	825	06	45	480	31094	053	533	28	73	396
PWE	258	37	824	04	40	478	31095	058	536	29	71	325
LMN	105	37	823	04	38	477	30087	056	533	12	29	296
POD	166	37	822	03	37	475	29053	046	521	19	45	251
–ORD	132						27030	028		23	21	230

A	– Flight plan check point

H	– Wind direction (26 = 260°) and speed in knots 045 = ⸳ 5kts

B	– Segment mileage (n. m.)

I	– Head or tail wind component (headwind "_")

C	– Flight level (thousands of feet)

J	– Ground speed

D	– Indicated Mach number

K	– Segment time (minutes)

E	– Deviation from standard temperature (all plus values)

L	– Segment fuel burnoff (hundreds of lbs.)

F	– Tropopause height

M	– Fuel remaining (hundreds of pounds)

G	– True airspeed (Kts.)

Note – total flight time 3 hrs. 13 mins.   Total fuel burnoff 50,600 lbs.

Figure 8.10 Flight plan—United Airlines. *(United Airlines)*

can be placed on permanent file with ATC. These are referred to in Great Britain as *stored flight plans* and are automatically printed out from ATC computer files in advance of flight departures. The airline flight plans, the operational or company flight plans, give a great deal of information including the en route consumption of fuel. Such details are not the concern of ATC, which requires altitudes and times in relation to the ATC system check points together with certain safety details (e.g., number of persons on board the aircraft and the detail of the instrument-flying aids and safety equipment carried by the aircraft). The international format for the ATC flight plan is shown in Figure 8.11.

FLIGHT PLAN   ATS COPY

PRIORITY

ADDRESSEE(S)

<<= FF →

<<=

FILING TIME   ORIGINATOR

→   <<=

SPECIFIC IDENTIFICATION OF ADDRESSESS(S) AND/OR ORIGINATOR

---

3. MESSAGE TYPE   7. AIRCRAFT IDENTIFICATION   8. FLIGHT RULES   TYPES OF FLIGHT

–

9. NUMBER   TYPE OF AIRCRAFT   WAKE TURBULENCE CAT   10. EQUIPMENT

–   /   /   <<=

13. DEPARTURE AERODROME   TIME

–   <<=

15. CRUISING SPEED   LEVEL   ROUTE

–   →

<<=

16 DESTINATION AERODROME   TOTAL EER HR. MIN   ALTN AERODROME   2ND ALTN AERODROME

–   →   →   <<=

18 OTHER INFORMATION/

–

)<<=

SUPPLEMENTARY INFORMATION (NOT TO BE TRANSMITTED IN FPL MESSAGES)

19   ENDURANCE   EMERGENCY RADIO

HR. MIN   PERSONS ON BOARD   UHF   VHF   ELBA

–E/   → P/   → R/ U   V   E

EURVIVAL

EQUIPMENT POLAR   DESERT   MARITIME   JUNGLE   JUNGLE   LIGHT   FLUDRES   UHF   VHF

S   P   D   M   J   → J   / L   F   U   V

DINGHIES

NUMBER   CAPACITY   COVER   COLDUR

→ D /   →   → C →   <<=

AIRCRAFT COLOUR AND MARKINGS

–A/

REMARKS

N   <<=

PILOT-IN-COMMAND

C/   )<<=

FILED BY

SPACE RESERVED FOR ADDITIONAL REQUIREMENTS

CA48/RAF F2919 (Revised November 1984)   Carts. D. O.8810

Figure 8.11 International flight plan.

Aircraft weight and balance

After the fuel required for a particular flight has been determined, it is possible to proceed to a calculation of the weight available for the carriage of passengers, mail, and cargo (payload). It should be noted that these calculations may be in either pounds (lb), which is the case in the United States, or in kilograms (kg). However, before any actual load calculations can be carried out, account must be taken of the physical weight limitations, the design limits, of the aircraft structure in the various operation phases.

Takeoff

There is a maximum takeoff weight (i.e., at brake release) that the available power can lift off the runway and sustain in a safe climb. The value is established by the manufacturer in terms of ideal conditions of temperature, pressure, runway height, and surface condition. Along with these values, the manufacturer will provide performance details for variations in any of these conditions.

In flight

There are limits on the flexibility of the wings of each aircraft design. These are imposed by the upward bending loads that the wing roots can sustain without breaking. The greatest load would be imposed if there was no fuel remaining in the wings (fuel cells), so this zero fuel weight is taken as a limitation on fuselage load.

Landing

Depending on the shock-absorbing capabilities of the aircraft undercarriage, there is a maximum landing weight that it can support on landing, without collapsing. Thus the three design-limiting weights are maximum takeoff weight, maximum zero fuel weight, and maximum landing weight. Typical examples of these values for a Boeing 747-300 are:

- Maximum takeoff weight 883,000 pounds (377,850 kg)
- Maximum zero fuel weight 535,000 pounds (242,630 kg)
- Maximum landing weight 574,000 pounds (260,320 kg).

The completed flight plan will provide two fuel figures:

Takeoff fuel. The total amount of fuel on board for a particular flight. This does not include taxi fuel but will include required fuel reserves for flight to an alternative destination or for holding or delay before landing.

Trip fuel. Is that fuel required for the trip itself, that is, between takeoff and point of first intended landing: also sometimes referred to as *burnoff.*

In order to arrive at the maximum permissible takeoff weight, we compare three possible takeoff weights:

- Takeoff weight' = maximum takeoff weight
- Takeoff weight'' = zero fuel weight + takeoff fuel
- Takeoff weight''' = landing weight + trip fuel

The lowest of these three values is the maximum allowed takeoff weight, and this value minus the operating weight will give the allowed traffic load. These and other values are used in relation to aircraft weight calculations and load, and they also appear on the load sheet, for which there is a format agreed by the International Air Transport Association (IATA). Together with the values for takeoff fuel and trip fuel, the following operational figures are included in a load-sheet calculation:

- *Dry operating weight.* The weight of the basic aircraft, fully equipped, together with crew and their baggage, pantry/commissary supplies, flight spares but not including fuel and payload.
- *Operating weight.* The sum of dry operating weight and takeoff fuel.
- *Takeoff weight.* The operating weight plus payload (traffic load).
- *Total traffic load.* The sum of the weights of the various types of load, that is, passengers, baggage, cargo, mail, and also the weight of any unit load devices (ULDs) (containers) not included in the dry operating weight. All these various weights appear on the load sheet together with a breakdown of the weight distribution.

Balance/trim

Having ensured that the aircraft load is within the permitted weight limitations, it is then necessary to distribute the load in such a way that the center of gravity is within the prescribed limits. This is calculated by means of a trim sheet, which might be a separate form or a part of a combined load and trim sheet (Figure 8.12). On the trim diagram, each of the aircraft compartments is given a scale graduated either in units of weight, for example 2200 lbs (1000 kg) or blocks of passengers (e.g., 10 passengers). Starting from the Dry Operating Index scale the effect of weight in each compartment is then indicated by moving the required number of units along the scale in the direction of the arrow and dropping a line down from that point to the next scale, where the process is repeated ending up with a line projecting down into the CG envelope, where its value is noted as a percentage of the wing mean-aerodynamic chord (MAC). The outer limits of the envelope are clearly indicated by the shaded areas. Certain sections of the load sheet side of the form are also shaded to indicate data that should be in-

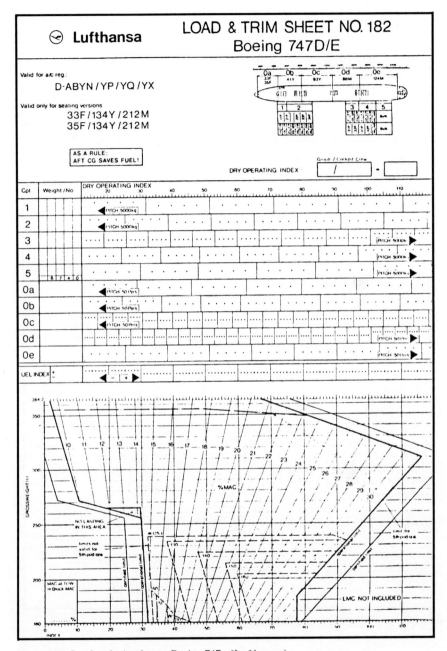

Figure 8.12 Load and trim sheet—Boeing 747. *(Lufthansa)*

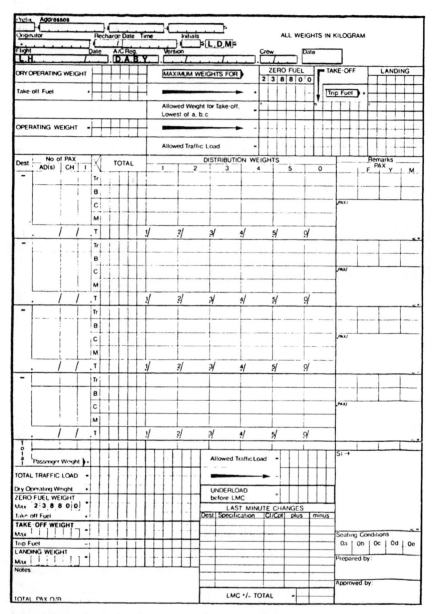

Figure 8.12 Load and trim sheet—Boeing 747. *(Lufthansa)*

cluded in a load message to be transmitted to the aircraft destination(s). These functions are now almost universally computerized.

Loading

The distribution of the load into various compartments must be detailed for the information of ramp loading staff, and this is achieved by the issue of loading instructions, usually in the form of computer drawn diagrams. In Figure 8.13, the details are given of the various container positions. Where containers are not used, it will be necessary at this stage to take into account limitations in respect to dimensions vis-à-vis the measurements of the hatch openings and also maximum floor loadings, and the loading instructions will be drawn up accordingly.

All matters relating to the load carried on an aircraft and the position of the C G have such a direct influence on flight safety that the documents used are of considerable legal significance, reflecting as they do the regulations of each country. For this reason, they have to be signed by the airline staff responsible for these various aspects.

Flight crew briefing

The purpose is to present to flight crew appropriate advice and information to assist them in the safe conduct of a flight. The information will include a

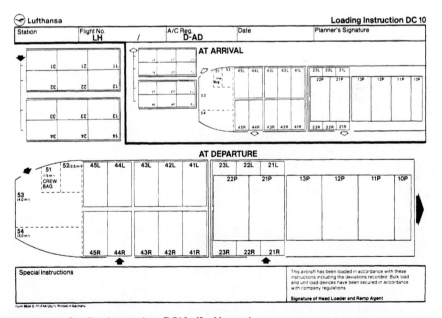

Figure 8.13 Loading instructions DC10. *(Lufthansa)*

flight plan and load details together with information regarding en route and destination weather and notices regarding any unserviceabilities of navigation or landing aids. This latter information is contained in Notices to Airmen (NOTAM), an internationally agreed system whereby the civil aviation authorities of each country exchange information on the unserviceability of any of the facilities in their country (e.g., navaids and airports). Airline flight dispatch staff will obtain NOTAM from the appropriate governmental agency, edit them, and, where necessary, add details relating to any company facilities. Weather information will also be obtained from the meteorological department at the airport and might be augmented by in-flight reports received from other flight crews. An example of the presentation of briefing information is given in Figure 8.14, for a flight from Los Angeles (LAX) to Chicago (ORD). Further details of the NOTAM system and of the various kinds of weather information available are given in Chapter 11.

```
LAXFO
..LAXFOUA 231828 2970/ROS
WBM 108-23      LAX--ORD       RT: 19            ALTNT MKE
MAP FEATURES WESTERN U.S. 231652Z-240600Z
SFC TROF XTNDS NWWD FROM THE GULF OF CALIF THRU CALIF TO WRN
OREG.HIGH CNTRD OVER XTRM NW MONT DRFTG SE.FOG AND ST OVT THE
PAC NW LIFTING.MARINE ST CONTRL CALIF COAST LIFTING AND
RETURNING TOWARD 06Z.


LAX 1748 150 SCT E280 OVC 4H 110/90/51/1704/986/ 802 1087 65
LAX FT23 231616  180 SCT 250 SCT. 20Z 180 SCT 250 SCT 2512. 02Z
    CLR. 10Z VRF..
LAX NO 9/4 7L/25R OPEN 1600
LAX NO 9/5 EFF 1600 THR 25R DSPLCD 962
LAX NO 9/10 VORTAC OTS 20-2200
LAX NO 13/1 GATE 80 TOW-IN GATE. GATE 72 D737 ONLY NO
    ACCU-PARK, WILL HAVE SIGNALMAN. GATE 74 IS TOW-IN GATE FOR
    DC10 WHEN 83A OR 83 OCCUPIED.
LAX NO 13/2 RWY 25R/07L OPEN FOR ALL UAL ACFT, NO STRUCTUAL.
    WT RSTN, WIDEBODY'S DO NOT APPLY T/O THRUST UNTIL TAXIWAY OJ


ORD 1750 E100 DKN 250 DKN 15 164/64/44/2016/001/ 717 1031 40
ORD FT23 231515 80 SCT 250 --DKN 1810 SCT V BKN. 18Z C80 DKN 250
    DKN 2112 LWR DKN V SCT CHC C30 DKN 3TRW AFT 21Z. 097 VFR..
ORD NO 9/98 14R--32L CLSD 02--1400
ORD NO 9/106 14L--32R CLSD 16--1800
ORD NO 9/108 9R--27L CLSD 241100--1600
ORD NO 9/109 14R--32L CLSD 02--1400 NIGHTLY THRU 11/24 EXCP
    SUN


UA571 /0V HCT 1708 F390/TA MS56/WV 300 TO 305/WV AT 105 TO 110
    KNOTS/TD SMTH/SK NO CLDS BLO CLR
UA235 /0V GCK 1711 F390/WX OVER OCK WIND 045125 TS PLUS3... OVER
    CIM 1743Z WIND 035100 CLEAR WEST OF GCK
UA709 /0V SLN 1643 F350/TA MS26/WV 30090/SK CLD TOPS FL360/TD
    LT TURBC IN CLDS
UA235 /0V SLN 1649 F390/TD FL 390 SMOOTH ACFT BLG REPORTING MOD
    CHOP.TTSM  ACTVTY 40 DME N.IRK TOPS 890 EASY DETOURABLE
```

Figure 8.14 United Airlines flight crew briefing sheet.

Flight watch (flight control)

This is a procedure by which flight dispatch/flight operations personnel monitor the progress of individual flights. For this reason, it is also sometimes described as *flight following.* (Not to be confused with flight following by ATC in the U.S. for VFR aircraft.) Due to the worldwide nature of air transport, it is carried out using Greenwich Mean Time (GMT), sometimes written "Z" time. Flight watch is not intended to be entirely passive: however, information of any unexpected changes in weather or serviceability or facilities is transmitted to aircraft in flight. Depending on the extent of an airline route network, the responsibility for flight watch might be divided into areas. In addition, most larger airlines have one centralized coordinating operations center equipped with comprehensive communications facilities providing the latest information on the progress of all its aircraft. The center for United Airlines is located at Chicago; for Air Canada, at Toronto International Airport; and for British Airways, at London (Heathrow) Airport. It is useful for airport operations management to know the location and telephone/telex addresses of such centers for airlines using their airport as well as the organization of flight watch responsibility.

8.7 Governmental Requirements

Most airports handling passenger movements of any reasonable scale will be required to provide office and other working space in the vicinity of the passenger terminal for the civil aviation authority and the air traffic control authority, if this is separately constituted. At major airports where international passengers are handled, it is also possible that up to four governmental controls must be accommodated:

- Customs
- Immigration
- Health
- Agricultural produce

In most countries, the facilities necessary for health and agricultural inspection are not particularly demanding. On the other hand, customs and immigration procedures can be lengthy, and the requirements in terms of operational space for the examining process can be very great. Figure 8.15 shows the layout of an immigration hall at a major international airport. Because of the filtering effect of immigration and the relatively speedy processing at most customs examination halls, customs facilities are not usually space extensive. The use of red/green customs, procedures, especially in Europe, has materially improved customs processing time without any apparent deterioration in enforcement. Some countries, however, still have

Figure 8.15 Arrival, immigration.

very time-consuming and involved customs examination procedures that require the provision of many desks and extensive waiting areas. In addition to their processing areas, most governmental agencies require office and other support space, such as rest, changing, and toilet areas.

8.8 Nonpassenger-related Airport Authority Functions

It is often convenient at smaller airports to locate within the terminal building for ease of intercommunication, all the airport authority's nonpassenger-related functions. These include:

- Management
- Purchasing
- Finance
- Engineering
- Legal
- Personnel
- Public relations
- Aeronautical services
- Aviation public services (e.g., noise monitoring)
- Plant and structure maintenance

At larger airports, it is customary to separate these authority functions into distinct buildings or buildings away from the terminal building to ease the level of congestion associated with busier terminals. At multiple airport authorities such as Aeroports de Paris and the Port Authority of New York and New Jersey and the privatized multifunction BAA, many of the management and staff functions are carried out entirely off-airport, only the line-operating functions being staffed by airport-based personnel. The detailed design of a terminal should take great account of the way in which the authority intends to operate its facility, since space requirements revolve around operational procedures.

8.9 Processing Very Important Persons

Air travel is still a premium method of travel, attracting important, famous, and very rich individuals. Some of the busier airports process a large number of very important persons (VIPs). For example, it is estimated that more than 6000 groups of VIPs pass through London Heathrow in a year. This requires special facilities and staff to ensure that the arriving and departing party can pass through the terminal with all necessary courtesies, sheltered from the conditions of the average traveler. Consequently, VIP facilities have separate landside access, a fully equipped and comfortable lounge in which the party can wait for either landside or airside transport, and a separate access to the apron. The facility must be capable of holding fairly large parties; often traveling heads of state have VIP parties in excess of 25 persons. In addition to the needs for sufficiently large and adequately equipped accommodation, the facilities must be safe from the security viewpoint, since they might become the target of unlawful acts. Figure 8.16 shows the VIP lounge at a large airport. At multiterminal airports it is not unusual to have either several VIP lounge facilities or one central facility (Changi), to minimize congestion and inconvenience.

8.10 Passenger Information Systems

Passengers move through airport terminals under their own power. They are not physically transported in a passive manner as is freight, although in larger terminals, mechanical means are used to aid in the movement through the facility (see Section 8.12). This, of course, does not refer to physically challenged persons who need special ramps and other necessities, which are beyond the scope of this book. Equally important, a large proportion of passengers reach airports in their own personal vehicles. There is, therefore, a need to ensure that the passenger has sufficient information in both the access phase of the journey and in passing through the terminal to reach the correct aircraft gate at the right time with a minimum of difficulty and uncertainty. Additionally the passenger requires information on the location of many facil-

Figure 8.16 VIP lounge facilities—Bahrain.

ities within the terminal, such as telephones, toilets, cafeterias, and duty-free shops. Information is therefore usually functionally classified into either directional guidance or flight information categories. Directional guidance commences some distance from the airport and will normally involve cooperation with some local governmental authority to ensure that suitable road signing is incorporated into the road sign system on all appropriate airport access roads (Figure 8.17). Often such signs include an aircraft symbol to help the driver to identify directions rapidly. Nearer the airport, terminal approach road signs will guide the passenger to the appropriate part of the terminal. It is essential that the driver be given large clear signs in positions that permit safe vehicular maneuvering on the approach road system. The driver must obtain information on the route to be taken with respect to such divisions as arrivals/departures, domestic/international flights, and often to airline specific locations (Figure 8.18). In multiterminal airports, there will be signage to each individual terminal, either by terminal designation or by airline groups. Within the terminal, departing passenger flows are guided principally by directional guidance signs, which indicate check-in, governmental controls, departure lounges, gate positions, and so on. Other terminal facilities that must be identified are concessionary areas and public service facilities such as telephones, toilets, and restaurants (Figure 8.19). It is essential that the signing is carefully designed. ICAO has a set of recommended pictograms for signing inside terminals. Many airports have adopted their own signing convention. In some

Figure 8.17 Road sign to airport in Arabic and English with pictogram.

Figure 8.18 Signage with directions to specific airline terminal areas.

Figure 8.19 Information signs in terminal.

cases the signing used falls short of acceptable standards. Sufficient signage must be given to enable the passenger to find the facility or the direction being sought; on the other hand, there cannot be such a proliferation of signs that there is confusion. It is essential that the signing configuration is designed to conform with available internal building heights, which itself must be set recognizing that overhead signing is essential. Once in the terminal, information concerning the status and location of departing flights is conveyed by the departure side of the flight information system. Conventionally, this information has been displayed on mechanical, electromechanical or electronic departure flight information boards. More recently, these have been supplemented and to some degree supplanted by cheaper visual display units (VDUs), which can be located economically at a number of points throughout the terminal. Figure 8.20 gives an example of a modern bank of VDUs.

The arriving passengers are given similar guidance information, which helps convey them to the baggage reclaim area and to the landside access area, stopping en route at immigration and customs in the case of the international arrival. It is necessary to have adequate exit signing within the terminal for all passengers, and on the internal circulation roadways for those passengers using the car mode. An example of an airport road exit signing is shown in Figure 8.21. Meeters who have come to the airport to greet a particular flight are informed of flight status and location either by a conventional arrivals board (Figure 8.22) or by VDUs. Arrival and departure VDUs have the

Figure 8.20 Bank of Visual Display Units.

Figure 8.21 Terminal roadway exit signs.

advantage that they are readily compatible with computerized information systems and can be easily updated. The units themselves, which are relatively inexpensive, are easily removed, replaced, and repaired in the case of failure.

Most airport operators supply at least one airport information desk per terminal on the departures side. This worker-staffed desk, an example of which is shown in Figure 8.23, supplies information that goes beyond that supplied by the visual systems. Also it is capable of assisting those unable to use the automatic system for one reason or another. In the case of failure of the automatic systems, the only means of providing flight status and location might be through the manned desk.

8.11   Space Components and Adjacencies

Earlier it was stated that the organization of a terminal must closely follow operational strategies and requirements if the terminal is to function adequately. Consequently, no hard-and-fast rules can be set down for the overall division of terminal space. However, Figure 8.24 provides a rough guide of the functional distribution of terminal space in a typical U.S. airport. More than half the terminal area is likely to be rented out if baggage rooms are included in the figure of rented areas. For detailed estimates of the terminal space requirements, it is suggested that the reader refer to design texts and guides (Ashford and Wright 1992, FAA 1996, FAA 1988, FAA 1980). However, there still remains the question of interrelationships of the provided spaces, that is, adjacencies that are operationally desirable. In a typical terminal layout, there are several facilities that ideally should be

Figure 8.22   Arrivals board.

Figure 8.23 Worker-staffed information desk.

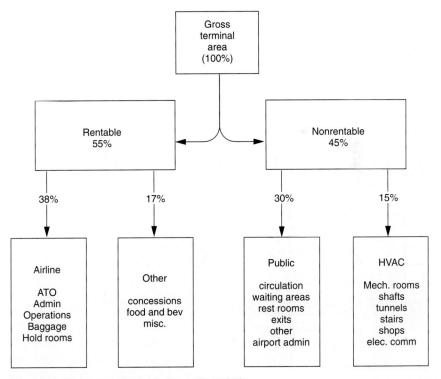

Figure 8.24 Terminal space distribution. *(FAA 1976)*

grouped in close proximity, whereas juxtaposition of other facilities is nonessential. For example, grouping is desirable for concession areas, but nonessential for airline administration areas. It is essential, however, that customs-check areas be in the immediate vicinity of the baggage claim. Figure 8.25 indicates a functional adjacency chart published by IATA to aid in the location of terminal facilities. These adjacencies are of as much interest to the authority that must operate a facility as they are to the designer (Hart 1985, Blow 1996).

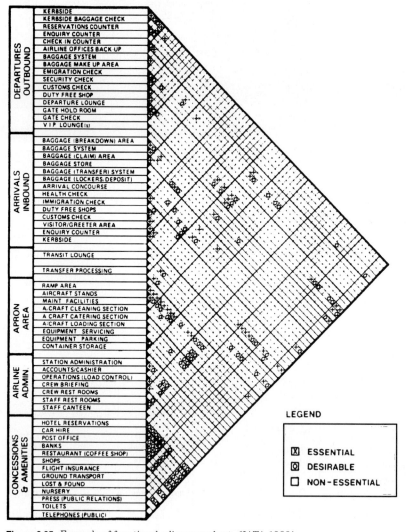

Figure 8.25 Example of functional adjacency chart. *(IATA 1989)*

8.12   Aids to Circulation

Large airport terminals with multiple gate positions for large transport aircraft necessarily involve large internal circulation distances. At some largely single-terminal airports such as Chicago O'Hare, the distance between extreme gate positions is close to 1 mile (1½ km), and the distance from the center of the car parking area to the extreme gate is about the same. To ease the burden of walking long distances, it is now becoming common for airports to install some form of mechanized circulation aid. In airports with multiterminal designs (e.g., Kansas City, Charles de Gaulle, Seattle, New York JFK, Houston), remote piers (e.g., Atlanta and Pittsburgh), and remote satellites (e.g., London Gatwick, Miami, Tampa, Orlando), the distances can be very large, and mechanized movement becomes essential. For example, if the ultimate construction of the New Seoul International Airport (NSIA) is built to the master plan there will be more than 5 miles (8.5 km) between the extreme terminals. In any case, it is now becoming common practice to provide mechanized assistance where practical when walking distances exceed 1500 feet (450 m).

Three main methods of movement assistance are used:

1. *Bus.* Used to link unit terminal in multiple terminal operations (e.g., Paris Charles de Gaulle, New York JFK, Los Angeles, London Heathrow)

2. *Pedestrian walkways.* Used within piers and to connect to remote satellites or railway stations (e.g., Paris Charles de Gaulle, London Heathrow, Los Angeles, Atlanta, Barcelona)

3. *Automatic people movers.* Used to make connections with remote piers, railway stations or between terminals (e.g., Miami, London Gatwick, Orlando, Tampa, Atlanta, Houston, and Singapore).

Pedestrian walkways are an older and now fairly widely used technology in which there is a great deal of experience. Their great limitation is their speed, which must for safety reasons of boarding and alighting be kept to approximately 1.5 mph (2.5 kmh). For very long distances, therefore, they are unsuitable. Another disadvantage is the fact that there are technical reasons that limit their length. There is also the likelihood that at least one in a chain of walkways will be inoperable due to failure as the devices age. They must also be operated in one direction, which means that unlike a two-track people-mover system, one direction cannot operate in the shuttle mode should there be a failure in the other direction. Under conditions of equipment failure, walking might be the only other option.

A number of larger airports now use people movers, automatic vehicles acting essentially as "horizontal elevators" and capable of moving passengers at top speeds of approximately 30 mph (45 kmh). Figure 8.26 shows the subterranean tunnel connecting the airside remote piers to the main terminal area in Atlanta. Passengers can connect to the piers by walking, by using the moving walkways or by using the loop people-mover system, which can be entered at one of the pier stations. One of the first such connector vehicles was the kind used to connect the terminal to the satellites at Tampa and at Miami, as shown in Figure 8.27 and terminal to terminal as shown in Figure 8.28. Such automatic systems reduce manpower, but require extensive control systems (Figure 8.29). It is usual to provide maintenance areas such as that shown in Figure 8.30 either within the terminal area or close to one of the satellite areas it feeds. Where such systems are used, it is necessary to provide station areas, track, control room, maintenance areas, appropriate emergency evacuation areas, and escape points in addition to alternative methods of travel in case of failure.

System reliability is extremely important, because without the people mover, the design of the terminal area is no longer coherent— the passenger would be subjected to intolerable walking distances. Therefore, the airport authority sets high performance standards for such equipment. It is usual to require several months of break-in operation prior to carrying passengers. Authorities then specify system availability of 98 percent during the first few months of operation and subsequent performance at 99.5 percent availability. From operating systems, it is apparent that 99.9 percent availability is possible with current systems. A common arrangement is for the equipment manufacturer to operate and maintain the system for a period of the first two years and to perform subsequent maintenance on contract.

Figure 8.26 Walkway tunnel connecting piers to terminal area in Atlanta.

Figure 8.27 People mover between terminal and satellites.

Figure 8.28 People mover, terminal to terminal.

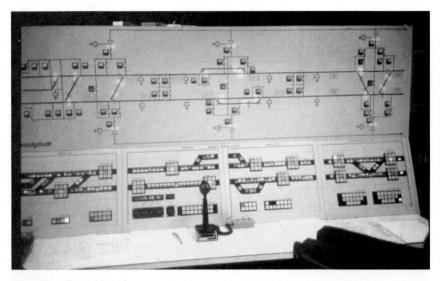

Figure 8.29 Control board.

8.13 Hubbing Considerations

In the last 15 years, particularly since deregulation, airlines have tended to set up hub and spoke operations to improve service frequency, load factors and available destinations. Consequently, a number of airports in the United States and elsewhere have become *hub* airports where passenger transfers are common and might amount to more than two-thirds of the total traffic (e.g., Dallas-Fort Worth and Atlanta). In some cases the hub operation is airline driven (e.g., Pittsburgh with USAir and Birmingham with British Airways). In other cases, the policy is airport driven where interlining as well as on-line transfers are encouraged (e.g., Manchester).

Hub terminals differ considerably from origin-destination terminals. They must accommodate large numbers of passengers moving between gates at the terminals rather than from the landside to the gate and vice versa. Similarly a large proportion of passenger baggage must be handled for on-line or interline transfer rather than being originating or destined baggage.

A hub terminal must be designed and operated to handle satisfactorily waves of passengers fed by banks of arriving and departing aircraft. During a single day at a major hub there might be as many as twelve such waves. Recognizing that the inter-gate transfers might require considerable distances to be covered in relatively short connection times, large hubs require mechanized aids to circulation which are speedy and reliable (Pittsburgh, Atlanta). Where the facility has to act as a hub between international and domestic flights (e.g., Birmingham) particular attention must be paid to customs and immigration facilities to ensure that connections can be made. International hub terminals (Singapore, Dubai, Bahrain) often develop extensive commercial facilities for tax-free and duty-free shopping with the knowledge that passengers are likely to have some free time for shopping during the connection. Even domestic hubs have developed extensive commercial facilities, which are designed to attract impulse buyers with time to spare (Pittsburgh and London Heathrow Terminal 1).

The requirements for baggage handling at hub terminals differ greatly from origin-destination airports. It is essential that there is a rapid and accurate on-line and interlining baggage transfer capability. The operational cost to airlines of mishandled baggage is unacceptably high where this cannot be guaranteed. The situation becomes even more complicated where domestic and international flights are concerned. ICAO regulations require that passengers and their baggage are reconciled to ensure that unaccom-

Figure 8.30 Subway train from front with rail visible.

panied bags of no-show passengers are not permitted on international flights. Should a passenger not make the connection, loaded bags should be unloaded from the aircraft, a costly and time consuming source of aircraft delay.

References

Ashford, N., and P. H. Wright. 1992. *Airport Engineering, 3rd edition.* New York: Wiley-Interscience.

Blow, C. 1996. *Airport Terminals, 2nd edition.* London: Butterworth.

Federal Aviation Administration (FAA). 1976. *The Apron and Terminal Building Planning Report, FAA-RD-75-191.* Washington, DC: Department of Transportation.

FAA. 1980. *Planning and Design of Airport Terminal Facilities at Non-Hub Locations, AC150/5360-9.* Washington, DC: Department of Transportation.

FAA. 1988. *Planning and Design of Airport Terminal Facilities, AC150/5360-13.* Washington, DC: Department of Transportation.

Hart, W. 1985. *The Airport Passenger Terminal.* New York: Wiley-Interscience.

International Air Transport Association (IATA). 1989. *Airport Terminals Reference Manual, 7th edition.* Montreal.

International Civil Airports Association (ICAA). 1979. *ICAA Manual on Commercial Activities, amended.* Paris.

Airport Security

"The wise man avoids evil by anticipating it."
PUBLILIUS SYRUS, 1ST CENTURY BC

9.1 Introduction

Airports, in common with other public facilities, have always been vulnerable to "conventional" crimes such as vandalism, theft, breaking and entering, and even crimes against the person. As part of a worldwide air transport system they have also become the focus of the particularly vicious crimes associated with terrorism. Such criminal acts have included exploding bombs aboard aircraft in flight, ground attacks on aircraft and on ground facilities, using firearms and missiles, and the hijacking of aircraft. This latter usually results in the taking of passengers and crew as hostages and the subsequent involvement of an airport in attempts to free the hostages and apprehend the hijacker(s). Table 9.1 shows that this particular threat shows no sign of abating. Even more alarming is the degree of careful planning that is seen to have taken place before many of these attacks, whether involving aircraft directly or airports—there were 17 attacks on the world's airports in 1993. Whatever the motivation for criminal attacks against civil aviation—very often it is political—there is an ever present need for airports to be on constant alert in order to frustrate any such acts. Nationally and internationally there is considerable concern over providing continuous protection against the possibility of attacks on civil aviation, and airports stand as the last line of defense. The occurrence of a security incident at an airport is as unpredictable as the probability of an aircraft accident, although both have the potential for leading to serious loss of life, injury and damage to

TABLE 9.1   Worldwide Breakdown of Highjacking Incidents 1989-1993

	1993	1992	1991	1990	1989	Totals
Hijackings						
Sub-Saharan Africa	4	5	3	2	1	15
Asia	17	1	2	4	3	27
Europe	2	. . .	2	. . .	3	7
Latin America	1	4	5	4	2	16
Middle East and North Africa	4	. . .	1	2	2	9
North America	. . .	. . .	1	1	1	3
Central Eurasia	3	2	10	27	3	45

property. A breakdown of incidents by category, worldwide, is given in Table 9.2.

Airport management, in common with others involved in the operation of elements of the civil air transportation system, are required to take measures that will provide a high level of protection of buildings and equipment (including aircraft), in addition to ensuring the safety and personal security of passengers and staff using the system. This must be done in a manner that disturbs the normal operating patterns as little as possible, while maintaining acceptable standards throughout the whole of the airport system. The achievement of this basic goal of a modern security operation requires the commitment and cooperation of central and local government agencies, airport authorities, airlines, other airport tenants, police and security staff, and the public itself. This chapter will discuss how security procedures affect airport operation and describes in general terms airport security requirements. For obvious reasons, descriptions of detailed procedural arrangements will be avoided, as will the identification of the procedures and arrangements at particular airports. This type of information is best regarded as restricted, and it is wisely made available only to those who need to know. A more com-

TABLE 9.2   Incidents by Category

Total Incidents, 1989–1993

	1993	1992	1991	1990	1989
Civil aviation					
Hijackings	31	12	24	40	15
Commandeerings	2	4	1	2	0
Bombings/attempted bombings/shootings	0	0	1	2	4
General/charter aviation	5	10	10	3	5
Airport attacks	17	15	27	6	7
Off-airport attacks	20	50	47	4	4
Shootings at aircraft	9	7	10	0	2
Totals	84	98	120	57	36
Incidents not counted	13	15	13	2	6

plete reference available on a restricted basis can be found in ICAO 1992 (Security Manual).

The gravity of the illegal acts against airlines and airports is such that it is now accepted that countermeasures must be of international concern. The report of the U.S. President's Commission on Aviation Security and Terrorism stresses the need for the necessary *will* on the part of all governments. "A consensus must be reached among law-abiding nations that terrorism is an act of aggression which can and must be destroyed." In seeking this consensus a series of conventions have addressed the issues:

1. *Tokyo 1963.* Convention on Offenses and Certain Other Acts Committed On Board Aircraft—concerned with the whole subject of crime on aircraft and particularly with the safety of the aircraft and its passengers.

2. *The Hague 1970.* Convention for the Suppression of Unlawful Seizure of Aircraft dealing with hijacking, specifically, recommending that it be made an extraditable offense.

3. *Montreal 1971.* Convention for the Suppression of Unlawful Acts Against Civil Aviation—enlarging the Hague convention and adding the offense of sabotage.

These conventions were followed in 1974 by the adoption by ICAO of Annex 17 to the original Chicago Convention. Entitled *Safeguarding International Civil Aviation Against Acts of Unlawful Interference,* the Annex established international aviation standard (40) and recommended practices (17). In addition, a supplement to the Montreal Convention was drawn up in 1988. The Montreal Protocol for the Suppression of Unlawful Acts of Violence at Airports Serving International Civil Aviation. This was intended to cover acts of violence against civil aviation that occur at airports and ticket offices, which were overlooked in the Montreal Convention of 1971. The need for such an addition was brought about by terrorist attacks on Rome and Vienna airports.

One of the most effective methods of combating the threats of sabotage, hijacking, terrorism and other acts against civil aviation is for individual governments to ratify such conventions and to enact laws that serve to deter and punish such acts. Although not all countries have satisfied the Conventions on the Protocol, most have enacted the necessary national legislation, which is administered through regulations of the governmental administrative departments and the courts. The need for the participation of all countries is illustrated by the worldwide spread of incidents and the varied nature of their acts of unlawful interference, as indicated by Figure 9.1. In order to further strengthen the force of its own legislation, the United States enacted the *Aviation Security Improvement Act—1990* creating new high-level security positions within the Department of Transportation

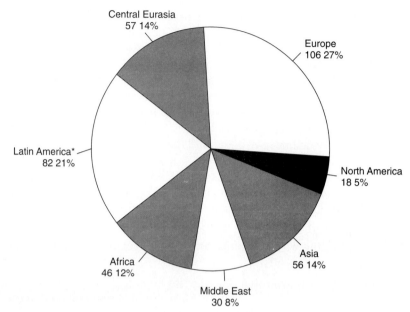

Incidents against aviation by geographic region 1989–1993
395 incidents * Also includes Central America and the Caribbean
Note: Total percentage exceeds 100% because of rounding

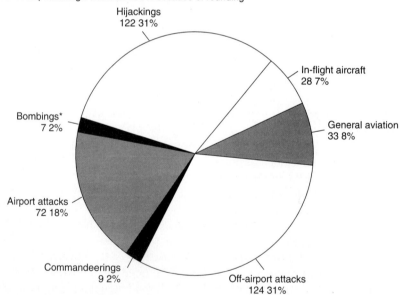

Incidents against aviation by category 1989–1993
395 incidents * Also includes attempted bombings and shootings on board aircraft
Note: Total percentage is less than 100% because of rounding

Figure 9.1   Incidents against aviation by geographic region and category, 1989 to 1993.

and the Federal Aviation Administration. The Canadian federal government, in strong legislation, has enacted an Act of Parliament setting terms of imprisonment from a minimum of 14 years to a maximum of life.

In Great Britain, the Tokyo Convention was satisfied by the Tokyo Convention Act of 1967, while the Aviation and Maritime Security Act of 1990 incorporates the provisions of the other conventions. The Airports Act 1986 empowers the Government Minister to issue such orders as he deems necessary to request airports to meet specific inter-governmental security agreements (e.g., U.S. FAA requirements).

9.2 Structure of Planning for Security

Clearly, the subject of security has wide implications that reach far beyond the jurisdictional limits of the airport to central government itself. Planning to meet the needs of a security emergency requires the involvement of a number of organizations, such as:

- The airport administration
- The operating airlines
- The National Civil Aviation Organization (FAA, CAA, etc.)
- Police
- Military
- Medical services
- Security services
- Labor unions
- Customs
- Government departments

Internationally, the ICAO recommends that each member state develop a national aviation security program that can be developed by a national aviation security committee formed from representatives of the organizations listed. If this body and the airports themselves are to be effective in countering security threats, there must be a clearly established process that starts with the issue of policy at national level and is operationally apparent in the procedures adopted by the individual airports. National policy is translated into a national security plan or overall security program, a necessity if airports and government wish to do other than react *post hoc* to a security incident.[1] The national plan is implemented by the provision of

---

[1]However, it is acknowledged that no security program can guarantee that incidents will not occur so *contingency plans* are prepared and exercised at both the national and airport levels.

staff, equipment and training at the airports and other sensitive aviation areas. Both as a whole, and at individual facilities, security operations are tested, evaluated, and modified to assure adequate performance standards.

Reviews of this nature must be carried out by qualified security officers and operations personnel, and assessments should include information on the severity of any deficiency and how it relates to airport security as a whole. In particular, efforts should be made to determine whether unsatisfactory conditions reflect individuals' carelessness or the existence of systematic problems. Only in this way, by employing an analytical approach, can a security system's strengths and weaknesses be evaluated. Alterations in major policy direction are by a continuous situation assessment of the changing security climate. Factors that can radically alter the security threat in a particular country or at a particular airport are political agitation or unrest and widespread publicity of other security incidents. Figure 9.2 indicates the conceptual structure of the security planning cycle. Reassessment of threat should take into account not just the level of threat, but also perceived trends, especially the type of weapons used and the techniques and tactics employed. Reassessment, if it is to be of any value in the preventative sense, should be based on accurate and timely intelligence concerning the intentions, capabilities and actions of terrorists before they reach the airport. Here too, international cooperation has a vital part to play, a fact un-

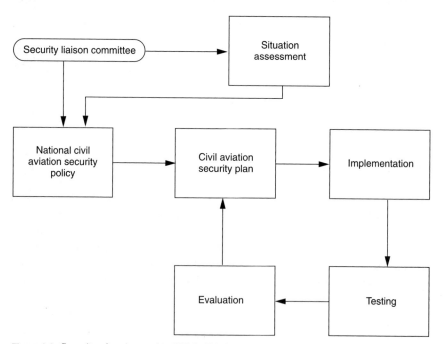

Figure 9.2 Security planning cycle. *(ICAO 1971)*

derlined when the United States signed into law the *Foreign Airport Security Act* as part of the International Security and Development Cooperation Act of 1985. The U.S. procedure for the Receipt, Assessment and Dissemination of Intelligence/Threat Information is outlined in Figure 9.3 (Report of President's Commission 1990). It was in order to support these intelligence requirements that the new inter-departmental offices were set up in the United States under the provisions of the *Aviation Security Improvement Act of 1990* (Figure 9.4) (Federal Research and Technology for Aviation 1994).

9.3 Responsibility and Organization

It is generally agreed that the nature of unlawful acts against civil aviation requires an appropriate immediate reaction. The safety of many innocent parties is frequently involved. Suitable and rapid responses to unlawful acts can be achieved only if there is a preestablished organizational structure with clearly assigned responsibilities. The general responsibility for maintaining law and order in a community is already established; threats to the security of civil aviation add another dimension to these responsibilities necessitating suitable additions and amendments to the laws. Governmental departments formulate and issue necessary orders and directives and also provide guidance material to enable each airport and airline to develop a security system appropriate to its own particular needs. While recognizing that all those concerned with the transport of passengers, cargo, and mail have a responsibility to take necessary safeguarding measures, the established practices of different countries will act as a modifying factor on the national procedures for aviation security. Nevertheless, there are broad principles of universal application:

1. *Feasibility.* The security program must be related to the resources available to the state, the airport, and the airlines, and must recognize real system constraints.

2. *Responsibility.* There must be an appropriate and unambiguous assignment of responsibilities to the central and local government authorities.

3. *Efficiency.* The efficiency of civil air transport must be retained as far as is feasible. Although central supervision of security matters is a function of the national government, there must be an appropriate delegation of powers to achieve the overall objective of transporting passengers efficiently, comfortably and economically.

4. *Coordination.* Appropriate security standards and practices can be established, maintained and updated only if there is a sustained level of coordination between the involved organizations. Good coordination

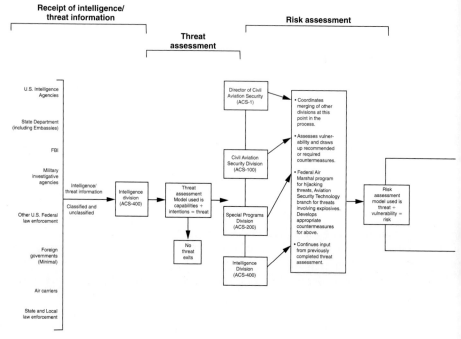

Figure 9.3   Receipt, assessment and dissemination of intelligence/threat information.

implies the creation of stable good relationships between the relevant parties.

5. *Resources.* Adequate resources must be supplied to attain specified security performance standards, and operators must ensure that the use of equipment and manpower is optimized.

Within the structure of airport management, the location of the director of security varies greatly between airports. One general rule should be observed: In an emergency, the chief security officer has immediate access to the executive director of the airport, who might have to make rapid and potentially difficult decisions. A direct reporting system is recommended to obviate the delay and indecision that can characterize a lengthy chain of command.

Undoubtedly one of the most difficult aspects of airport security, for management, is dealing with the public information/media implications of a serious incident such as a hijacking, both as the incident unfolds on the airport, and at its conclusion. Because civil aviation is international in nature, interest in such incidents is worldwide, facilitated by the availability of satellite television. As a result, reporting tends to be in real time. It is therefore essential that a very clear policy should be established in terms of what should (or should not) be said, as well as who should say it. In such circumstances

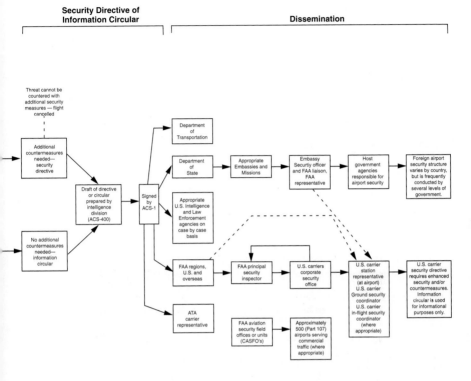

it would be usual for a senior governmental representative/Minister to be present on the scene, along with senior airport/civil aviation officers, and perhaps including a public relations/press officer. There should be unambiguous directives as to which of those present will make statements to the media, together with broad guidance on the line to be taken.

9.4 Airside Security Procedures

The airside of an airport is usually defined as the movement area[2] of the facility and all adjacent terrain and buildings to which access is controlled. Airport operators now recognize that the airside is likely to be the prime target area for those undertaking any unlawful action against either an aircraft and its crew or the passengers. It is therefore of paramount importance that the airside is protected against all unauthorized intrusion, which must be regarded as potentially dangerous. Specific action is therefore required in a number of areas.

---

[2] The movement area is defined as that part of the airport to be used for takeoff and landing of aircraft and for the surface movement of aircraft (ICAO 1990).

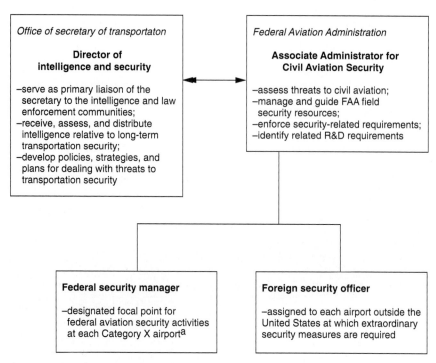

aCategory X airports (19) typically have a large number of passenger enplanements per year, along with departing international flights.

Figure 9.4 Aviation security positions created by the Aviation Security Improvement Act of 1990.

Fencing

The airside must have an adequate security fence that serves the multiple functions of clearly defining the protected area, providing a deterrent to the intruder, delaying and possibly inhibiting unlawful entry, and providing defined controlled access points at gates. These access points should be kept to a minimum. Where access is other than by key or an automatic control of entry system, gates must be manned, illuminated, and provided with alarms. Furthermore, the access points should be provided with communications to the central security unit. The fence itself must be a real deterrent to entry. Normally it is a high, possibly barbed wire-topped structure of nonscalable metal construction. Care must be taken to secure all conduits, sewers and other ducts and pipes that pass under the fence to ensure that entry to the airside is not possible in this way. However, one of the more alarming developments in recent years has been the incidence of attacks at airports. There were no fewer than 72 worldwide, during the period 1989–93. These have included 27 bombings; 17 attempted bombings and 28 shootings, shellings (such as mortar attacks). Shelling, which usually takes place from outside the airport boundary, is particularly difficult to thwart. Even a large international airport with good security fencing might still be vulnerable. London Heathrow airport was subject to a terrorist attack by mortars fired over the security fencing from positions on adjacent open land. Where such possibilities exist it will be necessary to extend the scope of security measures.

Identification of persons

Recognizing that the airside is potentially an area vulnerable to terrorist attack, access should be restricted to identified personnel. All airport and airline ground staff members needing access to perform their jobs and all necessary aircraft crews should have passes authorizing them to enter the airside. Identification cards bearing a photograph and details in tamper-proof covers should be issued by the security unit of the airport authority to staff. Control of the issue of cards is necessarily strict. Access by right should be given only where the operation requires it, and cards need not give universal access to the airside. Staff should be restricted to those areas in which they have a need to be. Visitors needing to visit the airside should have special passes. All passes should be subject to periodic checks. This is best achieved by having expiration dates that limit the validity to not more than one year. It is of particular importance that all staff display their passes permanently when airside. Strict control should be exercised over the identification cards/passes and they should be recovered from any employee who ceases to be employed or no longer has need to work in a restricted area. The grim result of not doing so was illustrated by the December 1987 crash of Pacific Southwest Airlines flight 1771 from Los Angeles to San Francisco. All 43 persons on board were killed when a recently dismissed employee shot the pilot and crew having reportedly brought the weapon aboard after bypassing preboarding screening by showing his company ID. A further precaution can be to issue an individual Personal Identification Number (PIN) to all holders of passes so that it is also necessary to key into a computer-controlled system the PIN associated with the pass before gaining entry to a restricted area. Even if the pass is not handed in upon cessation of employment, or reassignment to a non-restricted area, the PIN can be immediately deleted from the approved list.

Identification of vehicles

Similarly, airside access should be granted only to those vehicles that must be airside in performing their functions. Access can be restricted by the issue of individual vehicular passes, the control of which is maintained by central airport security. These also should be fixed-term passes and a firm rule that individuals within the vehicle require personal passes as well as their vehicle pass. The register of pass holders should be checked periodically to ensure continuing entitlement.

Protection of aircraft on aprons

If the airside is kept secure by restricting access, normally no special protection is required for aircraft while they are on the apron. General security precautions should be taken such as floodlighting the apron to deter intruders, locking aircraft doors, and removing steps and stairs from aircraft that are on the ramps for a long time with no service crews aboard, such as overnight layovers. Other general precautions include the use of temporary covers on all aircraft inlets where tampering is possible, and ensuring that there has been no tampering with vehicles, such as catering trucks and mobile stairs that will later approach the aircraft. Various intruder detection devices can also be used to ensure the integrity of the apron.

Control of general aviation

General aviation aircraft might arrive on the airside without having had any security check at their airfield of origin. Mixing of such "unsecured" aircraft and persons with security cleared air transport operations can be prevented by designating separate general aviation aircraft parking areas and separate taxiway routings and the provision of security checkpoints on general aviation taxiways and aprons. Where the general aviation aircraft will mix with other aircraft on the airside, air traffic control can alert central security of the arrival of a GA aircraft so that a security check can be carried out on the apron. Such clearance procedures require unobtrusive radio communication between central security, the security checkpoint, and air traffic control.

Transfer and interlining passengers, baggage, and cargo

The movement of transfer and interline passengers, baggage, and cargo through airports can give rise to security risks, especially in the case of flights from or to a high risk area of the world. Problems also arise where the transit movement is from an airport where security is lax for whatever reason. In most major international airports, transit passengers will be separated from other airside-security cleared individuals, or where this is not possible additional screening procedures will be introduced. Nonetheless, although it is probably correct to emphasize security risks in the international context, the possibility of domestic terrorism cannot be ignored.

Transfer and interline baggage and cargo have always been an operational concern because of the potential for the loss or misdirection of items.

Additionally, there are serious security implications that have been made only too clear in the recent past. One of the worst ever bombing outrages occurred with the destruction of Pan Am Flight 103 over Lockerbie, Scotland, on December 21, 1988. A bomb exploded on board shortly after the B747 reached its assigned flight level 310 (31,000 feet). All 243 passengers and crew and 11 residents of Lockerbie perished. It was subsequently determined that the bomb had been loaded on board in an unaccompanied transfer bag. Renewed efforts were made to devise a fool-proof method of preventing the re-occurrence of such an outrage. In this context transfer and interline baggage represents a potential weak link in the security chain, although the problem is by no means confined to these two possibilities. Recognizing this, the ICAO introduced a Standard in Annex 17: "Each contracting State shall establish measures to ensure that operators, when providing service from that state *do not transport the baggage of passengers who are not on board the aircraft* unless the baggage separated from passengers is subject to other security control measures" (paragraph 4.3.1). Although the "Standard" is binding on all nations that have signed the Convention on International Civil Aviation, the procedures for meeting it are not specified.

A variety of methods have been developed to meet the ICAO requirement. One of the most promising recent developments has been the introduction in 1994 at Frankfurt Airport of an automated *Passenger Baggage Reconciliation System*. By collecting and merging all flight-related passenger and baggage data, it enables a match to be made of the boarding data in the airlines' check-in system and the baggage data recorded in the system. The check is more usually made at the baggage make-up stage or, if necessary, right up to the last moment at the planeside before the hatches are closed (e.g., standby passengers' baggage). A bag whose owner is not onboard will either not be loaded or, if already loaded it will be unloaded again. Should this become necessary it is assisted by the fact that the system also records the number of load limit/container and the loading sequence number. The operator responsible for carrying out the check focuses a laser scanner on the baggage tag and receives a "YES" or "NO" indication on his screen (Figure 9.5). The "NO" indication can be qualified by the addition of the word "standby" indicating that the operator should recheck for possible later release of a standby passenger's baggage.

Air cargo, because it is "unaccompanied" inevitably gives rise to serious questions regarding security safeguards. There is a risk present because the original shipper might not be known to the air carrier and indeed a cargo consignment might pass through more than one shipping agency or so called "indirect air carrier" (e.g., trucker) before reaching the airline. Any one of these intermediaries could be lax in its observance of security controls. The only realistic solution is for the air carriers themselves to have in place their own systems of checks/screening because air cargo is as much a threat to an aircraft as checked baggage.

Figure 9.5 The part of baggage carousels not open to the public.

Aircraft isolated parking position and disposal area

An airport should designate an isolated aircraft parking position that can be used for parking an aircraft when sabotage is suspected or when an aircraft appears to have been seized unlawfully. This position should be at least 325 feet (100 m) from any other aircraft parking position, building, public area, or utility. Furthermore, a disposal area should be designated on the airport for disposing or exploding any device found in the course of sabotage or unlawful seizure. The disposal area should also be clear of all other used areas, including the isolated aircraft parking position, by at least 325 feet.

An airport might need several designated isolated positions to be used for different kinds of incidents. Some positions should be amenable to surreptitious approach.

9.5 Landside Security Procedures—Passenger Terminal

Perhaps more than in any other part of the airport, security measures in the passenger terminal are most effective in preventing subsequent unlawful acts in the air. If the public is made aware in general terms that a security program is in operation, the incidence of attacks is lowered, indicating a deterrent effect. It is still possible to publicize the fact that security systems are operating in a terminal without disclosing their nature. At least one major international airport makes a point of placing public notices to this effect in its terminal. The less well understood the security measures are, the greater the likelihood that the program will succeed in heading off all but the most determined attackers. Ideally a security system operates throughout the whole passenger facilitation process of ticketing, passenger and baggage check-in, and boarding. Abnormal behavior at the ticketing and check-in stage should alert staff to potential problems. In the boarding process, security procedures must ensure that no would-be assailant is able to convey any weapon to the aircraft. The mere presence of visible security systems is likely to reduce the likelihood of the occurrence of incidents.

Successful security necessitates that the airside-landside boundary be well defined and continuous through the passenger terminal, with a clear definition of security-cleared "sterile" areas. The number of access points to the airside must be very strictly limited; those that are available to passengers should be staffed by security personnel. Access to the airside through staff areas must not be direct and must be clearly signed as closed to the public. It should also incorporate the same security screening methods that are used for the passenger entry points to the airside or restricted areas. Doors from the passenger terminal to the apron must be locked. Where emergency exits give access to the apron, they should be fitted with alarms. Equal care must be taken to secure access to the apron via unattended loading bridges and other apron connectors. The security screening of passengers can be carried out in a decentralized way at the individual aircraft gates. Some operators believe gate screening achieves maximum security. However, it requires more screening equipment and staff, tends to cause longer boarding delays, and suffers the disadvantage that the challenge to any armed person or group is performed in the vicinity of the aircraft. Centralized security before entry to a central sterile departure lounge requires less staff and can utilize less but more sophisticated equipment. The main disadvantage is that unscreened individuals might be able to infiltrate the sterile areas from the apron or through staff routes. Other obvious security measures include the design of ticket and processing counters that prevent the public from obtaining unlawful access to the validators, tickets, and boarding passes. The advantages and disadvantages of centralized and decentralized search are sketched in Table 9.3.

Passenger screening is carried out by physical search and the use of electromechanical, electronic, and X-ray equipment. The screening process is significantly improved by limiting carry-on baggage to one piece. Baggage is either automatically machine checked or hand searched while the personal screening is carried out by walk-through machines, supplemented where necessary by a body search. The reader is referred to Reference 2 for procedures designed to ensure that material that could be used in an unlawful act on the aircraft is not conveyed into the cabin. In general, the security staff will take possession of all firearms; offensive weapons and imitations; explosives, inflammable, toxic, and corrosive substances. Where

TABLE 9.3 Advantages and Disadvantages of Centralized and Decentralized Search Procedures

Advantages	Disadvantages
Centralized Search	
Favored by passengers	Passenger segregation in a sterile departure lounge is difficult to achieve
Minimum personnel and equipment needed to process a given number of passengers	Requires staff search Control of food and merchandise
Encourages passenger spending in restaurants and duty-free and other shops	Passenger separation (arriving and departing) difficult to achieve
Easier to have policemen on duty in one place	Surveillance of passengers difficult at busy airports Only one standard of search is possible, whereas high risk flights may require more thorough search
Gate Search	
The separation and surveillance problem is eliminated	Requires earlier call-forward of passengers
The risk of collusion is minimized	Results in loss of revenue from restaurants, bars, shops, etc.
Allows special measures to be taken on high risk flights	Involves long waiting in crowded gate lounges with no facilities
	Requires more personnel and more equipment to process a given number of passengers
	Creates problems of search team availability of flight schedules go awry
	Makes a police presence difficult depending on number of gates in use at one time
	Allows passengers to get close to aircraft before search and access to the apron is always possible (emergency exists)
	Enables terrorists to identify specific passengers and lines them up for attack when queuing
	Current gate lounges inadequate for future aircraft
Pier Search	
Combines the advantages and disadvantages of the other systems	
Could be the best bet if space available in the right place to set up search points	

possession of these items is not against the law in the country of embarkation, they are taken into the custody of the airline for carriage in the hold and returned at the end of the flight.

Protection of the aircraft means that *all* avenues to any aircraft on the apron must be secure. Staff working on the airport must realize that terrorist organizations might be as well aware of airport operating procedures as the staff themselves. In the case of a planned attack, it is very likely that the particular configuration and operating procedures of an airport will have been examined. The systematic checking of the efficiency of security systems is therefore essential. One method might be to invite official specialist security personnel (e.g., police, military) to attempt to penetrate the system. Furthermore, the use of covert surveillance can be just as useful as overt screening devices. This introduces an element of insecurity in the minds of potential violators. Consequently, the full extent of the security system should be known to as few persons as possible from an operational viewpoint.

9.6 Landside Procedures—Cargo Terminal

The cargo terminal is a high-activity area in the airport that potentially gives illegal access to the airside areas and to passenger and cargo aircraft. Once this is grasped, it becomes obvious that there is the need for a security program in the cargo terminal that supplements that at the passenger terminal and is equally effective.

Because cargo terminals do not handle passengers, security procedures are relatively simple and can be described in general terms:

1. *Identification.* All staff with right to access to the airside should have security passes.

2. *Security of all doors and windows.* Audible warning signals can be placed on doors to indicate when they are left open. Where doors or windows must be left open for ventilation, grilles can be installed.

3. *Control of access.* Typically, no pedestrian access should be allowed through vehicular doors. Security staff should inspect the credentials of all persons entering the terminal. It is usual to control both vehicular and personal access to cargo terminals.

One excellent measure of the level of security within a cargo terminal is the level of theft and pilferage that is observed. Terminals with high levels of "conventional" crime are very likely to have insecure areas that are vulnerable to infiltration by terrorists and saboteurs.

9.7 Security Equipment and Systems

All airports providing air passenger transport service feed into the international system. Because of their particular vulnerability, large airports with heavy passenger volumes require sophisticated security precautions with elaborate equipment systems and substantial staffing. The requirements at very small community airports with only third-level operators is less stringent, but precautions are still necessary, given that the smaller airports can feed passengers into the worldwide system when their passengers transit to other airlines.[3] A major airport with an extensive security operation is likely to have most, if not all, the following devices. For further details, the reader is referred to ICAO 1992.

1. *Security fencing and manned barriers.* As discussed earlier, security fencing with manned barriers at points of access is essential to maintaining the integrity of the airside-landside concept of security.

2. *Intruder detection.* Electromechanical and electronic warning systems can be set up to detect intruders who have broken into buildings or gained access to the airside by illegal entry over or through the security fence.

3. *Lighting.* Aprons and other airside areas are lit to ensure that any unlawful activities are difficult to conceal on the airside areas.

4. *Metal detection.* For the detection of firearms and metallic explosive devices on the passengers, and in checked and cabin-carried-on baggage, various metal detecting devices are available. They fall into the following categories:

   (*a*) Flux gate magnetometers (passive)
   (*b*) Inductive loops (active)
   (*c*) Eddy current devices (active)
   (*d*) Hybrid inductive loop and eddy current devices (active)
   (*e*) Gamma spectrographs (active)
   (*f*) X rays

5. *Explosive and incendiary device detectors.* Explosive and incendiary devices that are not traced by their high metal content can be detected by

   (*a*) Vapor analysis detectors
   (*b*) Neutron activation equipment
   (*c*) Trained explosive detecting dogs

---

[3]Some countries have adopted very stringent security precautions that ensure that all enplaning passengers at their airports, whether transiting or originating, are subject to screening.

6. *Pressure chambers*. Pressure chambers, sometimes called rarified atmosphere simulation chambers, can be used to ensure against an explosive device with a barometric trigger. The chamber, because of its potential for an explosion, must be located well away (325 feet [100 m] minimum) from all facilities including the isolated aircraft parking position.

7. *Bunker*. For the disposal of bombs and incendiary devices and around any pressure chamber, a protection bunker, usually constructed of earth and timber, should be located well away from all other facilities.

8. *Office security equipment*. Protection is required for all security restricted documents, manuals, and plans. Secure metal cabinets should be used and strict measures adopted to ensure control of access by restricting the number of available keys.

9.8 Security Operations

Airports are responsible for maintaining security measures that might well operate for years without major incident but must respond very rapidly if a security emergency does arise. The fact that air transport is international only adds to the difficulties. In an emergency such as hijacking or bombing of an aircraft originating in one country, an airport many thousands of miles away in another country might have to finally deal with the incident. The hijacking of Lufthansa flight 592 shortly after takeoff from Frankfurt en route to Cairo on February 11, 1993, was finally dealt with at John F. Kennedy International Airport, 11½ hours later.

On a day-to-day basis, however, most security operations at an airport deal with activities in and around the terminal and the measures taken to ensure the safety of passengers and crews operating through the airport. One of the most effective deterrents to anyone likely to pose a threat would be the knowledge that security operations are not predictable. This uncertainty can be introduced by varying the operating procedures. This would apply to procedures at check-in, to the searching and screening of passengers and even to measures taken at planeside as passengers board. As discussed in the previous chapter, one of the major security concerns is the matching of individual boarding passengers with their baggage. Security operations in this area have been greatly assisted by the recent availability of sophisticated reconciliation systems. These are, however, not yet widespread and less elaborate methods have still to be used at many airports. One method is to call for last minute identification of checked baggage by passengers. After the passengers have passed through a screening procedure they are asked to identify their individual bags before final boarding. Any bags left over after this procedure are obviously suspect. These checks can be carried out on a sampling basis with random choice of flights so that there is no routine "pattern" of flights that are checked in this way. Occasionally, an airport authority might decide to carry out a search of the contents of checked baggage. Once this has been done, the searched baggage should be sealed or banded to maintain security and to identify the baggage that has been through the process.

In regard to the searching/screening of passengers themselves, this procedure can also be varied. Passengers might be picked at random for a personal search, even if they have already passed through a screening device. Again if this is done on a random sample basis, there can be no possibility of passengers being able to anticipate whether or not they will be subjected to extra search. It is, of course, necessary in these cases to have suitably screened areas where personal searches can be carried out with separate male and female facilities. Because of the concentration needed by the security staff examining passengers and baggage, it is desirable to also have someone standing by to oversee the security operation generally so that they can detect any attempt by individuals to circumvent the security procedures.

The actual scope and extent of these procedures will vary according to individual airport circumstances. An important factor in deciding the scope and extent of security procedures is the assessment of the likelihood and nature of any security threat. Accordingly, the procedures might vary from a relatively simple screening procedure where the level of threat is very low to a personal interview for each departing passenger together with detailed personal search and an examination of the contents of all carry-on and checked baggage when the threat is judged to be serious.

By way of general guidance for airport management in drawing up an airport security program, a suggested outline is offered here. It is emphasized, however, that this is not intended to be followed rigidly; individual airports will need to modify it to suit their own particular requirements.

Security program for (official name of airport)
1 General
    1.1 Objective—This security program has been established in compliance with Standard 3.1.1 of Annex 17 to the Chicago Convention, and in accordance with national legislation and regulations, namely laws—decrees—etc.

    The main purpose of the provisions and procedures contained in this program is to protect the safety, regularity and efficiency of international civil aviation by providing through regulations, practices and procedures, safeguards against acts of unlawful interference.

2 Organization of Security
    2.1 Name and title of the official(s) responsible for airport security.
    2.2 Organizational details of services responsible for the implementation of security measures, including:

- Airport Security Officers
- Police
- Government inspection agencies
- Airline operators
- Tenants
- Municipal authorities

3 Airport Security Committee

3.1 An airport security committee must be established to comply with Standard 3.1.11 of Annex 17 to the Chicago Convention. This committee is responsible for providing advice on the development and implementation of security measures and procedures at the airport. It must meet regularly:

- To ensure that the security program is kept up-to-date and effective
- To ensure that the provisions it contains are being satisfactorily applied
- To coordinate the activities of all the bodies concerned with security measures (police, gendarmerie, operators, airport management, etc.)
- To maintain liaison with the various security services outside the airport (responsible government departments, bomb disposal service, etc.)
- To give advice to the airport management on any reorganization or extension of the facilities.

Minutes must be kept for every meeting of the airport security committee, which, after approval by the members, are circulated to the main authorities concerned.

3.2 Composition of the Committee

The airport security committee should be made up of representatives of all the public and private bodies concerned with the operation of the airport. In addition, the airport manager will normally act as Chair with the chief of airport security as stand-in for those occasions when the chairperson is unable to attend.

The following would be appropriate members:

- Airport Manager
- Airport Security Chief
- Police
- Military
- Customs
- Immigration
- Air Traffic Services
- Fire Services
- Communications Representatives
- Health Service

- Postal Service
- Operators
- Cargo companies and forwarders
- Tenants

The names, titles and other useful details of all members should be included.

4 Airport Activities

4.1 Name, location, official address, telephone/fax number of the airport and identification code.

4.2 Hours of operation of the airport.

4.3 Description of the airport's location with respect to the closest town or province.

4.4 Attachments including a location map and plan of the airport with particular emphasis on the airside indicating the various security restricted areas.

4.5 Name of the airport owner(s)

4.6 Name of the airport manager

4.7 Airport operating services

4.8 Administration

4.9 Air traffic services

4.10 Maintenance

4.11 Others

4.12 Airline operators and route/traffic details

5 Security Measures at the Airport

5.1 Definition and description of restricted areas/airside, describe measures designed to safeguard the airside; (boundary fencing, guard posts, lighting, alarm systems, closed circuit television systems, walk-through units, patrols, etc.).

5.2 Restricted areas

5.2.1 Restricted areas in the air terminal
- Departure lounges
- Transit lounges
- Immigration area
- Customs area

5.2.2 Airside
- Air traffic services facilities
- Taxiways
- Maneuvering areas
- Parking/ramp access
- Cargo areas

5.2.3 Public areas
- Public areas of the air terminal
- Car parking areas

5.3   Access and movement control
    5.3.1 Identification procedures for persons:
        Attach the text(s) that regulate the movement of persons at the airport:
- Define the areas where access passes are mandatory
- Specify the access points where access passes are required
- Specify the criteria for granting access passes
- Specify how and by whom the access passes are issued
- Describe in detail the format and contents of the various cards, badges and signs used for identification
- Specify the procedures for checking the access pass and the penalties for not complying with the regulations

    5.3.2 Identification procedures for vehicles:
        Vehicles authorized to enter the restricted area shall be equipped with a pass. This pass might also specify in which particular sectors the vehicle is authorized to circulate, and during which hours. As in the case of personal access, passes specify the allocation procedures and describe the passes.

5.4   Security control for passengers and baggage
    5.4.1 Passengers
- Custody and control of flight documents (tickets, etc.)
- Identification of passengers at check-in or other identified locations (e.g., passport check at boarding gate)
- Agency implementing security controls
- Equipment and procedures for passenger screening

    5.4.2 Control of hold baggage
        Procedures: Searches using security equipment, percentage of hand searches required on a random basis, identification and disposition of removed articles; procedures for off-airport checked baggage; procedures for "short-shipped" and mishandled baggage.

5.5   Security control of cargo, mail, small parcels
- Assignment of responsibility for security control
- Screening of cargo; courier and express parcels and mail
- Nature of control procedures: hand searches, searches using security equipment
- Measures for the treatment of suspect cargo, mail, small parcels
- Airline responsibilities in relation to the control of flight catering and other stores

5.6   Security control of VIPs and diplomats
- National guidelines for special procedures

- Procedures for VIPs and diplomats
- Private or semi-private arrangements for special passengers
- Measures to limit arrangements to strict minimum
- Procedures for dealing with diplomatic bags and diplomatic mail

5.7   Security control of certain categories of passengers
- Staff members including crew members in uniform
- Facilities and procedures for disabled passengers
- Procedures for inadmissibles, deportees, escorted prisoners (notification to operator and relevant captain)

5.8   Security control of firearms and weapons
- National laws and regulations
- Carriage of firearms on national aircraft, foreign aircraft
- Authorized carriage in the aircraft cabin, prisoner escort, VIP escort, sky marshals
- Authorized carriage in checked baggage or as cargo

5.9   Protection of aircraft on the ground
- Responsibilities and procedures
- Security measures for aircraft not in service
- Positioning of aircraft
- Use of intruder detection devices
- Preflight security checks
- Special measures available to operators on request

5.10   Security equipment
- Responsibilities for operation and maintenance of equipment
- Detailed description
- X-ray equipment
- Walk-through metal detectors
- Hand-held metal detectors
- Explosion detectors
- Simulation chambers— location, type and construction

6 Contingency Plans to Respond to Acts of Unlawful Interference
6.1   Categories
- Reception of unlawfully seized aircraft
- Bomb threat to an aircraft in flight or on the ground
- Bomb threat to a facility at the airport
- Ground attacks—ground to air and ground to ground

6.2   Responsible organizations
- Operational command and control
- Air traffic services procedures
- Special services (location day/night)
- Explosives ordinance disposal unit(s)
- Armed intervention teams

- Interpreters
- Hostage negotiators
- Police authority
- Fire brigade
- Ambulances

7 Security Training Program
  - Training policy
  - Training objectives
  - Curriculum outline
  - Course syllabi
  - Procedures for evaluating training

  All personnel with direct responsibilities for security and all staffs at the airport should either attend a training course or a security awareness presentation adapted to the particular needs of the various levels.

8 Appendices to Security Program

  Organizational diagram showing the structure of the airport administration and security management map of the airport and peripheral area:
  - Detailed maps: air and landside, terminal, layout of all categories of areas.
  - Agreements/instructions to tenants
  - Instructions for air traffic services
  - Legislative and regulatory texts relating to aviation security including those in a national context, or any other document/reference which would help the program

Thorough though all these security procedures might be, it is important to also have in place measures to increase the motivation of security staff, especially those engaged in the repetitive tasks involved in screening passengers and their baggage. Here training has a continuing part to play in such areas as continuation and refresher training. There is an extensive series of standardized training packages (STPs) available from the ICAO in Montreal (Table 9.4). Although concerned with international civil aviation, they can be easily adapted to domestic airports that serve international facilities.

TABLE 9.4 Standardized training packages (STPs) of the ICAO training program for aviation security

Part	Short title	Full title
1	AVSEC 123/Basic	Basic Airport Security Personnel Training
2	AVSEC 123/Management	Aviation Security Management Training
3	AVSEC 123/Instructors	Aviation Security Instructors
4	AVSEC 123/Crisis Management	Aviation Security Crisis Management Training
5	AVSEC 123/Cargo Security	Aviation Cargo and Mail Security Training
6	AVSEC 123/Supervisor	Airport Security Supervisor

TABLE 9.4   Standardized training packages (STPs) of the ICAO training program for aviation security (Continued)

Part	Short title	Full title
7	AVSEC 123/Airline Security	Airline Security Training Programme
8	AVSEC 123/Equipment Maintenance	Aviation Security Equipment Maintenance Training
9	AVSEC 123/program Development	Aviation Security Equipment Design and Development Training—All Categories
10	AVSEC 123/Awareness	Airport Personnel Security Awareness Training
11	AVSEC 123/Refresher	Airport Security Refresher Training
12	AVSEC 123/In-service	Airport Security Personnel Continuation Training
13	AVSEC 123/Advanced	Advanced Aviation Security Training

Note: Each part confined in substance to preventing acts of unlawful interference.

References

International Civil Aviation Organization (ICAO). 1992. *Security: Safeguarding International Civil Aviation Against Acts of Unlawful Interference, ICAO Doc 8973/4, Annex 17 to the Convention on International Civil Aviation.* Montreal.

ICAO. 1995. *Aerodromes. Volume 1, Civil Aviation 2nd edition. Annex 14 to the Convention on International Civil Aviation.* Montreal.

Report of the President's Commission on Aviation Security and Terrorism, May 15. 1990. Washington, DC.

Office of Civil Aviation Security. 1993. *Criminal Acts Against Civil Aviation.* U.S. Department of Transportation, Federal Aviation Administration, Washington, DC.

Federal Research and Technology for Aviation. 1994. U.S. Congress, Office of Technology Assessment, Washington, DC.

10

Cargo Operations

10.1 The Air Cargo Market

For more than 40 years the air cargo market has been a steadily growing sector of the air transport market. During the late 1960s, the total world tonne kilometrage of freight doubled every four years, an average growth rate of 17 percent (ICAO 1995). At that time, the aviation world was replete with extremely optimistic forecasts of a burgeoning air cargo market. For example, McDonnell Douglas in 1970 projected that growth rates would increase and that the total market would grow from 10 billion route ton kilometers (6.2 billion route ton miles) in 1970 to 100 billion route kilometers (62 billion route ton miles) in 1980. In fact, this figure was not reached until 1995 due to recurrent economic recessions and steep fuel cost rises in the 1970s and 1980s. More recent forecasts reflect the steady annual growth of 7.8 percent for the 20 years after 1975, indicating a most likely average annual growth rate of 6.5 percent between 1995 and 2015 with high and low estimates of 8.6 percent and 4.3 percent annually (Boeing 1995). The historical and forecast growths are shown in Figure 10.1, which displays this anticipated strong growth over the long term. A closer examination of the development of air cargo indicates a number of factors are involved.

Gross domestic product

There is a very strong positive correlation between air cargo and gross domestic products (Figure 10.2). In times of economic buoyancy, air freight grows rapidly but the cyclic recessions of the last 20 years have retarded air freight growth in the Western industrialized nations. For a while air freight

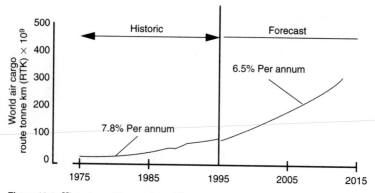

Figure 10.1 Historic and forecast world cargo traffic.

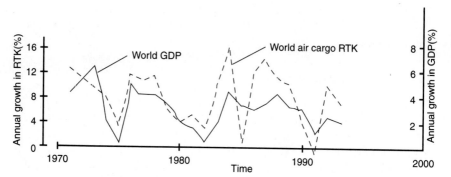

Figure 10.2 Relationship between world GNP and world air cargo revenue ton kilometer growth rates.

to the oil-producing companies of the Middle East was also buoyant but growth rates fell in the early 1990s as oil prices decreased in real terms. The area generating highest growth rates during the early 1990s was the industrializing area of the Western Pacific Rim nations.

Cost

In real terms, the cost of the air freight declined until 1974, aided by the decreasing real cost of fuel and technological improvements. Declining real costs ceased abruptly with oil price increases and the subsequent growth rates were significantly lower. For the 10 years following 1985, the real cost of oil again declined and inflation fell dramatically in the industrialized nations; these two factors combined to make the real cost of air freight tariffs fall once more and the demand for air freight to rise in a healthy manner.

Technological improvement usually manifests itself in terms of lower freight costs through improved efficiency. Improvements to technology have taken place in three principle areas: the *air vehicle*, with the introduction of wide-bodied large-capacity aircraft; the development of a *wide range of ULDs*, and the necessary subsidiary handling and loading devices on the aircraft, on the apron, and in the terminal; and finally in *facilitation* with the maturing of freight forwarding organizations and the development of computerized control and documentation. In the last 20 years, freight yield has declined at an annual rate of 2.9 percent reflecting productivity efficiencies and intense competition.

Other factors

Various other secular trends have contributed to the increasing demand in air freight. For example, *miniaturization* of industrial and consumer products has made items much more suitable for carriage by air. The expected growth of the silicon-chip market will continue this trend. Another factor is the increasing trend for industry to move away from regional warehousing and the high associated labor, construction, and land costs. Manufacturers find that centralized warehouses backed by sophisticated electronic ordering systems and air cargo delivery are as efficient and less costly to operate than decentralized regional warehouses. Since the mid 1970s the concept of *just-in-time* delivery has revolutionized many industrial production processes. This concept requires reliable delivery procedures that are capable of adjustment at short notice. The consequent reductions in industrial inventories produce savings that can pay for air freight charges. Finally, as real incomes rise in the industrial countries, more wealth falls into the bracket that can be designated as discretionary. Such income is less sensitive to transport costs for the goods purchased; consequently, as the real standard of living rises the higher cost of air transport is more easily absorbed into the price of goods in either an explicit or implicit trade-off between cost and convenience.

Air cargo is extremely heterogeneous in character. It is often convenient to categorize the freight according to the manner in which it is to be handled in the terminal[1].

1. *Planned.* For this type of commodity, the air mode has been selected as the most appropriate after analysis of distribution costs. It is either cheaper to move by air freight or the added cost is negligible when weighed against improved security and reliability. Speed of delivery is not of vital importance to this type of freight.

---

[1]The first three categories include shipments from the following four categories, which might be carried by air on either planned, regular, or emergency basis, due to their peculiar nature and the special attention that becomes necessary.

2. *Regular.* Commodities in this category have a very limited commercial life, and delivery must be rapid and reliable. Newspapers and fresh flowers are examples of regular commodities.

3. *Emergency.* Speed is vital and lives might depend on rapid delivery of emergency cargo such as serums and blood plasma.

4. *High value.* Very high value cargo such as gemstones and bullion require special security precautions in terms of staffing and facilities.

5. *Dangerous.* The carriage by air of dangerous goods is a topic of much concern with airlines because of onboard hazard. Even on the ground, dangerous chemicals and radioactive materials, for example, require special handling and storage in the terminal. Especially where containerized loads are used, it is important that personnel are adequately trained in the handling of dangerous shipments. IATA includes within its definition of hazardous goods the following: combustible liquids, compressed gases, corrosive materials, etiologic agents, explosives, flammable liquids and solids, magnetized materials, noxious and irritating substances, organic peroxides, oxidizing materials, poisons, polymerizable and radioactive materials.

6. *Restricted articles.* In most countries, arms and explosives can be imported only under the severest restrictions. Normally, restricted goods such as these can be transported only under very strict security conditions.

7. *Livestock.* Where livestock is transported, arrangements must be made for animals to receive necessary food and water and kept in a suitable environment. In a large terminal with considerable livestock movements, such as London Heathrow, the care of animals occupies a number of full-time staff.

Patterns of flow

The cargo terminal, like the passenger facility, experiences significant temporal variations in throughput. Unlike passenger terminals, freight facilities often demonstrate very large differences between inbound and outbound flows on an annual basis.

Cargo flow variations can occur across the year, across the days of the week, and within the working day. The pattern of variation differs quite noticeably between airports and might even vary quite remarkably between airlines at the same airport. Figure 10.3a shows the monthly variation of flows observed at four major airports: Kennedy, London Heathrow, Schiphol Amsterdam, and Paris Charles de Gaulle. Other than a general lowering of traffic volumes during the summer holiday periods, the patterns differ substantially. Figures 10.3b and 10.3c indicate observed daily and hourly variations for a particular airline's cargo terminal. Perhaps only one conclusion

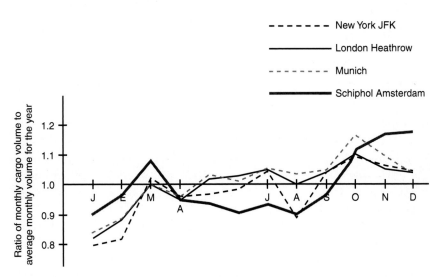

Figure 10.3a Variation of cargo volume throughout the year. *(PANYNJ, BAA, Flughafen München GmbH, and Schiphol Amsterdam Airport)*

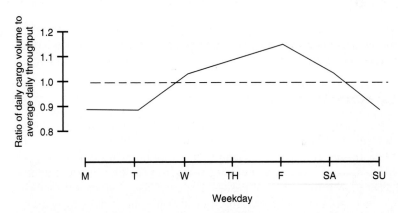

Figure 10.3b Daily variations of cargo throughput at British Airways freight terminal, London Heathrow. *(British Airways)*

can be drawn from these figures. From the operator's viewpoint, it is important that the traffic variations peculiar to the particular airport be known and understood. This requires careful data collection, because without adequate records the provision of adequate and economic facilities and staffing cannot be properly planned. Although terminal facilities are designed around peak rather than average conditions, they are not necessarily sized to cope with immediately processing the highest peak flows. Freight, unlike passengers, can be held over from the peak hour. It is not unusual when unloading arriving aircraft during an arrival peak for contain-

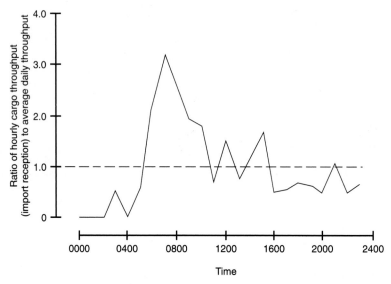

Figure 10.3c Hourly variations of cargo throughput at British Airways freight terminal, London Heathrow. *(British Airways)*

ers to be off-loaded to a bed of lazy rollers outside the terminal building. They remain there until they can be processed after the peak.

10.2 Expediting the Movement

Figure 10.4 shows diagrammatically the parties and organizations involved in the movement of air freight. Freight is moved from the shipper to the consignee, usually through the agency of a freight forwarder, by one or more airlines, using premises and infrastructure provided to some degree by the airports through which it passes. In many cases, on-airport facilities are provided not only for the airline, but also for the freight forwarders. Even large firms frequently use the facilities of a freight-forwarding agency because air cargo requires rather specialized knowledge, and air cargo might form only a small part of the firm's normal shipping operation. In order to provide an air shipping service, the freight forwarder performs several functions that are likely to be beyond the expertise or capability of the shipper. These are:

- To determine and obtain the optimum freight rate and to select the best mixture of modes and routes

- To arrange and oversee export and import customs clearances, including preparing all necessary documentation and obtaining requisite licenses (these are procedures with which the specialist forwarder is familiar)

- To arrange for the secure packing of individual consignments

- To consolidate small consignments into larger shipments to take advantage of lower shipping rates (the financial savings obtained by consolidation are shared between the forwarder and the shipper)

- To provide timely pickup and delivery services at both ends of the shipment

Most airlines see freight forwarders as providing a necessary and welcome intermediary service between themselves and the shipper and consignee. The freight forwarder, being familiar with the necessary procedures, permits the airline to concentrate on the provision of air transport and to avoid time-consuming details of the facilitation and landside distribution systems. Shippers with large air-cargo operations frequently use their own in-house expertise within a specialized shipping department.

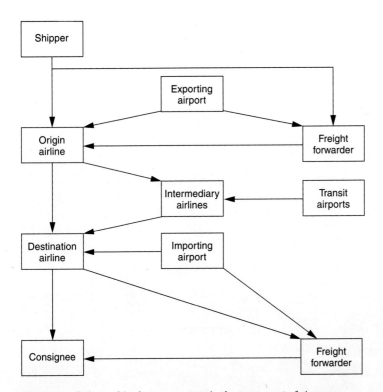

Figure 10.4 Relationships between actors in the movement of air cargo.

To encourage shipments that are more economical to handle, airlines have a complex rate structure, of which the main components are:

- **General cargo rates.** These apply to general cargo between specific airport pairs.

- **Specific commodity rates.** Often over particular routes, there are high movement volumes of a particular commodity. IATA approves specific commodity rates between specific airports. For *general cargo* and *specific commodity* rates, there are quantity discounts.

- **Classified rates.** Certain commodities because of their nature or value attract either a percentage discount or surcharge on the general commodity rate. Classified rates frequently apply to the shipment of gold, bullion, newspapers, flowers, live animals, and human remains.

- **ULD rates.** This is the cost of shipping a ULD container or pallet of specified design containing up to a specified weight of cargo. ULDs are part of the airline's equipment and are loaned to the shipper or forwarder free of charge, provided they are loaded and relodged with the airline within a specified period, normally 48 hours.

- **Consolidation rates.** Space is sold in bulk, normally to forwarders at reduced rates, because the forwarder can take advantage of quantity and ULD discounts. The individual consignee receives the shipment through a break bulk agent at destination.

- **Container rates.** Containers in this context are normally owned by the shipper rather than the airline. They are usually nonstructural, of fiberboard construction, and suitable for packing into the aircraft ULDs. If a shipment is delivered to them in approved containers, airlines make a reduction of air freight rates.

The very rapid movement of air cargo requires precise documentation. This is provided in terms of the *air consignment note* of the freight forwarder, an example of which is shown in Figure 10.5, and the airline's *airway bill*, which form the major essential documentation of carriage (Figure 10.6). The airway bill is a document with multiple uses. It provides:

- Evidence of the airline's receipt of goods

- A dispatch note showing accompanying documentation and special instructions

- A form of invoice indicating transportation charges

- An insurance certificate, if insurance is effected by the airlines

- Documentary evidence of contents for export, transit, and import requirements of customs

- Contents information for constructing the loading sheet and flight manifest
- A delivery receipt

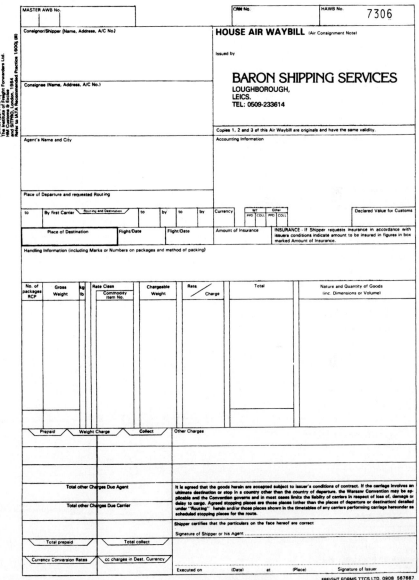

Figure 10.5 Consignment note.

Shipper's Name and Address		Shipper's Account Number	Not negotiable

Air Waybill*

⊘ **Lufthansa**

Issued by
Deutsche Lufthansa AG,
D-50664 Köln, Von-Gablenz-Straße 2—6

Member of International
Air Transport Association

Copies 1, 2 and 3 of this Air Waybill are originals and have the same validity

Consignee's Name and Address   Consignee's Account Number

It is agreed that the goods described herein are accepted in apparent good order and condition (except as noted) for carriage SUBJECT TO THE CONDITIONS OF CONTRACT ON THE REVERSE HEREOF. ALL GOODS MAY BE CARRIED BY ANY OTHER MEANS INCLUDING ROAD OR ANY OTHER CARRIER UNLESS SPECIFIC CONTRARY INSTRUCTIONS ARE GIVEN HEREON BY THE SHIPPER. THE SHIPPER'S ATTENTION IS DRAWN TO THE NOTICE CONCERNING CARRIER'S LIMITATION OF LIABILITY. Shipper may increase such limitation of liability by declaring a higher value for carriage and paying a supplemental charge if required.

Issuing Carrier's Agent Name and City   Accounting Information

Agent's IATA Code   Account No.

Airport of Departure (Addr. of First Carrier) and Requested Routing

To	By First Carrier	Routing and Destination	To	By	To.	By	Currency	CHGS Code	WT/VAL PPD COLL	Other PPD COLL	Declared Value for Carriage	Declared Value for Customs

Airport of Destination   Flight/Date   For Carrier Use Only   Flight/Date   Amount of Insurance

INSURANCE — If Carrier offers insurance, and such insurance is requested in accordance with the conditions thereof, indicate amount to be insured in figures in box marked 'Amount of Insurance'

Handling Information

These commodities licensed by USA for ultimate destination .... Diversion contrary to USA law prohibited.

No of Pieces RCP	Gross Weight	kg lb	Rate Class Commodity Item No.	Chargeable Weight	Rate / Charge	Total	Nature and Quantity of Goods (incl. Dimensions or Volume)

Prepaid	Weight Charge	Collect	Other Charges

Valuation Charge

Tax

Total Other Charges Due Agent

Total Other Charges Due Carrier

Shipper certifies that the particulars on the face hereof are correct and that insofar as any part of the consignment contains dangerous goods, such part is properly described by name and is in proper condition for carriage by air according to the applicable Dangerous Goods Regulations.

Total Prepaid	Total Collect

Signature of Shipper or his Agent

Currency Conversion Rates	CC Charges in Dest. Currency

Executed on   (date)   at   (place)   Signature of Issuing Carrier or its Agent

For Carrier's Use only at Destination	Charges at Destination	Total Collect Charges

* Luftfrachtbrief (nicht begebbar) — eine verbindliche Übersetzung dieses Frachtbriefformulars (einschließlich der Vertragsbedingungen) in die deutsche Sprache liegt bei allen Lufthansa Frachtbüros aus.

Figure 10.6 Air waybill.

10.3   Flow Through the Cargo Terminal

Figure 10.7 shows in conceptual terms the principal stages involved in freight flow through the terminal. The terminal operation on the input side is one that accepts over a very short period of time a large "batch" of freight (i.e., the aircraft payload). This batch is then sorted and that which is inbound and not direct transfer is checked in, stored, processed, stored again prior to delivery in relatively small shipments (i.e., up to container size). The export operation is the reverse process. Small shipments are received, processed, stored, and assembled into the payload for a particular flight that is then loaded by a procedure which keeps aircraft turnaround time to an acceptable minimum. Figures 10.8a and 10.8b show for import and ex-

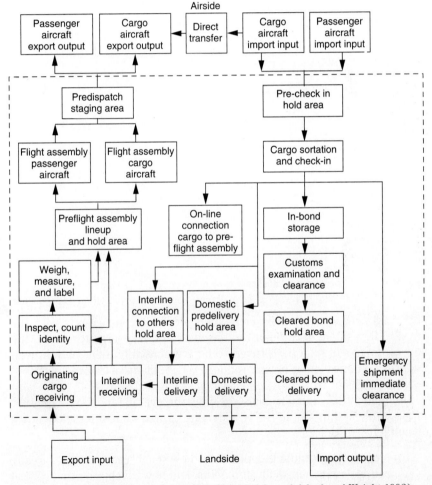

Figure 10.7   Flow through the cargo terminal. *(Adapted from Ashford and Wright 1992)*

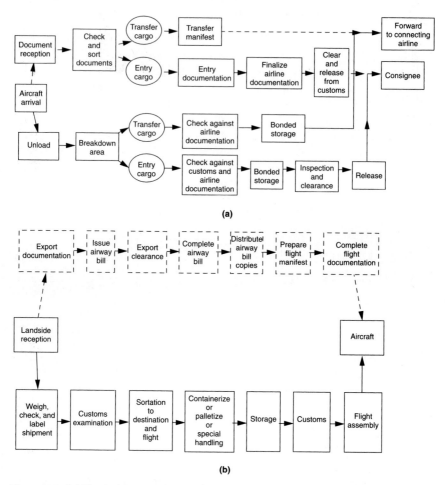

Figure 10.8 *(a)* Physical flow and documentation of *import* cargo; *(b)* physical flow and documentation of *export* cargo.

port cargo, respectively, how facilitation proceeds concurrently with the physical flow of cargo through the terminal. For very large terminals, the flow of documentation has proved to be a potential bottleneck; however, the use of on-line computers has reduced the problems of large paper flows, resulting in faster more efficient cargo processing.

Although most cargo terminals are similar in overall function, the nature of the traffic and therefore the actual mode of operation differs depending on a number of factors:

1. *Type of cargo.* Handling depends on how much cargo arrives already utilized, how much in small shipments, and how much requiring special handling.

2. *Type of shipper.* Terminal operations are simplified when receiving shipments through freight forwarders rather than private shippers. The forwarder will have already partially consolidated the handling and facilitation.

3. *Domestic/international split.* Domestic cargo requires less documentation and handling than international. At some U.S. and other airports, for example, suitably containerized domestic shipments physically bypass the terminal, the only handling being the facilitation. Bypassing is also permitted with international traffic with special arrangements with customs.

4. *Transfer.* Some terminals, such as Frankfurt, London, Bahrain, and Chicago, have very high levels of transfer cargo, which transfers between flights *across the apron.* This special requirement should be reflected in terminal design, since it requires an apron-handling capacity that is not needed in routine terminal processing. Many modern designs effect the across the apron transfer within a special area within the car ;o terminal building. So-called across-the-apron movements are therefore frequently conducted within the cargo terminal building, to provide shelter during the transfer period.

5. *Surface shipments.* In many countries, especially in Europe, the major air cargo terminals receive a large amount of "air" cargo by road. For example, cargo taken into the air-cargo terminal at Lille bound for Singapore is likely to be assigned to a "flight" from Lille to Paris that is in fact a road vehicle. After the first leg by surface, the leg from Paris to Singapore is achieved by air. London Heathrow provides a similar hub service for many of the cargo services of British regional airports.

6. *Interlining.* Interlining traffic to other airlines is unlikely to require the same processing as other import cargo.

7. *All-freight operations.* Where cargo is carried by all-freight aircraft, the airside operation is characterized by dramatic peaks of cargo flow. Severe overloading of the cargo apron is much more likely. Combined passenger/freight operations lessen the total peak flows and transfer them to the peak-passenger hours, which are not likely to coincide with freighter movements. However, the use of passenger aircraft implies that cargo will be more remotely loaded at the passenger terminal apron.

10.4 Unit Load Devices (IATA 1994)

Cargo, which is freight, mail, and unaccompanied baggage, was originally carried, loosely loaded in the cargo holds of passenger aircraft and if small all-cargo aircraft. Until the mid 1960s, all air cargo was carried in this loosely loaded or "bulk" cargo form. The introduction of large all-cargo aircraft, such as the DC8 and the B707, meant very long ground turnaround times due to the lengthy unloading and loading times involved with bulk cargo. The ground handling process was substantially speeded up by combining loads into larger loading units on pallets. Pallets, in combination with suitably designed aircraft floors and loading equipment, permitted rapid loading and unloading times. However, pallets must be carefully contoured to prevent damage to aircraft interiors where the aircraft is used also for passengers. Goods on the pallets are themselves vulnerable to damage and pilferage in carriage as well as being unprotected from the weather while on the apron. *Igloos*, which are nonstructural shells or covers to the pallets, were introduced to overcome these drawbacks. Even with a cover, a pallet is a relatively unstable device, which might shift during the handling process. A more substantial structural unit, the ULD air-freight container, gives considerably more support to the cargo during the handling and carriage stages but the device must still be lifted from below, unlike the intermodal container. With the introduction of wide-bodied aircraft, belly containers also were available in shapes that could not be built from pallets. Figure 10.9 shows how ULD containers can be used both on the main deck and the belly holds of a wide-bodied aircraft.

IATA recognizes a set of standard ULDs in the form of dimensional pallets, igloos, and ULD containers (IATA 1994 and IATA 1992). These ULDs are each compatible with a number of different aircraft types and are generally compatible with the terminal, the apron, and the loading equipment. However, as Table 10.1 indicates, there is serious incompatibility among aircraft types, which can cause considerable rehandling at freight transfer airports.

Because all ULDs do not give an optimum fit to all aircraft, without some degree of standardization aircraft will suffer from poor utilization of space and cargo rates will be unnecessarily high. Maximum space efficiency is achieved by use of the aircraft's optimum ULDs throughout. This, however, is likely to mean a break bulk operation at any transfer terminal and at the export and import ends of the journey. It is hoped that in the long run this might be overcome by the use of intermodal ULD containers that can be carried by surface modes. Intermodal containers will necessitate some loss of aircraft space efficiency. At the moment, most air ULD containers are too fragile for surface modes, and surface ULDs are too heavy for efficient aircraft use. Projections indicate that in the future, intermodal ULDs are likely to be developed and they might capture an increasing share of the market (Figure 10.10). This figure also indicates that by the mid 1990s manually loose-loaded bulk cargo had already shrunk to less than five percent of the total cargo market.

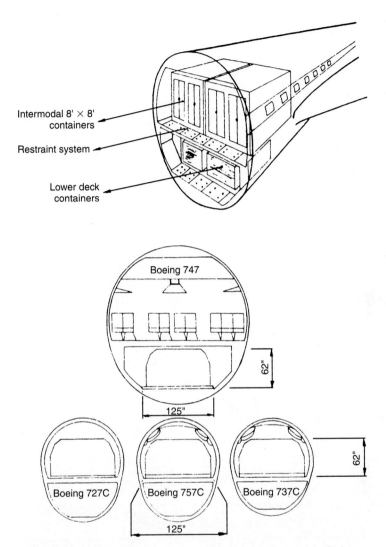

Figure 10.9 Container arrangements in wide-bodied and narrow-bodied aircraft. *(Ashford and Wright 1992)*

10.5 Handling Within the Terminal (IATA, 1992)

Unlike passengers, who merely require information and directions in order to flow through a terminal, cargo is passive and must be physically moved from landside to airside or vice versa. The system to be used for achieving

TABLE 10.1 Compability of Container Standard Contours with Aircraft Loading Envelopes

Aircraft Type/ Configuration (see ULD Technical Manual, Chapter 2, for loading envelope)	IATA Standard Container Contours (see ULD Technical Manual) Specification 50/0, Appendix 'E')											
	A B	C	D N	E	F	G H	K	M	P	U	Y	Z
B707C						•				•		•
B727C						•				•		•
B737C						•				•		•
B747 main deck	•	•				•	•	•		•		•
B747 lower deck			•		*			*				
L100	•					•	•	•		•	•	
L1011 lower deck			•	•		•		•				
DC8F						•			•			
DC10 main deck	•					•	•	•		•		•
DC10 lower deck			•	•		•		•				
A300 lower deck			•	•		•		•				
A310 lower deck			•	•		•		•				
A300C main deck	•					•	•	•		•		•
A310XC main deck	•					•	•	•		•		•
A320 lower deck					•							
AN12	•					•	•	•		•		•
IL76	•					•	•	•		•		•
IL86 main deck			•	•		•		•				

- Contour compatible with all aircraft (see Note).
- Contour compatible with large optional door only (check with carrier).

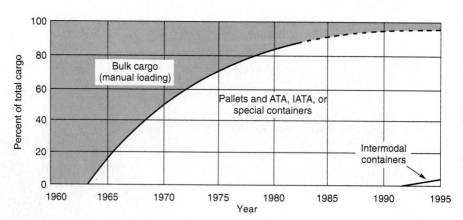

Figure 10.10 Air cargo unitization trend.

this physical movement will depend partly on the degree of mechanization to be used to offset manpower costs. The range of terminal designs is encompassed by three main types. Any particular terminal is likely to be made up of a combination of these types.

Low mechanization/high worker handling

Typically, in this design, all freight within the terminal is handled by workers over unpowered roller systems. Forklift trucks are used only for building and breaking down ULDs. On the landside, freight is brought to the general level of operation in the terminal by a dock-leveling device. This operational level, which is maintained throughout the terminal, is the same as the level of the transporting dollies on the airside. Even heavy containers are fairly easily handled by the workers over the unpowered rollers. This system is very effective for low- to medium-volume flows in developing countries where unskilled labor is cheap, where mechanizatio1 is expensive, and where there might be a lack of skilled labor for servicing equipment. The system is space extensive and is unsuitable for large flows simply from the number of unskilled workers in the terminal. Figure 10.11a shows the VARIG cargo terminal in Sao Paulo, Brazil, which has extensive roller beds and rail transfer tables.

Figure 10.11a Air cargo interior - Varig terminal Sao Paulo.

Open mechanized

The open mechanized system has been used for some time in developed countries at medium-flow terminals. All cargo movement within the terminal is achieved using forklift vehicles of various designs that are capable of moving fairly small loads or large aircraft container ULDs. Moreover, these forklift vehicles can stack up to five levels of bin containers. Many older terminals operated successfully with this system, but the mode is space extensive and forklift operations incur very high levels of ULD container damage. As pressure has come on cargo terminals to achieve less costly, higher volume throughputs in existing terminal space, many open mechanized terminals are being converted to fixed mechanized operation. A typical open mechanized system is shown in Figure 10.11b.

Fixed mechanized

The very rapid growth of the use of ULDs in aircraft has led to cargo terminal operations in which extensive fixed mechanical systems are capable of moving and storing the devices with minimum use of workers and low levels of container damage in handling. These fixed-rack systems are known as *transfer devices* (TV) if they operate on one level and *elevating transfer devices* (ETV) if they operate on several levels. Because they have very large ULD storage capacities, they can level out the very high apron throughput peaks that can occur with all freight wide-bodied aircraft. ETV rack storage can absorb for several hours incoming ULD freight and con-

Figure 10.11b INFRAERO terminal, Sao Paulo.

Figure 10.11c Lufthansa terminal, Frankfurt.

versely can provide departure flows on the airside greatly in excess of the terminal's ULD capacity. New and renovated terminals at New York JFK, Tokyo Narita, Frankfurt, Paris Charles de Gaulle, and London Heathrow, all include ETV and TV systems. A typical ETV layout is shown in Figure 10.11c. Fixed mechanized systems also have a great advantage over open mechanized systems from the viewpoint of container damage. Forklift operations can cause millions of dollars damage to containers at high volume open mechanized terminals each year.

10.6 Cargo Apron Operation

During the 1960s, there was a widely held belief that the next 20 years would see a general trend to virtual separation of passenger and cargo transport and the rapid development of all-cargo fleets. Two principal factors combined to ensure that this did not take place. First, wide-bodied passenger aircraft were introduced very rapidly in the 1970s to achieve crewing and fuel-efficiency objectives. The new wide-bodied aircraft had substantial and under-utilized belly space that was suitable for the movement of containerized cargo. Second, as indicated in Section 10.1, while exhibiting healthy growth rates, air cargo did not achieve the explosive growth expected at that time. By the early 1980s, the air-cargo operation had changed so much in character that a number of major airlines, which previously operated all-cargo aircraft, abandoned this form of operation in the short term in favor of using lower-deck space on passenger aircraft. This position is likely to continue while such space is available, although by the mid 1990s several major carriers had reintroduced all-cargo aircraft. All-cargo aircraft are common at airports served by all-cargo airlines. There are many small airlines that are all-cargo operations and a very few major operators if the integrated carriers such as Federal Express (now known officially as FedEx), UPS, and DHL are excluded. These operations are discussed briefly in Section 10.9.

Even though much freight is carried by other than all-cargo aircraft, very large volumes are moved by such operations through the air-cargo apron. All-cargo aircraft are capable of very high productivity, provided that there is a sufficient level of flow to support these productivity levels. The maximum payload of the B747F is more than over 270,000 pounds (122,000 kg). Figure 10.12 indicates that with containerized cargo, the aircraft manufac-

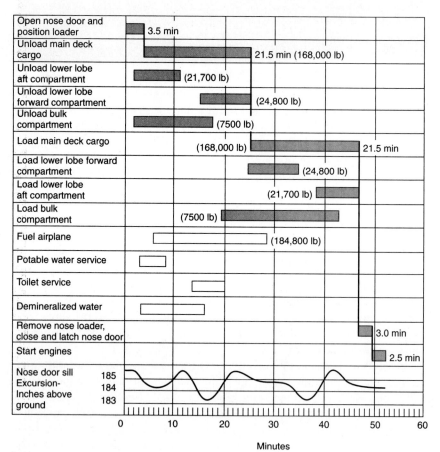

Figure 10.12 GANTT chart for turnaround of a 747F all-freight aircraft. *(Boeing)*

turer estimates that it is possible to off-load and on-load 220 total tons in slightly less than one hour. A more typical operational time would be considerably longer.

The times given by the manufacturer must be regarded as being ideal times where the load is immediately available and sequenced for loading. Real-world apron operations often mean that load control of the aircraft seriously inhibits total loading time. For a 100 series 747 with only side-door loading , a turnaround time of 1½ hours would be considered very good, and for a 200 series aircraft only nose-loaded, 2½ hours is more likely. The latter time can be reduced by simultaneously loading the nose and side doors of the main deck, but this ties up two high-lift loaders. Minimum total turnaround time might well be seriously affected by outside considerations such as off-loading equipment availability or the necessity to wait for customs or agricultural inspection before any off-loading can be started. Average turnarounds are much greater than minimum times because frequently aircraft are constrained by schedules that give a total ground handling time much greater than these quoted minima.

Using containers, the payload is decreased, but considerable gains are made in operational efficiency. Typically, the total payload of a B747F would be constituted in the following way:

Main deck cargo	168,000 lb (containerized)
Lower lobe aft	21,700 lb (containerized)
Lower lobe forward	24,800 lb (containerized)
Bulk compartment	7500 lb
Total payload	222,000 lb

Figure 10.13 shows the location of ground handling and servicing equipment required for simultaneous upper- and lower-deck unloading and loading sequence assumed in the Gantt chart shown in Figure 10.12. Such an operation places a very heavy load on apron equipment and apron space. In all, 45 containers will be off-loaded and a similar number loaded. Two lower deck low-lift loaders and one upper deck high-lift loader will be required, each requiring at least one and possibly two container transporters. Figure 10.14 shows a range of apron equipment used in loading air cargo. The bulk cargo hold is operated with a bulk loader fed by an apron trailer unit. Additionally, for general turnaround servicing, two fuel trucks are required if hydrant fueling is unavailable, a potable water truck, a truck to supply demineralized water for water injection, a sanitary truck for toilet servicing, a ground power unit, a compressed air start unit, and a crew access stair. In the immediate vicinity of the cargo building, it is usually necessary to provide a bed of "lazy rollers" or slave pallets to accept and temporarily store the off-loaded containers that might arrive from the apron at a peak rate be-

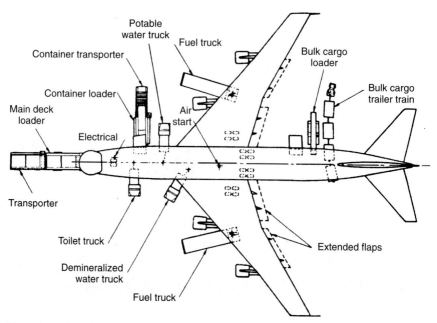

Figure 10.13 Location of servicing and loading/unloading equipment on cargo apron. *(Boeing)*

Figure 10.14a Upper deck freight container loader.

Figure 10.14b Lower deck freight container loader.

Figure 10.14c Nose loader for freighter in position.

Figure 10.14d Cargo container transporter.

yond the terminal capacity. Peak apron capacity is normally in excess of terminal throughput capacity. An ETV system is equally capable of absorbing apron peaks by storing received ULDs until they can be processed in the terminal.

10.7 Computerization of Facilitation

The secure and efficient shipment of a consignment can take place only when documentation keeps pace with the physical movement of cargo. This is a fairly straightforward matter with low volumes moving through a single air-cargo terminal building. At major airports with high volumes, numerous airlines, and multiple processing facilities, the control of facilitation becomes extremely complex and very necessary. This has been achieved for the last 25 years through computerization. Initially main-frame systems, such as the LACES and ACP 80 systems at Heathrow, were utilized. These connected Customs with the inventory-control systems of a number of airline operators and the airlines connected their own computers to the central bureau in order to transmit inventory control data to a communal file. Information from the communal file was made available to airlines, agents, and customs provided that the information sought is within the authority of the operator requesting it. Many main-frame systems still operate but are being steadily replaced by cargo community systems.

Main frame systems

Figure 10.15 shows the major events in the computerized life of a typical consignment, first on export then in import. A computer record of the consignment is initially created either at the time of space reservation or at reception. The initial file contains all details of the consignment from the airway bill, such as weight, contents, destination, carrier, shipper, and consignee. As the consignment moves through the system, numerous subsidiary files are created. Additionally, files are created that create flight records up to seven days in advance of departure date, indicating maximum weights and volumes to be allocated to each particular destination. When the consignment is completely received (i.e., the number of packages agrees with the airway bill), the consignment is either allocated to a flight or to an ULD, which itself will be allocated to a flight. The computer then provides a flight-tally file that indicates shed storage location, ULD details, and any instructions on handling. Once all consignments have been tallied, a flight manifest, the working document reporting the movement of cargo, is produced. The manifest is used on the load control of the aircraft itself. The final input to the information system is the statement that the flight and the goods have departed. The data system can be interrogated at any point to determine the current location and status of the cargo. There is also the possibility of modifying initial input data to allow for consignment splitting, off-loading and for coping with short shipments where a partial shipment is made when part of the consignment is held up.

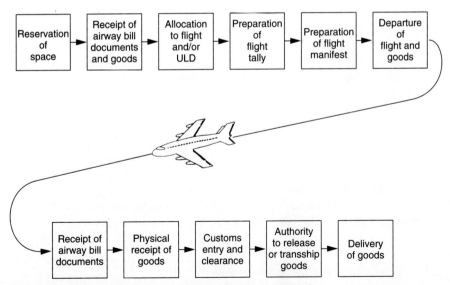

Figure 10.15 Principal stages in exporting and importing freight.

On the import side, a file record is created on receipt of the waybill documentation. This computerized record identified the receiving shed, the airline, the airway bill number, and the necessary data contained on the airway bill. Status reports are made on the physical receipt of the goods; first that all goods have been received and second that the consignee is creditworthy. After customs entry and clearance, either a release note is produced after clearance, or removal authority to permit transshipment to another airport for clearance, or for removal to another import shed. At this stage, a transfer manifest is produced for through-transit consignments. The final input in the life of the record is that the goods have been delivered. Several reports are automatically available that record discrepancies, such as the receipt of more or less packages than expected. As a standard inventory control measure, a number of daily operational reports are produced. These give the current status of problem areas such as:

- Consignments not received within 24 hours of the receipt of the airway bill
- Consignments not delivered with 2 days of customs clearance
- Through transits not delivered to onward carrier in 6 hours
- Transshipments and interairport removals not achieved within 7 days

At all stages, the status files may be interrogated by operators with necessary authority.

Cargo community systems

With the widespread introduction of personal-computer (PC) technology to business, PC hardware has become generally available. Cargo community systems enable freight operations with PCs to join systems which, using *electronic data interchange* (EDI), enable computers to exchange data directly without human intervention. The ICARUS system by the mid 1990s, links more than 50 major airlines with in excess of 1500 forwarders, carriers and other EDI-based systems, including the CCS-UK system used by British Airways and others at London Heathrow. The worldwide ICARUS system connects with the CCN system at Changi Singapore, the TRAXON systems in France, Japan and Hong Kong, Avex in the United States, Cargonet in Australia, and other CCS systems in Austria, Italy, South Africa, and the United Kingdom. The user is able to exchange information between computers without the need for paper-based documentation and can:

- access flight space availability
- obtain electronic booking/reservation
- transmit electronic documents (e.g., airway bill)

- receive electronic documents
- provide consignment tracking and status
- community among forwarders
- provide schedule information

Although the initial investment to join such a CCS system is relatively expensive, for airports handling in excess of 100,000 tons of freight each year the system is likely to be seen as an essential element of the operation and is much cheaper and more flexible than a main-frame system.

10.8 Examples of Modern Cargo Terminal Designs

In 1995, Lufthansa redesigned its cargo terminal at Frankfurt with a handling capacity of approximately 1 million tons/year. Figure 10.16 shows a schematic layout of the facility which, in common with most modern terminals, is extensively mechanized and is designed around the continued use of ULDs. Frankfurt has a very high proportion of transfer freight, much of which requires reconsolidation within the cargo terminal itself. The system adopted within the terminal recognizes that workers and floor area can be saved by extensive use of mechanical systems. Given that little can be done to standardize the world's range of containers, this operation is designed to standardize the handling of the various container types. Prior to consolidation, bulk cargo shipments are assigned to rollboxes in a continuous handling section where vertical stackers are able to assign the rollboxes to vertically racked rollbox slots. An ETV feeds a vertical container racking system, which can hold 10-foot containers and 20-foot units. The very high level of mechanization is designed around a concept of centralized storage rather than separate storage for import and export. The layout permits lateral, transverse, and round-the-corner movements by conveyor systems without the use of mobile wheeled equipment. This form of operation has three principal objectives:

- Minimization of accidents to personnel and cargo
- Minimization of damage to ULDs
- Maximum utilization of cargo terminal space

10.9 Freight Operations for the Integrated Carrier

By the mid 1990s, ten of the world's airports were handling more than one million tons of cargo. Of these ten, Memphis ranked as the world's busiest freight airport and Louisville as the sixth. The former is the base of Federal Express and the latter of UPS, both so-called *integrated cargo carriers*

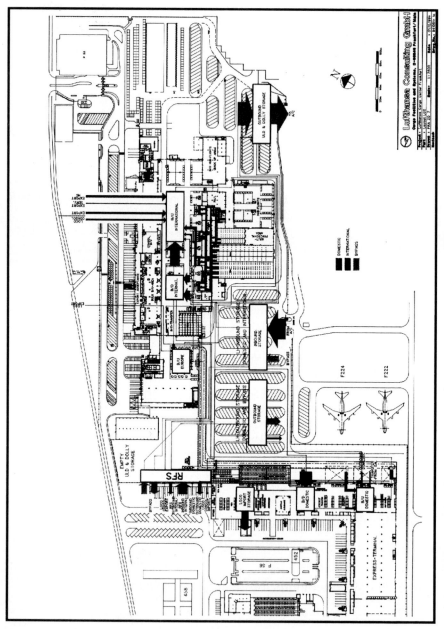

Figure 10.16 Schematic of Lufthansa cargo terminal. (*Lufthansa Consulting GmbH*)

(ICs). The growth of the integrated carriers over the last 20 years has been extraordinarily buoyant. Such carriers offer door-to-door service usually within a stated time limit. Federal Express, the largest of the integrated carriers, is also the world's largest all-cargo airline serving more than 200 airports in 180 countries, owning in excess of 30,000 vehicles and employing approximately 100,000 persons.

The operation of the integrated carrier's air freight terminal is quite different from the conventional air freight terminal, and it is difficult to draw comparisons between the two operations even at the same airport. The terminals of the integrated carriers have very high daily peaks and are characterized by the lack of storage space required because very little freight dwells in the terminal for any significant time. The principal characteristics that differentiate this type of terminal from the conventional operation are:

- The freight is under the physical control of one organization throughout the whole of its journey
- Delivery to the exporting airport and pick-up at the importing airport is by the integrated carrier itself, the landside pick-up area is more easily controlled and organized
- No freight forwarders or clearing agents are involved
- No other airline is involved
- All-cargo aircraft are used and the characteristics of the fleet are well known to the terminal operator
- The freight is given a guaranteed delivery time; this implies rapid clearance through the airport facility
- Documentation and facilitation is through one company's system
- High security and limited access to the terminal is possible because no outside organizations are involved other than customs.

Typically, the large global integrated carriers set up intercontinental hubs such as Memphis, Newark, and Paris Charles de Gaulle. Additionally, there are regional hubs. Federal Express, for example, has six auxiliary hubs in Europe: Prestwick, Stansted, Frankfurt, Cologne, Basle, and Milan. Trucks from these facilities serve the remainder of Europe.

The location of an integrated carrier's intercontinental or regional hub at an airport will have little relationship with the operations of the passenger facilities. The IC hub terminal tends to operate in isolation, with peaking characteristics quite independent of the activities on the passenger apron. Often such facilities see banks of arrivals around midnight and banks of departures in the early hours of the morning. As such they have much in common with air-mail terminals, although in scale they can be very much larger.

References

Ashford N. J., and P. H. Wright. 1992. *Airport Engineering, 3rd edition.* New York: Wiley Interscience.

Boeing. 1995. *World Cargo Forecast.* Seattle.

International Air Transport Association (IATA). 1991. *Principles of Aircraft Handling.* Montreal.

IATA. 1992. *Principles of Cargo Handling.* Montreal.

IATA. 1993. *Airport Handling Manual.* Montreal.

IATA. 1994. *Unit Load Devices Handling Guide.* Montreal.

International Civil Aviation Organization (ICAO). 1995. *Civil Aviation Statistics of the World.* Montreal.

11

Airport Technical Services

11.1 The Scope of Technical Services

Various operational services found at air carrier airports can be conveniently grouped together under this general heading. They are concerned with the safety of aircraft operations in terms of control, navigation and communications, and information. Clearly these matters are of interest to all aviation nations and they form the subjects of four of the technical annexes (ICAO 1991, ICAO 1991, ICAO 1992, ICAO 1993) to the International Convention on Civil Aviation (Chicago Convention). The services are:

- Air traffic services
- Aeronautical telecommunications (including navaids)
- Meteorology
- Aeronautical information services

A general aviation airfield might need only some of these services to fulfill its particular type of operations (e.g., flight training, executive aviation, personal flying).

In addition to the four preceding services, emergency service also has to be provided at all airports and general aviation airfields in order to provide fire-fighting and rescue capabilities in the event of aircraft accidents. This service is dealt with separately in Chapter 12.

11.2 Air Traffic Control

Function of ATC

Air traffic control has, as its primary purpose, the prevention of collision between aircraft in flight and also between aircraft and any obstructions either moving or stationary on an airport. Additionally, it is concerned with promoting an efficient flow of air traffic.

In most countries, the central administration and management of air traffic services is vested in a governmental or quasi-governmental agency. This is usually a civilian organization but might be the military authority in some countries. In the case of a civilian organization, for example, United States legislation contained in the Federal Aviation Act 1958, Title III, Section 307, provides the FAA with powers relating to the "Use of Airspace, Air Navigation Facilities, and Air Traffic Rules." The FAA is also the central authority for issuing airman[1] certificates, including those for air traffic controllers. It is not, however, responsible to the economic regulation of the air transport industry.

In Great Britain, the CAA is the designated agency set up by the Act of Parliament (Civil Aviation Act 1971) with specific powers relating to air traffic control as set out in Air Navigation Orders. As in the United States, this is also the agency responsible for the issue of airman licenses, but unlike the FAA it is also responsible for the economic regulation of the air transport industry.

In West Germany, the air traffic services agency is the Bundesanstalt fur Flugsicherung (Federal Administration of Air Navigation Service.)

At most airports, therefore, air traffic control and its associated departments will be under the immediate management of a central government agency official and not a member of airport management. At the moment there are only a few exceptions to this where air traffic control staffs are employees of the airport authority, although licensed by the government authority.

In addition to airports, a large part of any air traffic service is responsible for en route airspace exercising control from regional centers; in the U.S., Air Route Traffic Control Centers (ARTCC); in Europe, Air Traffic Control Centers (ATCC). Although most of these staffs are governmental employees, there are some areas of the world where this service is contracted out to specialist organizations (e.g., IAL, International Aeradio Limited, a company active in some areas of the Middle East). In all cases, however, governments maintain ultimate policy control in line with the central concept of national sovereignty over airspace.

In the United States and Great Britain there is current discussion regarding proposals to privatize the air traffic control services. The basis for the

---

[1]A comprehensive term for holders of pilot, flight engineer, and other specialized qualifying licenses, including air traffic controller, aircraft dispatcher, and so on.

proposals is the perceived need to assure the predictability of funding rather than have the scope and speed of technological improvements jeopardized by the vagaries of governmental/political budgetary processes. There is general agreement that governments should still have overall responsibility for policy control of safety standards.

Types of airspace

Due to the variations in the density of air traffic and to the constraints imposed by weather conditions, air traffic control authorities apply more stringent control in some areas than in others. As a result, there are several different types of airspace. Those in the vicinity of busy airports, for example, are designated for an intensive level of control while other areas with only light traffic might not have any positive control at all; this would be *uncontrolled* airspace. The basic geographical division of airspace is that which occurs at national boundaries. Within the national boundaries there might be one or more Flight Information Regions (FIRs). The profusion of FIRs in Europe is brought about mainly by the comparatively large numbers of national frontiers (Figure 11.1). For air traffic purposes, the division of airspace within the continental United States is by Flight Advisory Areas (Figure 11.2). There are 20 of these areas, although it is planned to reduce this number to 16. Each is the responsibility of an Air Route Traffic Control Center (ARTCC). Those areas adjoining international airspace are:

- EAST AND SOUTH
  ~New York Oceanic FIR
  ~Miami Oceanic FIR
  ~Houston Oceanic FIR

- WEST
  ~Oakland Oceanic FIR
  ~Anchorage Oceanic FIR
  ~Honolulu Oceanic FIR

There are occasions when FIR boundaries between countries do not coincide with national geographic boundaries. In these cases, boundaries are mutually agreed between the nations involved as occurs over the high seas or in airspace of undetermined sovereignty (e.g., the rim of the North Atlantic, where the United States, Canada, and Portugal agree on the boundaries between the New York Oceanic FIR, Gander Oceanic FIR, and Santa Maria Oceanic FIR).

Those parts of a FIR that are uncontrolled are sometimes referred to as the *open FIR* because no restrictions in terms of air traffic control are placed on aircraft in these areas—no separation is provided by ATC. It is usual, however, for another air traffic service to be available on a notified

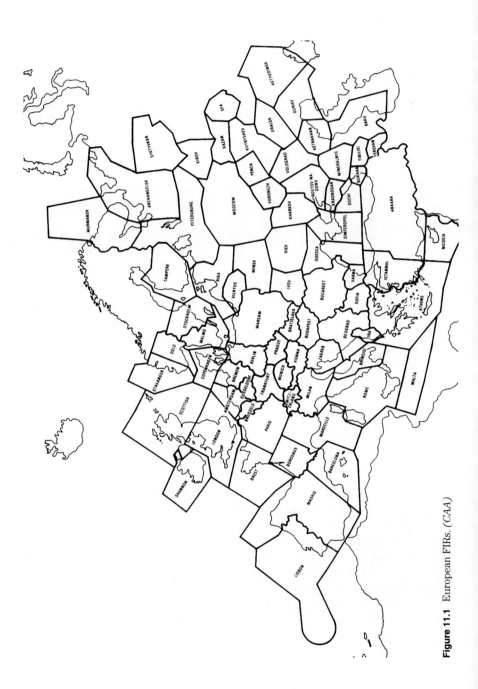

Figure 11.1 European FIRs. *(CAA)*

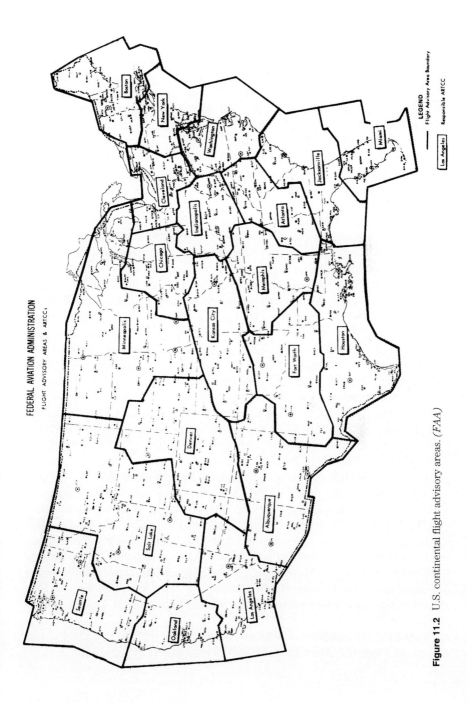

FEDERAL AVIATION ADMINISTRATION

FLIGHT ADVISORY AREAS & ARTCCs

Figure 11.2 U.S. continental flight advisory areas. *(FAA)*

LEGEND

—— Flight Advisory Area Boundary

Los Angeles   Responsible ARTCC

frequency via flight information service. It might be broadcast at notified times or supplied on request to pilots by ground radio stations. Flight information service can, of course, also be provided by any control unit including airport or approach control. The basic operational units of airspace are *control zones* and *control areas,* the difference being that zones start from the surface to a given altitude while control areas are controlled airspaces extending upwards from a specified limit above the earth including airways. Controlled airspace, as defined by Annex 2, ICAO 1990a, is an "airspace of defined dimensions within which air traffic control service is provided to IFR flights and to VFR flights in accordance with the airspace classification."

Due to the differing requirements for IFR and VFR operations in various types of airspace there is now an international classification of airspace that is divided into various classifications: A, B, C, D, E, F, G, not all of which are allocated.

The intention is that worldwide it will be sufficient to know the classification of airspace to know exactly what the ATC conditions are in respect of IFR and VFR, and the type of air traffic service available in the particular class of airspace. Not all countries will immediately implement all classes; Great Britain, for example has not yet allocated Class C, while the United States has not allocated Class F. All have agreed, however, that Class A is for IFR operations only, though it is interesting to note that unlike the United States, which classified only upper airspace as Class A, Great Britain has included its major terminals, Heathrow and Manchester. The controlled airspace found in the vicinity of busy airports generally consists of a surface area and two or more layers often resembling an upside-down wedding cake. In a typical example the layers surrounding a surface area might have lower levels gradually lifting in steps from 7000 feet to 8000 feet to 10,000 feet. Figure 11.3 illustrates this effect as it relates to the Cleveland-Hopkins Airport; the figures represent hundreds of feet, referring to the base and tops of the controlled airspace. In the United States such airspace includes primary civil airports such as Atlanta Hartsfield, Boston Logan, Chicago O'Hare International, Los Angeles International, Miami International, Newark International, New York Kennedy, New York La Guardia, San Francisco International, Washington National, Dallas-Fort Worth International, among others; all are now classified as Class B airspace. Class A airspace in the United States is generally the airspace from 18,000 feet msl up to and including FL 600 in which only IFR operations are authorized. Secondary airports (U. S. definition) in the United States are designated Class C airspace, and include less busy locations such as Fort Lauderdale-Hollywood International, Spokane International, Lincoln Municipal (FAA 1994).

Details of the airspace classifications employed by the U. S. are given in Table 11.1 and the international ICAO classifications in Table 11.2.

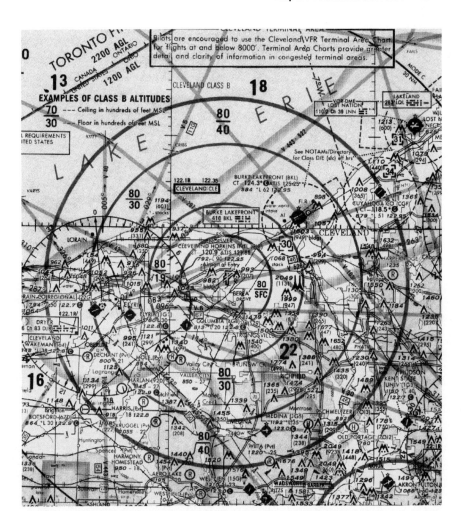

Figure 11.3 Class B airspace around the Cleveland-Hopkins primary airport.

Flight rules

There are three sets of flight rules depending on the circumstances listed:

General flight rules	Observed by all aircraft in whatever type of airspace
Visual flight rules (VFR)	Observed by aircraft flying in weather conditions above prescribed limits
Instrument flight rules (IFR)	Observed by aircraft in weather conditions below visual limits and/or in certain categories of airspace

General Flight Rules. As the name implies these rules refer to the conduct of flight in such general matters as the safeguarding of persons and prop-

TABLE 11.1 United States Airspace Classifications.

Airspace Features	Class A	Class B	Class C	Class D	Class E	Class G
Entry requirements	ATC clearance	ATC clearance	Prior two-way communications	Prior two-way communications	None	None
Minimum pilot qualifications	Instrument rating	Private or student certificate location dependent	Student certificate	Student certificate	Student certificate	Student certificate
Two-way radio communications	Yes	Yes	Yes	Yes	Not required	Not required
Special VFR allowed*	No	Yes	Yes	Yes	Yes	N/A
VFR visibility minimum	N/A	3 statute miles**	3 statute miles**	3 statute miles**	3 statute miles**	1 statute mile**
VFR minimum distance from clouds	N/A	Clear of clouds	500 feet below, 1000 feet above, 2000 feet, horizontally**	500 feet below, 1000 feet above, 2000 feet horizontally**	500 feet below, 1000 feet above, 2000 feet horizontally**	Clear of clouds**
VFR aircraft separation	N/A	All	IFR	Runway operations	None	None
Traffic advisories	Yes	Yes	Yes	Workload permitting	Workload permitting	Workload permitting
Former airspace equivalent	Positive control area (PCA)	Terminal control area (TCA)	Airport radar service area (ARSA)	Airport traffic area and control zone	General controlled airspace	Uncontrolled airspace

*Authorized by an ATC clearance and conducted within the lateral boundaries of the surface area.
**Flight visibility and cloud clearance requirements differ for operations below 1200 feet AGL, above 1200 feet AGL but below 10,000 feet MSL, day, night, or student pilot. See FARs 61.89 and 91.155 for specifics.
NOTE: IFR operations in controlled airspace require filing an IFR flight plan and an appropriate ATC clearance.

TABLE 11.2   I.C.A.O ATS Airspace Classifications

Class	Type of flight	Separation provided	Service provided	VMC visibility and distance from cloud minima*	Speed limitation*	Radio communication requirement	Subject to an ATC clearance
A	IFR only	All aircraft	Air traffic control service	Not applicable	Not applicable	Continuous two-way	Yes
	IFR	All aircraft	Air traffic control service	Not applicable	Not applicable	Continuous two-way	Yes
B	VFR	All aircraft	Air traffic control service	8 km at above 3050 m (10,000 ft) AMSL 5 km, below 3050 m (10,000 ft) AMSL Clear of clouds	Not applicable	Continuous two-way	Yes
	IFR	IFR from IFR IFR from VFR	Air traffic control service	Not applicable	Not applicable	Continuous two-way	Yes
C	VFR	VFR from IFR	1) Air traffic control service for separation from IFR; 2) VFR/VFR traffic information (and traffic avoidance advice on request)	8 km at and above 3050 m (10,000 ft) AMSL 5 km below 3050 m (10,000 ft) AMSL 1500 m horizontal; 300 m vertical distance from cloud	250 kt IAS below 3050 m (10,000 ft) AMSL	Continuous two-way	Yes

TABLE 11.2 I.C.A.O ATS Airspace Classifications (Continued)

Class	Type of flight	Separation provided	Service provided	VMC visibility and distance from cloud minima*	Speed limitation*	Radio communication requirement	Subject to an ATC clearance
D	IFR	IFR from IFR	Air traffic control service including traffic information about VFR flights (and traffic avoidance advice on request)	Not applicable	250 kt IAS below 3050 m (10,000 ft) AMSL	Continuous two-way	Yes
	VFR	Nil	Traffic information between VFR and IFR flights (and traffic avoidance advice on request)	8 km at and above 3050 m (10,000 ft) AMSL 5 km below 3050 m (10,000 ft) AMSL 1500 m horizontal; 300 m vertical distance from cloud	250 kt IAS below 3050 m (10,000 ft) AMSL	Continuous two-way	Yes
E	IFR	IFR from IFR	Air traffic control service and traffic information about VFR flights as far as practical	Not applicable	350 kt IAS below 3050 m (10,000 ft) AMSL	Continuous two-way	Yes
	VFR	Nil	Traffic information as far as practical	8 km at and above 3050 m (10,000 ft) AMSL 5 km below 3050 m (10,000 ft) AMSL 1500 m horizontal; 300 m vertical distance from cloud	250 kt IAS below 3050 m (10,000 ft) AMSL	No	No

Class		Separation provided	Service provided	VMC minima	Speed limitation	Radio communication requirement	Subject to ATC clearance
	IFR	IFR from IFR as far as practical	Air traffic advisory service; flight information service	Not applicable	250 kt IAS below 3050 m (10,000 ft) AMSL	Continuous two-way	No
F	VFR	Nil	Flight information service	8 km at and above 3050 m (10,000 ft) AMSL 5 km below 3050 m (10,000 ft) AMSL 1500 m horizontal; 300 m vertical distance from cloud. At and below 900 m AMSL or 300 m above terrain whichever is higher — 5 km **, clear of cloud and in sight of ground or water	250 kt IAS below 3050 m (10,000 ft) AMSL	No	No

erty on the ground, avoidance of collision, right-of-way rules, and aircraft navigation lights. Details are listed in the various regulatory documents of each country. For example, in the United States, this is Part 91 of Federal Aviation Regulations; the appropriate regulation in Great Britain is achieved by means of the Air Navigation: Order and Regulations.

Visual Flight Rules. In addition to observing the basic rules of the air, each flight has to be conducted according to either *visual flight rules* (VFR) or *instrument flight rules* (IFR). In the case of VFR, the flight is conducted on a see-and-be-seen basis in relation to terrain and other aircraft. It is, therefore, necessary for a pilot to have certain minimum weather conditions known as *visual meteorological conditions* (VMC). Anything worse than these conditions is referred to as *instrument meteorological conditions* (IMC). United States usage is *VFR conditions* and *IFR conditions*. The weather criteria for visual flight are intended to provide to pilots an adequate opportunity to see other aircraft or obstructions in time to avoid collisions. For this reason, limits for lower and slower aircraft are less than for those flying at the higher speeds and some allowance is also made for areas where traffic is less dense (i.e., uncontrolled airspace).

ICAO set out VFR minimums for the various classes of airspace, which countries by and large have adopted with some slight variations to suit their own circumstances. In the case of the United States the conditions are very similar to those of ICAO (Table 11.3), although not identical. The VFR minimums adopted by Great Britain are also very similar (Table 11.4).

Instrument Flight Rules. When visibility and/or proximity to cloud are less than the quoted VFR or VMC limits, flight has to be conducted under IFR. In respect to flight under instrument flight rules, the rules require that ATC must be notified of flight details in advance by what is known as an *ATC flight plan*. Thereafter the flight has to conform to the plan or to any other instructions issued by ATC. To do this, a continuous watch has to be maintained on the appropriate radio frequency by the pilot and reports made, as required, to ATC regarding the aircraft's position. In the United States, there is also a requirement for weather reports from pilots when unforeseen conditions are encountered. Another of the rules requires that an instrument flight must be conducted at a minimum height of 1000 feet (300 m) above the highest obstacle within 5 miles (8 km) of the aircraft's position, except while landing or taking off. Elsewhere in the legislation of all countries, there is a requirement that an aircraft must be suitably equipped for the type of flight being undertaken and the pilot suitably qualified for that flight.

Flights may, of course, be conducted under IFR outside controlled airspace and therefore not receive a specific altitude assignment from ATC. In order to provide some safeguard for such flights and for VFR flights, there

TABLE 11.3 U.S. Basic VFR Weather Minimums

Airspace	Flight Visibility	Distance from Clouds
Class A	Not applicable	Not applicable
Class B	3 statute miles	Clear of clouds
Class C	3 statute miles	500 feet below 1000 feet above 2000 feet horizontal
Class D	3 statute miles	500 feet below 1000 feet above 2000 feet horizontal
Class E Less than 10,000 feet MSL	3 statute miles	500 feet below 1000 feet above 2000 feet horizontal
At or above 10,000 feet MSL	5 statute miles	1000 feet below 1000 feet above 1 statute m le horizontal
Class G 1200 feet or less above the surface (regardless of MSL altitude) Day, except as provided in section 91.155(b)	1 statute mile	Clear of clouds
Night, except as provided in section 91.155(b)	3 statute miles	500 feet below 1000 feet above 2000 feet horizontal
More than 1200 feet above the surface but less than 10,000 feet MSL Day	1 statute mile	500 feet below 1000 feet above 2000 feet horizontal
Night	3 statute miles	500 feet below 1000 feet above 2000 feet horizontal
More than 1200 feet above the surface and at or above 10,000 feet MSL	5 statute miles	1000 feet below 1000 feet above 1 statute mile horizontal

are basic rules for a simplified form of vertical separation that is self administered by pilots. Under this system, the altitude to be flown depends on the magnetic course (ground track) being followed by the pilot. This system assures vertical separation which increases with increasing altitude/FL resulting in a separation of 2000 feet (600 m) for an aircraft in the upper airspace (intervals of 4000 feet (1200 m) in the "semi-circular" system (Figure 11.4.) In Great Britain a quadrantal split of tracks is used for the lower airspace (i.e., 3000 feet [900 m] msl to 24,500 feet [7500 m]) with a similar "semi-circular" split of altitudes and flight levels above 24,500 feet,

TABLE 11.4 VFR Weather Minimums—Great Britain

Airspace	Flight Visibility	Distance from Cloud
Class B	8 km at and above FL 100	Clear of cloud
	5 km below FL 100	Clear of cloud
Class C,D,E	8 km at and above FL 100	1500 m horizontally
		200 ft vertically
	5 km below FL 100	Clear of cloud
	5 km at or below 3000 ft MSL	Clear of cloud
	at an airspeed 140 kts or less	and in sight of
		the surface
Class F,G	8 km at and above FL 100	1500 m horizontally
		1000 ft vertically
	5 km below FL 100	1500 m horizontally
		1000 ft vertically
	5 km at or below 3000 ft MSL	Clear of cloud and
		in sight of the surface
	1500 m at or below 3000 ft MSL	Clear of cloud and
	at an airspeed 140 kts or less	in sight of the surface

(Figure 11.5). In the U.S. IFR traffic below FL 180 will be assigned altitudes on the even thousands of feet (e.g., 4000, 5000, 6000, etc.) depending on magnetic course being followed while VFR traffic operates 500 feet above or below the even thousands (e.g., 2500, 3500, 4500, etc.) to provide 500 feet separation between IFR and VFR traffic.

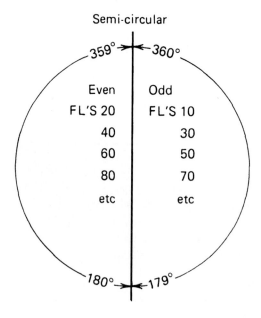

Figure 11.4 U.S. semicircular flight levels for IFR traffic. VFR traffic flies at these altitudes plus 500 feet.

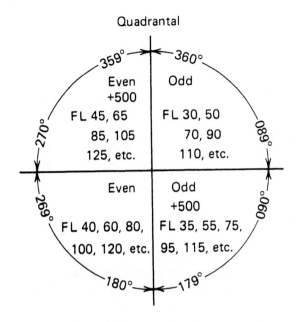

Figure 11.5 Quadrantal flight levels.

Separation standards

The criteria used by ATC to determine the required spacing between aircraft to achieve safety are known as *Separation Standards*. These are specific criteria relating to distance apart in the vertical or horizontal plane. In the vertical plane, IFR aircraft are separated by requiring them to fly at different altitudes or levels so that they are 1000 feet (300 m) apart up to FL 290 and 2000 feet (600 m) apart from FL 290 upwards. Horizontal separation is classified into three groups: lateral, longitudinal, and radar. Aircraft are *laterally* separated if their tracks diverge by a minimum angular amount with reference to a radio navigational aid, for example: 20, 30, or 45° or if they report over different geographical locations, and thereafter they continue to move farther apart.

ATC prescribe intervals of time or distance between aircraft to provide *longitudinal* separation to those at the same altitude or departing from the same airport. When ATC are able to see for themselves on radar where aircraft are in relation to each other and are thus not dependent on aircraft position reports, they are able to substantially reduce these intervals. Typically a 15-mile (24 k) standard longitudinal separation might be reduced to a 3-mile (5 k) *radar* separation. The widespread use of radar has been a major factor in helping to increase airspace capacity.

In relation to arriving aircraft, the radar separation of 3 miles might need to be increased to take account of the phenomenon known as *vortex wake*. This is the turbulence created in the wake of an aircraft, particularly a wide bodied, commonly referred to as a "heavy" jet and the associated hazard to a smaller, lighter aircraft following behind.

Operational structure

Before any authority is able to provide positive air traffic control, especially in the present busy environment around most air carrier airports, it must establish a complex system of navigation aids and communications facilities to delineate the pattern of routings for arriving and departing aircraft and to provide information to pilots and controllers on the position and progress of flights.

The short-range ground-based radio aids used include nondirectional beacon (NDB), VHF omnidirectional radio range (VOR), distance measuring equipment (DME), and fan markers (FM). For precision approach and landing guidance, there is the Instrument Landing System (ILS). Further information on these aids is to be found in Chapter 5.

The improved technology now being used in aircraft equipment provides better instrumentation and enhanced instrument flight capabilities. There is a requirement for this to be matched on the ground by ILS installations and associated ground aids capable of providing for automatic landing in fog. Landing requirements are categorized as shown in Table 11.5 according to two criteria:

Runway visual range (RVR). The maximum distance in the direction of takeoff or landing at which the runway or the specified lights or markers delineating it can be seen from a position above a specified point on its centerline at a height corresponding to the average eye level of pilots at touchdown.

Decision height. A specified height at which a missed approach must be initiated if the required visual reference to continue the approach to land has not been established. All air transport aircraft are operated into airports in compliance with company weather minima. These may conform to specific guidance from government, as in the United States, or be established by the company and then passed on to the government agency as is the case

TABLE 11.5 ILS Categories

Category	Decision Height	RVR
I	200 ft (60 m)	ICAO 2500 ft (800 m), FAA 1800 ft (600 m)
II	100 ft (30 m)	1200 ft (400 m)
III		
A	0	700 ft (200 m)
B	0	150 ft (50 m)
C	0	0

with the CAA in Great Britain. They are expressed in terms of visibility and decision height or cloud base.

When aircraft are approaching and landing in minimum weather conditions, ATC requires especially precise information on the positions of all aircraft. To assist with this, ATC relies in the first instance on primary surveillance (search) radar, which generally gives good coverage out to some 30 miles (48 k) from the airport. Such radars do not, however, give two other vital pieces of information: identification of the radar echoes and height. To provide this information requires secondary surveillance radar (SSR). This has now become so vital to ATC in busy terminal areas and other types of controlled airspace that its carriage has become mandatory in such areas. It will almost certainly supplant surveillance radar as the primary display system used by ATC to give position and related information on aircraft.

Finally, once the aircraft is on the ground in fog conditions, ATC requires a "mapping" radar ASMI or ASDE. Such radars are also useful for departing taxiing traffic, especially at airports subject to fog.

Clearly, the extent of the radar requirement will depend very much on the prevailing weather at an airport and the density of traffic and will range from a simple search radar to the more complex operating modes of SSR.

Operational characteristics and procedures

The characteristics of an airport with the full range of ATC equipment described above will clearly differ greatly from the smaller GA airfield. At the GA airfield, the balance of traffic will lean toward relatively small aircraft, possibly including training flying, mainly under VFR conditions. In the U.S. however, nonprecision instrument approaches are available at hundreds of GA airports, most of which do not have air traffic control towers. ATC services are adjusted to suit the differing characteristics. Some might have the full range of air traffic services including airport, approach, and radar while others might have only airport. The two main categories of air traffic control units associated with airport operations are airport control (tower) and approach control.

Airport control

This is the more familiar part of air traffic control carried out from the glass-topped part of the tower building that affords controllers a panoramic view of the airport and its immediate surroundings. There has been a tendency for tower structures to become higher as airports have expanded, as shown in Figure 11.6. The extension of terminal buildings and other ground facilities has resulted, on occasions, in critical points on the airfield, runway touchdown point, apron, taxiway, and parking areas being hidden from the

Figure 11.6 Old and new ATC towers, Schiphol Airport, Amsterdam.

view of the tower. Remote TV cameras with display units in the tower have assisted in solving such problems. Airport staffs responsible for stand/gate assignment can face similar problems of finding a suitable vantage point and some tower buildings now also make provision for them. The essential characteristic of airport control is that it controls only that which can be seen from the tower itself—the "visual" control room. Control is therefore exercised over aircraft moving on the ground on operational areas or flying in the immediate vicinity of the airport either in a specified traffic pattern known as the *circuit* or in the U.S., the *pattern*, or in the final stages of approaching to land or departing. Because it controls aircraft movement on the airport, airport control also exercises authority over all other traffic, including vehicles using the same operational areas. For this reason, vehicles using these areas are usually fitted with radio equipment to provide immediate voice communications with the tower. At some airfields that are not dealing with many air transport operations or a large amount of GA activity, there might not be a control tower. Thus it might not be a requirement that all vehicles are fitted with radio. If not, all users of the operational areas will need to be familiar with the internationally agreed light signals used by air traffic control, as set out in Table 11.6. In the U.S., of the more than 5000 airports open for public use, many of which have heavy concentrations of GA traffic, fewer than 700 have air traffic control towers. Most of these airports without a control tower, however, operate a unicom radio, which cannot be used for air traffic control but over which pilots may announce their positions and intentions for others to hear.

At airports with high volumes of traffic, however, it has become necessary to split airport control into "ground" and "air." Ground control is responsible for all movement on the operating areas *except* the runways and approaches to the runways. Air control (in the U.S. usually referred to as *local* control) is responsible for the runways and immediate approaches/turnoffs. This results in departing aircraft coming under ground control once clear of the stand/gate until the holding point prior to moving onto the runway itself. The split of control responsibility in an airport control tower at busy airports calls for extreme vigilance, especially at night, as demonstrated by the runway collision at Los Angeles between a B737 and a metroliner on February 1, 1991 (NTSB 1991). At busy airports, ground control will also be responsible for implementing any departure flow-control regulations that might be in force. This is usually originated by a central control unit, for example in Great Britain it is at the main ARTCC near London. The U. S. equivalent is the FAA Command Center Operations room in Washington, DC. The implementation of flow control means that aircraft may be delayed on the ground. In such circumstances ground control will also provide "start up' clearances once a "slot" is known to be available that will allow the departing aircraft to join the general flow of air traffic over the area. In most of the cases where there are separate frequencies for ground and air control they are so busy that it has become necessary to provide a third communications channel devoted solely to passing clearances to aircraft or "clearance delivery." In common with all elements of split responsibility, coordination between all elements of airport control is vital, particularly between air and ground control to ensure that there is no obstruction to the landing and departing aircraft on the runways.

TABLE 11.6 Aerodrome Control Light Signals

Color and Type of Signal	On the Ground	In Flight
Steady green	Cleared for takeoff	Cleared to land
Flashing green	Cleared to taxi	Return for landing (to be followed by steady green at proper time)
Steady red	Stop	Give way to other aircraft and continue circling
Flashing red	Taxi clear of landing area (runway) in use	Airport unsafe—do not land
Flashing White	Return to starting point on airport	
Alternating red and green	General warning signal—exercise extreme caution	

Approach control

Approach control has similar problems of coordination. It deals with IFR traffic approaching the airport and departing IFR traffic once it is handed over by airport control, usually almost immediately after takeoff. In the U.S. approach control also handles arriving and departing VFR traffic. The area of responsibility of approach control extends out, typically to a distance of some 20 miles (32 k) from the airport. Although approach controllers are situated usually in the same building as airport control, they are not usually in the same room but in a separate room, sometimes called the IFR room. Also, approach control might be responsible for more than one airport if the airports are adjacent to each other. If this is feasible, there are obvious savings in resources although there is some increased need for careful coordination. Approach control has to coordinate not only with airport control but also, on the other hand, with the appropriate airways or area-control center because its area of responsibility lies in between these two.

The use of radar, both primary and secondary, has proved an essential aid to dealing with the difficult conflicts of departing climbing traffic and arriving descending traffic. In this regard, the European practice is to employ compulsory IFR in busy traffic areas round airports, even in VMC. Furthermore, it is not the practice to mix VFR and IFR traffic in the same airspace. The reasoning behind this is that the critical factor for safe visual separation by pilots themselves is not so much prevailing weather in terms of visibility and distance from cloud, but the pilots' ranges of vision from the flight deck. In the U.S. VFR operations are permitted in all but Class A airspace, although there are certain equipment and pilot qualification requirements in some of the classes of airspace.

At all busy IFR airports, the dominant factors for approach control are the established instrument procedures. These take several forms but all have the common factor of providing pilots and controllers with known, predictable patterns that will be flown by arriving and departing aircraft. Instrument approaches may be flown using any of the short range and radio aids already mentioned, including ILS and radar and in some few cases by the use of PAR. A typical precision approach and landing on R/W 28R at San Francisco is shown in Figure 11.7. This defines the plan view of the path to be flown in the upper part of the diagram and the heights in the lower vertical cross section.

At particularly busy airports, procedures have also been standardized for departing and arriving routes:

- Standard instrument departure (SID)

- Standard terminal arrival (STAR)

One of the obvious advantages of SIDs and STARs is that their use reduces very considerably the load on radio frequencies as a result of the

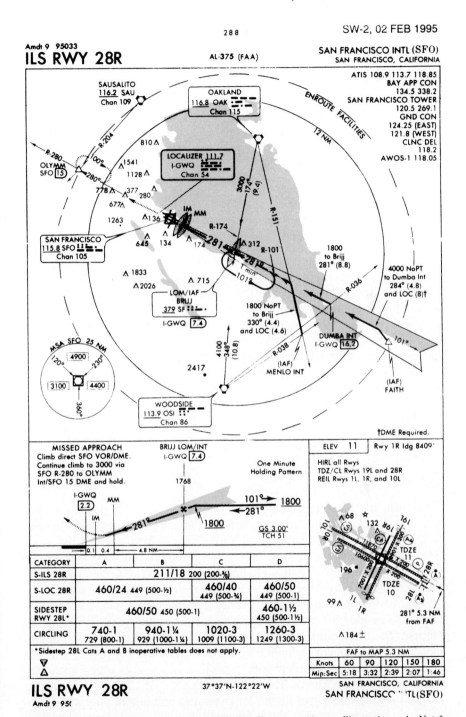

Figure 11.7 Approach plate, ILS runway 28R, San Francisco Airport. Illustration only. Not for navigation. *(FAA)*

"shorthand" descriptions that can be used for the complicated patterns of tracks and altitudes to be flown. The complex pattern to be flown for one Los Angeles procedure illustrated in Figure 11.8 can be transmitted to the pilot as "Gorman One Departure".

In similar fashion, the intricacies of the arrival routing shown in Figure 11.9 for Oakland can simply be transmitted as "Panoche One Arrival."

When it is appreciated that these and the many other procedures associated with ATC must be followed by pilots often thousands of miles away from their home base, the need for internationally agreed-on standards and procedures is clearly evident, as is the necessity for such information to be made available on a world-wide basis.

11.3 Telecommunications

The provision and maintenance of suitable aviation communication and navigation equipment and facilities are worldwide requirements for civil aviation and as such is another of the technical services subject to international agreement and standardization through ICAO. Details are contained in Annex 10 to the ICAO Convention: *Aeronautical Telecommunications* in two volumes. The standardization of communications equipment and systems is dealt with in Volume 1 while Volume 2 deals with communication procedures. International aeronautical telecommunications services are formally classified as:

- Fixed services
- Mobile services
- Radio navigation services
- Broadcasting services

All these are the responsibility of member states of ICAO although some of these facilities might be provided by commercial companies. In the Unites States there is Aeronautical Radio Inc., (ARINC) which has progressed from providing airlines with voice communications to a comprehensive data network service, including ACARS—Aircraft Communications Addressing and Reporting System. ARINC is airline-owned as also is another airline cooperative effort the telecommunications network, Societe International de Telecommunications Aeronautique (SITA) based in Europe. The main difference between the commercial and governmental agencies is that the governmental agencies restrict the traffic they will accept to certain types of messages, essentially those concerned with the safety of civil air transport. Thus the government circuits are in frequent use for the transmission of flight plan messages and urgent operational information between member states while commercial channels may be used

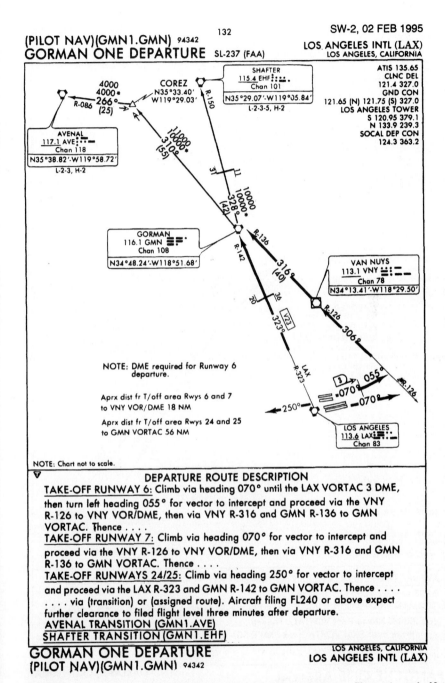

LOS ANGELES, CALIFORNIA
LOS ANGELES INTL (LAX)
GORMAN ONE DEPARTURE
(PILOT NAV)(GMN1.GMN) 94342

Figure 11.8 Gorman One SID routing, Los Angeles International Airport. Illustration only. Not for navigation. *(FAA)*

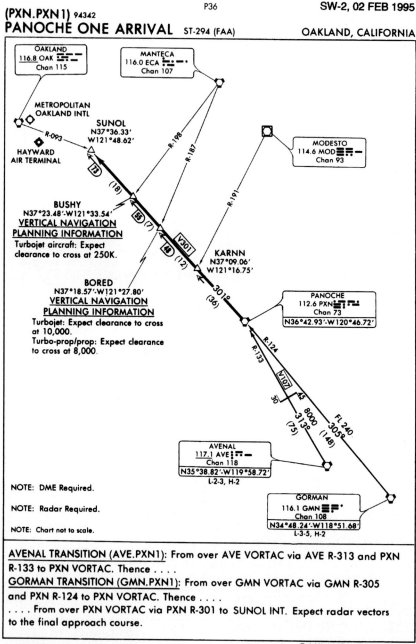

PANOCHE ONE ARRIVAL   ST-294 (FAA)   OAKLAND, CALIFORNIA

OAKLAND
116.8 OAK
Chan 115

MANTECA
116.0 ECA
Chan 107

METROPOLITAN OAKLAND INTL

SUNOL
N37°36.33'
W121°48.62'

R-093

HAYWARD AIR TERMINAL

MODESTO
114.6 MOD
Chan 93

BUSHY
N37°23.48'-W121°33.54'
VERTICAL NAVIGATION PLANNING INFORMATION
Turbojet aircraft: Expect clearance to cross at 250K.

KARNN
N37°09.06'
W121°16.75'

BORED
N37°18.57'-W121°27.80'
VERTICAL NAVIGATION PLANNING INFORMATION
Turbojet: Expect clearance to cross at 10,000.
Turbo-prop/prop: Expect clearance to cross at 8,000.

PANOCHE
112.6 PXN
Chan 73
N36°42.93'-W120°46.72'

AVENAL
117.1 AVE
Chan 118
N35°38.82'-W119°58.72'
L-2-3, H-2

NOTE: DME Required.

NOTE: Radar Required.

NOTE: Chart not to scale.

GORMAN
116.1 GMN
Chan 108
N34°48.24'-W118°51.68'
L-3-5, H-2

AVENAL TRANSITION (AVE.PXN1): From over AVE VORTAC via AVE R-313 and PXN R-133 to PXN VORTAC. Thence . . . .
GORMAN TRANSITION (GMN.PXN1): From over GMN VORTAC via GMN R-305 and PXN R-124 to PXN VORTAC. Thence . . . .
. . . . From over PXN VORTAC via PXN R-301 to SUNOL INT. Expect radar vectors to the final approach course.

Figure 11.9 Ocean One STAR routing for Los Angeles International Airport. Illustration only. Not for navigation. *(FAA)*

for passing company messages (e.g., requirements for crew transport, catering, supplies, etc.). Precise formats for various types of messages exchanged via the telecommunications and other governmental communications channels have been agreed on at international level, and they are usually coded so that problems of language differences are much reduced. Coding has the additional advantage of lending itself to computer techniques, and it is possible for teleprinter/teletype messages to travel around the world passing through several ground stations en route by means of automatic switching exchanges, without the need for human intervention.

Fixed services

Fixed service communication fills the need for a rapid means of point-to-point ground communications between "fixed" points (either by cable or radio link) to pass messages relating to safety and to the regular, efficient, and economical operation of air transport and general aviation. The basis of the worldwide service is the Aeronautical Fixed Telecommunications Network (AFTN). It is in effect a dedicated network confined to the following categories of messages:

- Distress messages and distress traffic
- Flight safety
- Meteorological
- Flight regularity
- Aeronautical administration
- NOTAM Class I distribution (see also Section 11.19)
- Reservations
- General aircraft operating agency

The first six in the preceding list are broadly classified as Category A traffic; the last two as Category B. U. S. policy is to not accept Category B traffic for U. S. government operated AFTN circuits. Priority for all AFTN messages is indicated by a two-letter priority code in the following order: SS, DD, FF, GG, JJ and KK, LL. Distress messages, for example, would bear the highest priority indicator, SS and "General aircraft operating agency" messages the lowest, LL. There is also a prescribed format for AFTN messages and this is set out in detail in Volume 2 of Annex 10. A typical international AFTN network is shown in Figure 11.10.

Mobile services

In the context of telecommunications, the term *mobile* refers to the service being provided for aircraft (moving vehicles) although the facilities pro-

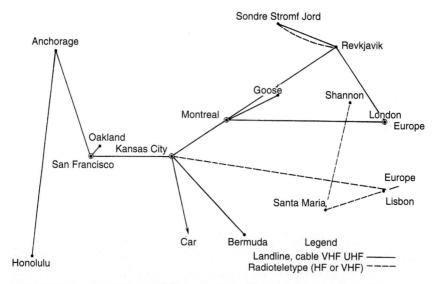

Figure 11.10 Aeronautical Fixed Telecommunications Network—International USA.

vided by the individual government agencies are primarily fixed installations on the ground. The mobile service covers two vital aspects of aircraft
movement:

- Communications
- Navigation

The major users of the communications facilities for air/ground and vice
versa contact are aircraft, but to a very small extent they may also be used
by ground vehicles moving on the airport (e.g., vehicles towing aircraft).
While the vast majority of communications facilities in the mobile service
are devoted to radio voice communications, there are still a few areas in the
world where the medium of air/ground communications is wireless telegra-
phy (WT). For some aircraft, there is also a need for the now largely outdated manual direction finding by means of which a ground-based direction
finding (D/F) installation is able to give an aircraft navigational guidance. A
mobile service is required to provide aircraft with:

- Flight information service
- Alerting service
- Air traffic advisory service
- Air traffic control service
- Area/airways control service

- Approach control service
- Airport control service

At the level of individual airport facilities, the main concern is with the voice communications used by airport and approach control and for this purpose very high frequencies (VHF) are used to avoid the risk of interference. The band of frequencies 118.0–136.0 MHz is used, and this provides 358 channels at 50-MHz spacing or 714 channels with 25 kHz spacing . A 100-kHz block is used to protect either side of the international emergency frequency of 121.5 MHz. These frequencies and others in the VHF range 30–200 MHz are characterized by line-of-sight reception range. Representative reception distances versus aircraft height are: 39 nm at 1000 feet, 55 nm at 2000 feet, 122 nm at 10,000 feet, 200 nm at 25,000 feet. At greater ranges remote receivers/transmitters are used with microwave or cable links back to the ATC unit involved. In cases where it is not possible to provide the remote facilities, for example, ocean or desert areas, then a different frequency band, high frequency (HF) 3, MHz, (3000 kHz), to 30 MHz is used to provide longer range propagation. "Families" of such frequencies are provided by governmental agencies on a worldwide basis for international flights in areas such as the North Atlantic, Europe, Mediterranean, Pacific. These are all part of the ICAO Major World Air Route Area HF Network (MWARA). Typical frequency groups for the European/Mediterranean Region are 2910, 4689, 6582, 8875 kHz. A commercial HF facility has been set up by ARINC in the USA at the major international ports such as New York, Miami, San Juan, San Francisco, and Honolulu. This provides HF SSB channels with each station having at least three frequencies for company operational messages. Communication can be established in this way with airline offices on the so-called *Phone Patch* System, where the transmissions go first to the ARINC ground station and then by telephone to the office being called. There is, however, a slow move towards the use of satellite communications, and although this is at the moment confined to data link use, it will inevitably come to be used also for voice.

Although developments in technology have allowed closer spacing of frequencies and with this an increased number of channels, there is still heavy congestion, particularly at the peak hours characteristic of air transport operations. For this reason, radio messages must be as precise and succinct as possible. Certain standard phraseology and abbreviations are therefore used in air/ground exchanges. Even so, misunderstandings can arise as was so tragically demonstrated at Tenerife in 1977 with the collision and destruction of two B747 aircraft—one just about to become airborne. Depending on the volume of traffic at a particular airport and any adjoining airports, there might be as many as six or seven different channels used by ATC for various communications purposes. A typical example is San

Francisco International Airport, California, which has separate frequencies for each of the following:

- Information service
- Approach control (Bay Area)
- Airport (air)
- Ground control
- Clearance delivery
- Helicopter

The channels/frequencies used by ATC are not necessarily operated from the airport itself. In the preceding example, approach control serves not only San Francisco, but also three other airports (Oakland, Alameda, and Hayward) and is in fact located at Oakland. It is normally the practice in Europe for the larger airports used by air carriers to have their own approach frequency operated from the individual airports. London (Heathrow) and London (Gatwick), only 15 miles (24 k) apart, have, however, adopted a similar system to that of the New York Area and now have a common IFR facility for all approaches. Even without this, there are still six frequencies in use at Heathrow airport.

Radio navigation services

There are internationally agreed standards for radio navigation equipment laid down by ICAO (ICAO, 1991a).

The basic short-range navigation aid to be found in the vicinity of airports is the VOR which transmits a radial pattern through 360° producing 360 separate tracks or radials that can be used by the pilot to fly *to* or *from* the VOR station. By use of one of the cockpit instruments, a pilot is able to select any desired track and by reference to a deviation indicator needle fly that track. While VORs provide a means of delineating airways and jet routes it is a lower powered version that is used near and on airports, the Terminal VOR (TVOR), usually limited to a range of about 25 miles (40 k). Because VORs provide directional guidance only, they are usually associated with *distance measuring equipment* (DME) to give an accurate position in terms of bearing and distance from the facility. More recent developments, which include an onboard computer, provide the means for a pilot to use the VOR/DME combination to set up his or her own "offset" track so that it is not then necessary to fly overhead the actual VOR installation. VORs and TVORs operate in a band, close to the VHF communications band of frequencies, between 108.0 and 118.0 MHz. DME operates at much higher frequencies between 960 and 1215.0 MHz, which is within the group of ultrahigh frequencies (UHF).

Another, somewhat older type of radio aid used in the vicinity of airports is the *nondirectional beacon* (NDB). It radiates a nondirectional signal to which an aircraft receiver can be tuned to obtain a bearing. The NDB is a much earlier development than the VOR and offers no precise track guidance or selection. Because it is in the low frequency band (200–1750 kHz) it is subject to interference in bad weather. Low-powered NDBs are frequently used on the approaches to an airport as an aid to locating the airport itself and are then described as *compass locator beacons*. On occasions such a "locator" might be sited on the approach course of the ILS, usually at the same site as the outer marker of the ILS.

A VHF beacon in common use at airports is the *fan marker.* Operating on a frequency of 75 MHz, it radiates a vertical pattern somewhat similar to an inverted cone with an elliptical cross section. On flying through its pattern, the pilot receives an audio/visual indication in the cockpit. The primary purpose of fan markers is to indicate position along a specified track.

Radio navigation guidance approach and landing at an airport requires a precision-approach aid. Over a period of years the ILS has become the universal precision approach aid. It has been developed to the stage where it can now provide the pilot with a "blind" landing capability. The ILS radiates two intersecting sloping pathways upward from the touchdown point of the runway. One in the horizontal plane is called the *localizer* and provides azimuth/center line guidance to the runway. For details, refer to Chapter 5.

The other—the glideslope path—provides vertical/descent guidance. Normal slopes are to the order of 3° with slightly higher angles, 3.5° for example if the approaches are over built-up areas. The pilot, if flying a manual approach, uses horizontal and vertical needle indicators on the ILS instrument to fly a precise approach/landing profile. It is more usual in modern jets for the ILS signals to be fed into the aircraft's flight guidance system for an automatic (coupled) approach. Distance information on the approach is provided by marker beacons—usually an outer marker (OM) some 4 to 5 miles (6 to 8 k) from touchdown and a middle marker (MM) situated between 0.5 and 0.9 miles (0.8 and 1. 4 k) from touchdown. A small number of ILS installations also have an inner marker (IM) situated 1500 to 1700 feet (457 to 548 m) from touchdown. Passage over the beacons is indicated in the cockpit by a flashing light—blue, OM; amber, MM; white, IM—as well as by an audio signal. There are two different frequency bands used for each ILS installation. The localizer transmits on VHF 108.1–111.9 MHz. The glideslope operates on UHF between 328.6 and 335.4 MHz. The Microwave Landing System (MLS) also has a localizer and a glideslope, as in the ILS, but radiates at frequencies above 5000 MHz, providing a multipath approach rather than a single track as with ILS. It was originally planned to replace ILS with MLS, but the perceived potential for satellite global positioning has raised the possibility of leap-frogging MLS development and proceeding directly to a satellite system.

The United States is a strong advocate of such a policy. Already the FAA—with some limitations—has approved GPS for IFR and VFR operations and to make approaches. In February 1994, the FAA approved GPS as a primary IFR flight guidance and to fly "overlay" approaches. The overlay allows the FAA to place a GPS approach over an existing nonprecision approach that uses VOR, VOR/DME, NDB, or NDB/DME for position fixes. The necessary positioning for GPS users are noted and added to the approach chart. Through the overlay program, nearly 5000 nonprecision GPS approaches are quickly becoming available in the U.S. with an additional 2000 GPS approaches to be added. The first "stand alone" approaches were published by the FAA in July 1994. The equipment and installation must meet FAA standards, and to use GPS under IFR condition, the aircraft must have an approved and operational alternate means of navigation.

Less precise approaches can of course be flown using VOR and/or VOR/DME or in some cases even a NDB, but certainly the latter could not be used for approaches in extremely bad weather conditions. With all three, the vital missing link is descent indication in the cockpit. With these less precise aids the glideslope becomes a matter of mental calculation rather than positive visual indication. At airports where precision approaches are necessary but local conditions make the installation of ILS technically impractical, ICAO recommends a PAR installation. This uses two radar pictures, azimuth and elevation, to give the air traffic controller a vertical and a plan picture of the approach path. Depending on the aircraft's indicated position relative to the centerline and the glideslope, the controller passes heading and descent corrections to the pilot, thus the so-called "talkdown". The use of a plan position radar (search) indication alone is not considered sufficient for a precision approach, which by definition also requires a glideslope.[2] It must always be accompanied by a vertical/height display to constitute a full PAR.

Broadcast services

A great deal of information relating to air navigation is required by aircraft in flight or about to depart. Such information on weather and airport and radio aids serviceability is of particular importance. As the requirement is universal, the telecommunications agency of each country makes available suitable broadcast facilities and is required by international agreement to publish details of the frequencies used and times of broadcasts. These channels are separate from those used for normal control purposes and no acknowledgment is required from aircraft receiving the broadcasts. There is an increasing tendency for information to be prerecorded and a tape used for the actual broadcast.

---

[2] Note that a nonprecision approach is a standard instrument approach procedure in which no electronic glideslope is provided (e.g., VOR, NDB).

Automatic Terminal Information Service (ATIS) is the most common type of broadcast concerned with airport operations. It is a tape-recorded broadcast on either the voice facility of a nearby VOR or a discrete radio frequency of its own. Each broadcast is individually identified by a phonetic letter of the alphabet commencing the day with "alpha" then "bravo," and so on. Details given in each broadcast include surface wind (magnetic direction) and prevailing weather. The purpose of the broadcast is not only to inform but also to reduce the amount of traffic on the vital control frequencies. Thus on initial voice contact with the destination airport the pilot advises that the ATIS broadcast has been received by quoting its identifying letter thus: "Information Sierra Received." The controller then knows that is it is not necessary to pass this information on the control frequency. At some airports, there might be two ATIS broadcasts: one for arriving aircraft and one for those departing. Contents of ATIS messages are continuously repeated until such time as a change takes place in any of the items reported. The message is then changed and assigned a new letter identification.

VOLMET is meteorological information for aircraft in flight and comprises both reports of actual weather conditions at specified airports and also landing forecasts. Broadcasts are made via the "mobile" service on both VHF and HF frequencies. HF frequencies have been assigned to New York for broadcasts to North Atlantic flights (ICAO, 1990b). Broadcasts of five minutes duration are twice hourly by New York for four areas in the Eastern U. S. and Bermuda, for example:

- H + 00–05 for Detroit, Chicago, Cleveland
- H + 30–35 for Niagara Falls, Milwaukee, Indianapolis

In Europe, broadcast for long-range flights are also made on HF relating to European airports. There is in addition an extensive VOLMET broadcast system on VHF frequencies from major airports based on half-hourly reports, for example, Amsterdam broadcasts on 126.20 MHz providing weather information for Amsterdam, Rotterdam, Brussels, Dusseldorf, Hamburg, Bremen, Copenhagen, and London Heathrow. Broadcasts via the "fixed" service include meteorological data and also charts, the latter received by the addressee in identical graphic form (facsimile). Details of the information sent in this way are given in the following section.

11.4 Meteorology

Function

Although meteorological service is a government responsibility in all countries, on some occasions the aviation service is augmented by airlines themselves employing meteorologists to interpret weather data and forecasts in light of their own particular operational plans. They will, however, still use the basic information supplied by the government agency. In addition, private companies are now entering the area of meteorological service using information from government sources to provide a service tailored to particular customer needs (e.g., providing a flight planning service). Nor is distance a problem with the availability of satellite and fax communications, which enables a company on the west coast of the United States to provide flight plans and briefing documents in minutes to crews in London Gatwick Airport preparing to depart for the Middle East. The main functions of a meteorological service are to observe, report, and advise on weather conditions on the ground at airports and in the upper atmosphere. In order to do this, a worldwide network of various categories of meteorological offices gather information, exchange and analyze information and provide an appropriate reporting and forecast service to airport authorities, airlines, and aircrews.

Reports of surface weather

The most common feature of the meteorological service is the routine observation and subsequent reporting of surface weather, made either hourly or half-hourly. In the United States, these are *surface aviation weather reports*. Until recently there were two main types: record observations or sequence reports (SR), and special reports (RS or SP), which are observations taken when required by a significant change in the weather. The ICAO designations for corresponding types of messages and their acronyms are:

- Aviation routine weather report (METAR)
- Aviation selected special weather report (SPECI)

However, many of the differences that had existed disappeared July 1, 1996 , when the United States adopted these ICAO designations and the associated message formats. However, the U.S. did not go to the metric system or to kilometers. In the U.S. visibility is stated in statute miles, RVR in feet, altitude in inches of mercury, and winds in knots.

With large amounts of data that have to be circulated quickly it is necessary to use various types of codes. This shortens transmission times, thus enabling the information to be used while it is still current. Due to the large amounts of weather data that have to be collected and disseminated, it is necessary to have extensive communications networks in each country, connected to international networks, dedicated exclusively to use by the meteorological services. In the United States, these are given individual network identification for example: *Service C* is the primary meteorological network; *Service A* collects and distributes hourly surface aviation observations and also disseminates NOTAMs. The exchange of data with other countries takes place via the International Exchange System with its gateway station at Washington, DC. It utilizes radio teletype, long-line, and satellite communications channels. In Europe, the network is *Meteorological Operational Telecommunication Network Europe* (MOTNE), which is divided into two main circuits or "loops." The airports in the following groups of countries are in each loop:

Loop 1   Algiers, Belgium, France, Germany, Great Britain, Iberia, Ireland, Luxembourg, Morocco, Netherlands, Switzerland.

Loop 2   Austria, Denmark, Greece, Italy, Scandinavia, Tunisia, the Middle East countries, and those of Eastern Europe.

The internationally agreed METAR and SPECI reports contain the following information:

- Identification
- Surface wind direction and speed
- Visibility
- RVR—when appropriate
- Present weather
- Cloud—amount and type
- CAVOK (either "visibility" or "CAVOK" must be present)[3]
- Temperature and dew point
- QNH—altimeter pressure setting
- Supplementary information
- Trend forecast

The identification which precedes the weather information is comprised of: type of report (METAR/SPECI), location identification and the time of observation. An example of the format is shown in Figure 11.11, while a detailed key to the METAR/SPECI code is at Figure 11.12.

Forecasts

Most official references to forecasts are hedged with provisos regarding their reliability and the need to regard them as "probability" statements. Even taking into account the inherent weakness of all forms of forecasting, aviation meteorology has demonstrated an increasing degree of reliability in recent years due to improved communications for the transmission of in-

---

[3] The United States does not use the term CAVOK.

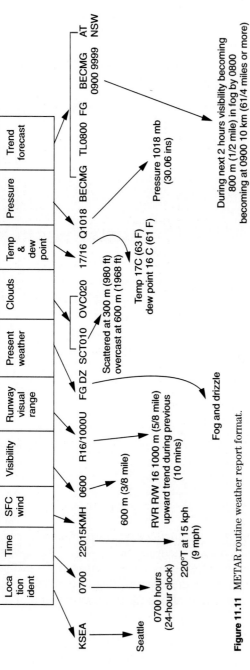

Figure 11.11 METAR routine weather report format.

TREND FORECAST (TWO HOURS FROM TIME OF OBSERVATION)

CAVOK — Replaces visibility, weather and cloud if these are forecast to be OK (see earlier definition)

FORECAST CLOUD — $N_sN_sN_sh_sh_sh_s(cc)$
- Cloud type - only **CB**
- Forecast height of base of cloud — Replaced when sky expected to be obscured and vertical visibility forecasts are undertaken by: VVh_sh_s; Indicator of Vertical Visibility; Vertical visibility in units of 30 metres (100 feet)
- Forecast cloud amount — **NSW** No Significant Weather; Replaced when significant weather ends by: **NSW**; Replaced when a change to clear sky forecast by: **SKC** Sky Clear

FORECAST WEATHER — $w'w'$ — Forecast significant weather. See Table.

FORECAST VISIBILITY — VVVV — Forecast surface visibility in metres 9999 = 10 km or more

FORECAST WIND — $dddff(G f_m f_m) \frac{KMH}{KT}$ or MPS
- Wind speed units
- Forecast maximum wind speed (gust)
- Indicator of Gust
- Forecast wind speed
- Forecast wind direction in degrees true, rounded to nearest ten degrees (VRB = VARIABLE) — 00000 = calm

CHANGE AND TIME — TTGGgg — Associated time group in hours and minutes UTC. Can be **AT** or **FM** = FROM or **TL** = TILL

CHANGE INDICATOR — TTTT or NOSIG —
- **BECMG** = BECOMING or **TEMPO** = TEMPORARY } see over for definitions
- **NOSIG** = NO SIGNIFICANT CHANGE

SUPPLEMENTARY INFORMATION

WIND SHEAR — WS and/or $RWYD_RD_R$ LDG / TKOF
- Runway designator - for parallel runways, may have **LL, L, C, R** or **RR** appended. (L=left; C=centre; R=right)
- RUNWAY
- TAKE OFF and/or LANDING
- Wind Shear

RECENT WEATHER — $REw'w'$ — REcent weather since last routine report of operational significance. Indicator of REcent weather. See Table.

PRESSURE — $Q P_HP_HP_HP_H$
- QNH in whole hectopascals or inches, tenths and hundredths of an inch depending on indicator
- Indicator of QNH in hectopascals.
- If **Q=A** then QNH in inches

TEMP AND DEW POINT — $T'T'/T'_dT'_d$
- Dew-point temperature in whole degrees Celsius (if below 0°C preceded by **M**)
- Temperature in whole degrees Celsius (if below 0°C preceded by **M**)

AERODROME ACTUAL WEATHER - METAR - SPECI DECODE

CAVOK — Cloud And Visibility OK. **Replaces** visibility RVR, present weather and cloud if: 1) Visibility is 10 km or more. 2) No cumulonimbus cloud and no cloud below 1500 metres (5000 ft) or below the highest minimum sector altitude whichever is greater, and 3) No precipitation, thunderstorm, sandstorm, duststorm, shallow fog or low drifting dust, sand or snow

CLOUDS — $N_sN_sN_sh_sh_sh_s(cc)$
- Cloud type - only **CB** (Cumulonimbus) or **TCU** (Towering cumulus) indicated
- Height of base of clouds in units of 30 metres (100ft) — Replaced when sky is obscured and information on vertical visibility is available by: VVh_sh_s; Vertical visibility in units of 30 metres (100 feet); /// = Vertical visibility unavailable; Indicator of Vertical Visibility
- Cloud amount: **SCT** = SCATTERED (half or less than half the sky covered); **BKN** = BROKEN (more than half but less than OVC); **OVC** = OVERCAST (sky totally covered) — Replaced when there are no clouds and CAVOK is not appropriate by: **SKC** Sky Clear

PRESENT WEATHER — $w'w'$ — Present weather. See Table. May be followed by two figure designator for reporting present weather in synoptic code which should be disregarded for aviation purposes

RUNWAY VISUAL RANGE — $RD_RD_R/VV V V_i$
- RVR tendency over past ten minutes. **U**=upward; **D**=downward; **N**=no change. Omitted if impossible to determine.
- Runway Visual Range in metres (10 minute mean) P1500= more than 1500 metres, M0050= less than 50 metres
- Runway designator - for parallel runways may have **LL, L, C, R** or **RR** appended. (L=left; C=centre; R=right)
- Indicator of RVR
— Replaced when there are significant variations in RVR by: $RD_RD_R/V_nV_nV_nV_nV_xV_xV_xV_xi$; RVR tendency; Runway visual range in metres (one minute mean maximum over last ten minutes); Indicator of significant Variation; Runway visual range in metres (one minute mean minimum over last ten minutes); Runway designator - for parallel runways may have **LL, L, C, R** or **RR**; Indicator of RVR

VISIBILITY — $VVVVD_v$
- Direction of lowest visibility (eight points of compass) where required
- Minimum horizontal visibility in metres 9999 = 10 km or more
— Followed when min. vis. < 1500 metres and max vis. > 5000 metres by: $VVVV D_v V_x V_x V_x V_x D_x$; Direction of maximum visibility (eight points of compass); Maximum horizontal visibility in metres 9999 = 10 km or more.

SURFACE WIND — $dddff(Gf_mf_m)\frac{KMH}{KT}$ or MPS
- Wind speed units used
- Maximum wind speed (gust) - if necessary
- Indicator of Gust - if necessary
- Mean wind speed (ten minute mean or since discontinuity)
- Mean wind direction in degrees true rounded off to nearest ten degrees (VRB = VARIABLE) — 00000 = calm
— Followed when there is a variation in wind direction of 60° or more and wind speed > 3 KT by: $d_nd_nd_nV d_xd_xd_x$; Other extreme direction of wind (measured clockwise); Indicator of Variability; Extreme direction of wind

IDENTIFICATION — GGggZ
- Indicator of UTC
- In individual messages, time of observation in hours and minutes UTC. In bulletins, time of observation in bulletin header instead.

— CCCC — ICAO four-letter location indicator

— METAR or SPECI — **METAR** - Routine weather report; **SPECI** - Selected special weather report

---

w'w' – SIGNIFICANT PRESENT, FORECAST AND RECENT WEATHER

QUALIFIER		WEATHER PHENOMENA			
INTENSITY OR PROXIMITY 1	DESCRIPTOR 2	PRECIPITATION 3	OBSCURATION 4	OTHER 5	
- Light	**MI** Shallow	**DZ** Drizzle	**BR** Mist	**PO** Well developed dust/sand whirls	
Moderate (no qualifier)	**BC** Patches	**RA** Rain	**FG** Fog	**SQ** Squalls	
+ Heavy	**DR** Low drifting	**SN** Snow	**FU** Smoke	**FC** Funnel cloud(s) (tornado or water spout)	
VC In the vicinity	**BL** Blowing	**SG** Snow grains	**DU** Widespread dust	**SS** Sandstorm	
	SH Shower(s)	**IC** Diamond dust	**VA** Volcanic ash	**DS** Duststorm	
	TS Thunderstorm	**PE** Ice pellets	**SA** Sand		
	FZ Supercooled	**GR** Hail	**HZ** Haze		
		GS Small hail and/or snow pellets			

NOTES:
1. The w'w' groups are constructed by considering columns 1 to 5 in the table above in sequence, that is intensity, followed by description, followed by weather phenomena. An example could be «SHRA (heavy shower(s) of rain)
2. A precipitation combination has dominant type first
3. **DR** (low drifting) less than two metres above ground, **BL** (blowing) two metres or more above ground
4. **GR** used when hailstone diameter 5 mm or more. When less than 5 mm, **GS** used
5. **BR** visibility at least 1000 metres but not more than 5000 metres. **FG** visibility less than 1000 metres
6. **VC** – within 8 km of the aerodrome but not at the aerodrome

Figure 11.12 Detailed key to the METAR/SPECI code.

formation, greatly enhanced data processing capabilities, and the availability of satellite observations.

Forecasts are exchanged between meteorological centers on worldwide meteorological communications network (fixed service). The forecast weather for airports is issued in terminal forecasts. In the United States, they are issued by Weather Service Forecast Offices (WSFO). There are 51 of these offices in the United States, one for each state except California, which has two—Los Angeles and San Francisco. Together the 51 offices prepare and distribute a total of 452 terminal forecasts, three times daily for specific airports in the fifty states and the Caribbean.

As with METAR and SPECI, the United States is to adopt the international format, TAF for terminal/airport forecasts. Many of the METAR code groups are also used in TAF. It is perhaps interesting to note that the international format has been in use previously for U. S. domestic military locations. An example of TAF coding for St. Louis, Missouri, is given in Figure 11.13 and the detailed key to the code as Figure 11.14.

Terminal/airport forecasts should normally be issued four times daily (24 hours) beginning at one of the main synoptic hours (00, 06, 12, 18 UTC). The period of validity should be of at least 18 hours or of 24 hours duration. When a TAF is issued on request, rather than routinely, for destination and alternate airport the period of validity should commence one hour before ETA (earlier if requested) and cover a period up to the ETA at the farthest alternate plus two hours.

In the event of it being necessary to issue any warning of hazardous weather phenomena, the information is transmitted under the acronym SIGMET—Significant Meteorological Phenomena, which might affect the safety of aircraft operations. A SIGMET message gives a concise description in abbreviated plain language of the occurrence and/or expected occurrence of specified en route weather phenomena. Table 11.7 gives a list of the approved ICAO abbreviations to be used in such messages. A typical example of how a message could be composed is given in Figure 11.15. Prior to departure, a considerable amount of meteorological information is used by airlines for flight planning purposes. Among the data used are forecast winds and temperature at various altitudes. Although this type of information can be supplied in tabular form (Figure 11.16), or in chart form (Figure 11.17), for flight crew information it is also particularly suited to computer processing, which considerably improves the efficiency of airline flight planning. In this instance, the meteorological service computer can transfer the data automatically to airline flight-planning computers, where it is programmed into flight-planning calculations along with details of aircraft performance. Airline operations personnel can then obtain from the computer a completed flight plan for a specified route and altitude. All this can be done at a distance if necessary, and at least one major European airline with

```
KSTL   161212   33025/35   0800   71SN   9//005
(a)     (b)        (c)       (d)    (e)     (f)

     BECMG   1215   0000   39BLSN   9//000
                     (g)

  BECMG   1618   33020   4800   38BLSN   7SC030
                  (h)

     PROB30   TEMPO   1920   85SNSH
                       (i)

BECMG   2123   33015   9999   WX   NIL   3SC030
                (j)

        FM   00   VRB05   9999   SKC
                   (k)
```

(a) Location Identifier St.Louis, Missouri (ICAO four-letter identification code)
(b) Time group Forecast valid for 24 hours from 1200Z(UTC) to 1200Z the following day
(c) Wind speed from 330° at 25 knots gusting to 35 knots
(d) Visibility 800 metres (approximately 1/2 mile)
(e) Significant weather Code for, Snow,Continous fall of snow flakes - slight at time of observation.
(f) Clouds Code group - (9) Sky obscured,(//) Clouds not observed, (150)Vertical visibility 500 ft.(units of 30 metres).Forecast may include several cloud groups.
(g) Expected between 1200Z and 1500Z to become, Visibility zero, blowing snow, sky obscured, clouds not observed, vertical visibility zero.
(h) Becoming between 1600Z and 1800Z , Wind 330°20 knots,visibility 4800 metres(3 miles), blowing snow with 7 oktas of strato cumulous at 900 metres(3000 ft)
(i) Weather fluctuating with a 30% probability of temporary or brief snow showers between 1900Z and 2000Z.
(j) Between 2100Z and 2300Z changing to, Wind 330°at 15 knots ,visibility 10 km or more (more than 6 miles), no significant weather, 3 oktas of strato cumulous at 900 metres
(3000 ft)
(k) Conditions changing at 0000Zto - Wind variable at 5 kts no significant weather, sky clear.

Figure 11.13 Example of TAFOT for St. Louis, Missouri.

intercontinental routes obtains its flight plans from a computer in California. Details of flight planning procedures are given in Chapter 8. However, certain calculation relating to takeoff performance can be carried out only immediately prior to departure by the flight crew. At major airports with their own forecasters, it is possible for pilots to obtain airport forecasts for specific departure times, giving the data needed to compute takeoff performance taking into account surface wind, temperature and pressure. The following is an example of such a forecast for Munich, West Germany, for a 1400Z takeoff:

```
EDDM 1400Z 240/15 26°C QFE 998mb
```

AERODROME FORECAST - TAF - DECODE

IDENTIFICATION

TAF	TAF - Name for an Aerodrome Forecast
CCCC	ICAO four-letter location indicator
YYGGggZ	Date and time of origin of forecast in UTC; Indicator of UTC
$G_1G_1G_2G_2$	Beginning G_1G_1 and end G_2G_2 of forecast period in hours UTC

FORECAST SURFACE WIND — $dddff(f)G\,f_m f_m$ KMH or KT or MPS
- Mean wind direction in degrees true rounded to nearest ten degrees. (VRB = VARIABLE)   00000 =calm
- Mean wind speed
- Indicator of Gust
- Maximum wind speed (gust)
- Wind speed units used

FORECAST VISIBILITY — VVVV   Minimum horizontal visibility in metres. 9999 = 10 km or more

FORECAST SIGNIFICANT WEATHER — w'w'   Forecast significant weather (see table on other side). Replaced when significant weather phenomenon obscured and forecast to end by: **NSW** No Significant Weather

FORECAST CLOUD AMOUNT AND HEIGHT — $N_sN_sN_sh_sh_sh_s(cc)$
- Cloud type - only **CB** (cumulonimbus)
- Height of base of cloud in units of 30 metres (100 feet)
- Cloud amount: **SCT** = SCATTERED (half or less than half the sky covered) **BKN** = BROKEN (more than half but less than OVC) **OVC** = OVERCAST (entire sky covered)
- Replaced when sky is expected to be obscured and information on vertical visibility is available by: $VVh_sh_sh_s$ — Vertical visibility in units of 30 metres (100 feet); Indicator of Vertical Visibility
- Replaced when clear sky is forecast by: **SKC** SKy Clear
- Replaced if agreed regionally, when no CB and no cloud below 1500 m (5000 ft) or below the highest minimum sector altitude whichever is greater, are forecast and CAVOK and SKC are not appropriate by: **NSC** No Significant Cloud

CAVOK — Cloud And Visibility OK. **Replaces** visibility, weather and cloud if:
1) Visibility is forecast to be 10 km or more
2) No cumulonimbus cloud and no other cloud forecast below 1500 metres (5000 ft) or below the highest minimum sector altitude whichever is greater, and
3) No precipitation, thunderstorm, sandstorm, duststorm, shallow fog or low drifting dust, sand or snow forecast

THESE GROUPS ONLY USED IF AGREED REGIONALLY

FORECAST TEMPERATURE — (T_TT_F/G_FG_FZ)
- Indicator of forecast Temperature
- Forecast temperature at G_FG_F. Temperatures below 0°C preceded by **M**
- Time UTC to which forecast temperature refers
- Indicator of UTC

FORECAST ICING — $(6I_ch_ch_ch_ct_L)$
- Indicator of forecast icing
- Type of icing (see below)
- Base of layer of icing in units of 30 metres (100 feet)
- Thickness of layer (thousands of feet) with code figure 0 = up to top of clouds

FORECAST TURBULENCE — $(5Bh_Bh_Bh_Bt_L)$
- Indicator of forecast turbulence
- Type of turbulence (see below)
- Base of layer of turbulence in units of 30 metres (100 feet)
- Thickness of layer (thousands of feet) with code figure 0 = up to top of clouds

SIGNIFICANT CHANGES IN FORECAST CONDITIONS INDICATED BY:

PROBABILITY — $PROBC_2C_2$
- PROBability
- Only 30 or 40 used, indicating 30% or 40%
- Probability is used to indicate the probability of occurence of: a) an alternative element or elements b) temporary fluctuations

TIME — GGG_eG_e Beginning GG and end G_eG_e of forecast period in hours UTC

CHANGE — TTTT
- Type of significant change: **BECMG** -BECOMING, used where changes are expected to reach or pass through specified values at a regular or irregular rate. **TEMPO** - TEMPORARY fluctuations of less than one hour and in aggregate less than half the period indicated by GGG_eG_e
- **Both replaced** if one set of weather conditions is expected to change more or less completely to a different set of conditions, thus indicating the beginning of another self-contained part of the forecast, by TTGG
- This takes the form **FMGG** where FM is the abbreviation for from and GG is time to the nearest hour UTC. All conditions before this group are superseded by conditions indicated after the group.

TIME — GGG_eG_e Beginning GG and end G_eG_e of forecast period in hours UTC

Icing (I_c) and Turbulence (B) codes

Code	Icing (I_c)	Turbulence (B)
0	no icing	no turbulence
1	light icing	light turbulence
2	light icing in cloud	moderate turbulence in clear air, infrequent
3	light icing in precipitation	moderate turbulence in clear air, frequent
4	moderate icing	moderate turbulence in cloud, infrequent
5	moderate icing in cloud	moderate turbulence in cloud, frequent
6	moderate icing in precipitation	severe turbulence in clear air, infrequent
7	severe icing	severe turbulence in clear air, frequent
8	severe icing in cloud	severe turbulence in cloud, infrequent
9	severe icing in precipitation	severe turbulence in cloud, frequent

Figure 11.14   Detailed key to TAF code.

TABLE 11.7   Abbreviations to be Used in SIGMET Messages

A) at subsonic cruising levels:

Thunderstorm

obscured	OBSC TS
embedded	EMBD TS
frequent	FRQ TS
line squall	LSQ TS
obscured with heavy hail	OBSC TS HVYGR
embedded with heavy hail	EMBD TS HVYGR
frequent with heavy hail	FRQ TS HVYGR
line squall with heavy hail	LSQ TS HVYGR

Tropical Cyclone

tropical cyclone	TC (+ cyclone name)

Turbulence

severe turbulence	SEV TURB

Icing

Severe icing	SEV ICE
Severe icing due to freezing rain	SEV ICE (FZRA)

Mountain Wave

Severe mountain wave	SEV MTW

Duststorm

heavy duststorm	HVY DS

Sandstorm

heavy sandstorm	HVY SS

Volcanic ash

volcanic ash	VA (+ volcano name)

B) at transonic and supersonic cruising levels:

Turbulence

moderate turbulence	MOD TURB
severe turbulence	SEV TURB

Cumulonimbus

isolated cumulonimbus	ISOL CB
occasional cumulonimbus	OCNL CB
frequent cumulonimbus	FRQ CB

Hail

hail	GR

Volcanic ash

volcanic ash	VA (+ volcano name)

Although not of immediate relevance to airport operators, there are a whole range of forecasts issued to provide en route planning data to airlines and aircrews. These include area and route forecasts both of surface conditions and upper air data such as pressure, temperature, and winds. The use of chart-type presentation is widespread for these forecasts, though tabular formats may be used. En route "significant weather" is of vital importance, and charts showing this for various levels are generally available. In the United States, there is also available a type of weather chart known as a *radar summary chart* on which are displayed the configuration, location,

EGTT SIGMET 4 VALID 251315/251700 SEV ICE LONDON FIR CBTOPS FL280 OCNL TS ISOL HAIL. INTSF

The fourth SIGMET message issued for the London flight information region by EGTT(London Area Control Center) with a warning of severe turbulence and severe icing in the London FIR with cumulonimbus tops at flight level 280. Occasional thunderstorms and isolated hail. Intensifying.

Figure 11.15 Example of SIGMET for London FIR.

and movement of precipitation echoes (Figure 11.18). A typical significant weather forecast (prog)[4] for low level over the North Atlantic is shown in Figure 11.19.

Significant weather charts are issued every three or six hours. They contain the forecast position of fronts and the position of clouds and weather. Areas are outlined on the chart with base and tops noted, in hundreds of feet, preceded by amount and type. In Figure 11.19, the cloud area in the vicinity of Gander (CYQX) is given as 7 oktas of layered cloud between 1500-foot base (457 m) and 20,000 feet tops (6000 m). Significant weather, as indicated by the "fork" symbol, is "moderate icing" between 11,000 and 16,000 feet (3300 and 4800 m). Surface weather charts are also available showing either current (actual) weather or forecast conditions. As a result of international agreement, further changes will take place in the way significant weather charts are drawn up. Facsimile broadcasts of charts are made in Europe and the United States. In the United Kingdom, they are broadcast by a land-line facsimile network Civil Aviation Operational Meteorological Facsimile (CAMFAX). The corresponding system in the United States is the National and Aviation Meteorological Facsimile Network (NAMFAX). In Germany, the Federal Meteorological Service provides a similar facility from its Frankfurt Center.

Among other charts of special interest to aircraft operators are fixed time prognostic charts for constant pressure levels. The equivalent altitude levels for constant pressure levels are shown in Table 11.8. The example in Figure 11.20 is for the 300-mb level equivalent to flight level 300 or approximately 30,000 feet (9000 m) and is drawn on a polar stereographic projection.

Recent years have seen the increasing availability of satellite photographs, which greatly assist the analysis and forecasting of weather for aviation, one is shown in Figure 11.21. Further details of various types of

---

[4]A prog or prognostication is a forecast for a set point in time and this will be stated on the chart in this case 1800 GMT (UTC). It is not a forecast for a period of time.

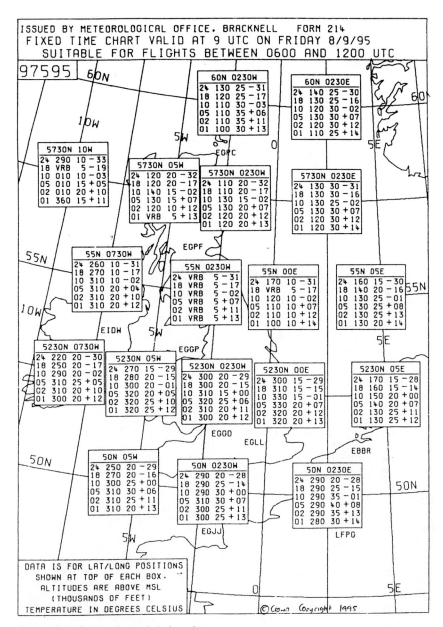

Figure 11.16 Tabular form of winds and temperatures.

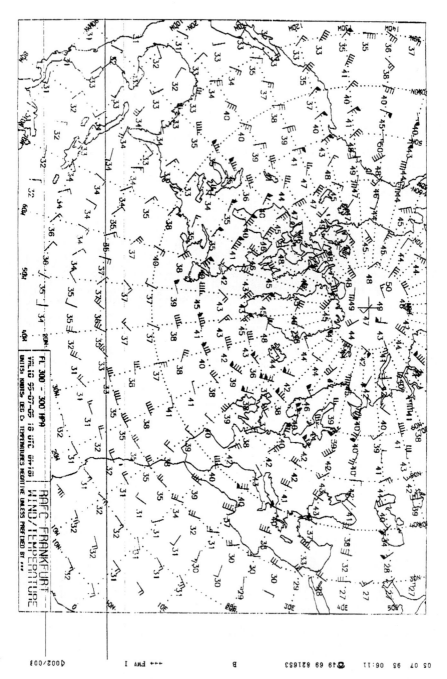

Figure 11.17 Chart format of winds/temperatures at various altitudes (North Atlantic). *(Deutscher Wetterdienst, Frankfurt Airport)*

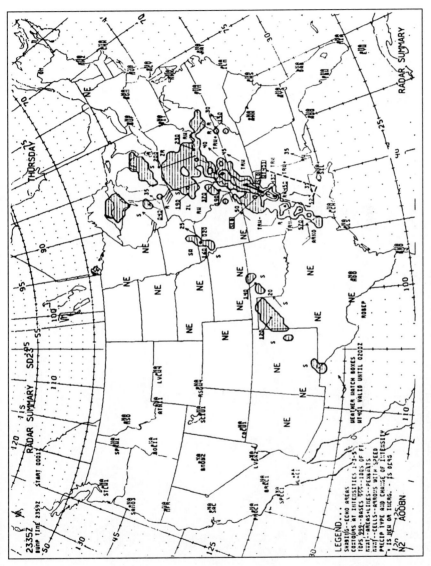

Figure 11.18 U.S. radar summary chart—significant precipitation. *(FAA and NOAA 1979)*

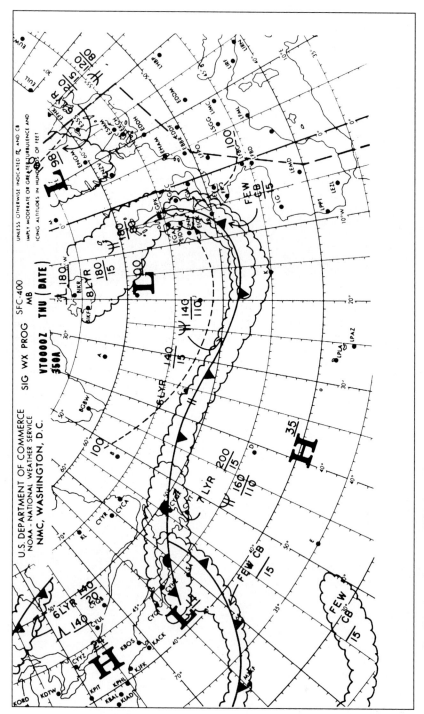

Figure 11.19 Significant weather prog. *(FAA and NOAA 1979)*

TABLE 11.8 Equivalent Altitudes for Constant Pressure Levels

Pressure (mb)	Altitude (ft)	Altitude (m)
850	5000	1500
700	10,000	3000
500	18,000	5500
300	30,000	9000
200	39,000	12,000

presentation of meteorological information are contained in national and international documents. The scope of meteorological information available is extensive and is of vital importance to all those concerned with operations including for example, airport management, air traffic control, airline, and GA flight operations.

Meteorological services for aviation have taken great strides in recent years towards a truly global service with the establishment of two World Area Forecast Centers (WAFCs)—Washington (Suitland) in the U. S. and London (Bracknell) in Great Britain—backed up by 15 Regional Forecast Centers. Both World Centers utilize the latest technology for the worldwide analysis of weather and for the production and transmission of operational meteorological information. Digital techniques permit the information to be transmitted speedily directly to users' own computer systems. Airlines in particular are able to take advantage of this in their flight planning procedures where meteorological information is only one of the required inputs.

In Great Britain SADIS—Satellite Distribution of Aviation Products—sponsored by the Meteorological Service and offered through WAFC London, can provide global coverage across a satellite footprint that extends from the eastern Atlantic (20° W) to central Australia (140° E). A complementary service the *International Satellite Communications System* ISCS, is provided by WAFC Washington using satellites covering the rest of the globe (except polar regions), which provides an overlap with SADIS to give worldwide coverage. The operational meteorological information available includes; METARs, TAFs, SIGMETs, charts of the standard ICAO designated map areas, wind and temperature charts at ICAO agreed flight levels, and significant weather (SIGWX) charts.

General aviation has benefited from the emergence of independent weather/flight planning companies with the capability of providing pilots/crews with the same computerized service as that provided for the crews of major carriers. Increasingly there is also a move to more automation in the delivery of weather information by telephone and by fax. The system, first developed in France, of a remote-controlled automatic weather briefing—Interrogation/Response Automatique (IRA)—for GA pilots has since spread

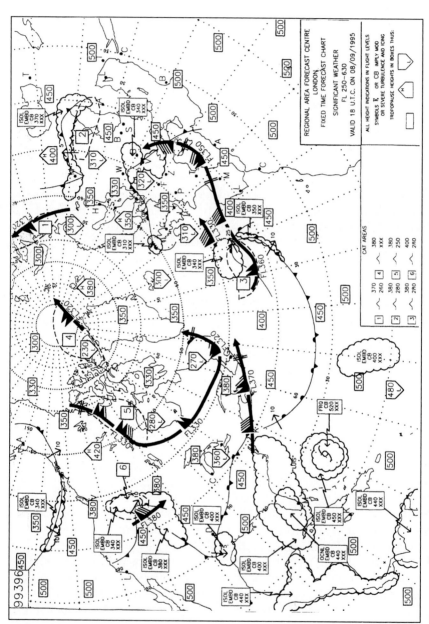

Figure 11.20 Example of a fixed-time prog. chart for approximately FL 300.

Figure 11.21 Satellite photograph, east and midcontinent United States. *(NASA, USA)*

and expanded. In the United States there is an Automated Weather Observing System as well as various delivery options: Telephone Information Briefing Service (TIBS) and Pilots Automatic Telephone Weather Answering Service (PATWAS). In the U.S. general aviation pilots also may obtain weather information using personal computers through a service called DUAT—Direct User Access Terminal—provided by the FAA.

11.5 Aeronautical Information

Scope

The complexities of civil aviation are such that it is almost impossible to conduct a flight of any kind, even a short GA flight, without having recourse to a considerable amount of aeronautical information such as air traffic control requirements (including airspace restrictions), airport layout, hours of operation, and availability of fuel. The requirement is multiplied many times over if an international flight is involved. Most states have acknowledged this and comply with the standards laid down by ICAO regarding an international format for production of aeronautical information. It is interesting to note, however, that a state may delegate the authority to provide such service to a nongovernmental agency, and there are several instances where this has been done. It would then be the responsibility of that agency to comply with the specifications for such a service as laid down by ICAO. These specify that an aeronautical information service is responsible under international agreement for:

- The preparation of an aeronautical information publication (AIP)
- The origination of NOTAMs
- The origination of aeronautical information circulars

In preparing an *AIP* (not to be confused in the U. S. with Airport Improvement Program AIP) there is a laid-down format and content for the AIP that comprises the following eight main sections listed by the agreed-on three-letter identifiers:

Section	Contents
GEN (general)	Description of aeronautical information services provided Summary of national regulations Abbreviations Units of measurement Time system Nationality and registration marks Special equipment to be carried on aircraft
AGA (airports)	Description of provisions for facilities, services, and ground aids for international air navigation International airports Airports available for use by international commercial air transport Airport directory Airport ground lights
COM (communications)	Description of provisions for communications services for international air navigation List of AFTN Special navigation systems, e.g., Long Range Navigation (LORAN) Time signals Aeronautical fixed service
MET (meteorology)	Description of provisions for meteorological service for international air navigation Airport meteorological observations and reports Meteorological service provided at airports VOLMET broadcasts
RAC (rules of the air and air traffic services)	Description of air traffic services provided and, where necessary, graphics of FIRs and divisions of airspace Air traffic services system Holding, approach, and departure procedures Prohibited, restricted and danger areas Information on bird migration Automatic terminal information (ATIS) broadcasts

Section	Contents
FAL (facilitation)	Description of provisions for facilitation of international air navigation Entry, transit, and departure regulations concerning airport use Fees and charges
SAR (search and rescue)	Provisions for search and rescue Search and rescue system Procedures and/or signals employed by rescue aircraft
MAP (aeronautical charts published)	Description of aeronautical chart series published A list of charts Sources for large-scale topographical maps of airport area if no airport obstruction charts are available

An amendment service is provided in each country to keep their AIP up to date. Amendments are usually in the form of replacement pages, particularly urgent amendments that cannot be held up for the normal amendment procedure may be sent by telex/fax or circulate by mail. The U. S. FAA through its Flight Services Division fulfills its international obligations by the publication of the Unites States of America Aeronautical Information Publication (AIP). This meets the need of the international operators flying to the United States. But with some 650 certificated (FAR 139) airports, and more than 16,500 general aviation airports there is a much greater requirement for U.S. domestic aeronautical information. This is met by a basic document—the Aeronautical Information Manual (AIM) which is re-issued every six months with amendments provided during the intervening months. The FAA, along with other government agencies, also publishes aeronautical charts, U.S. Terminal Procedures for individual airports, Airport Facility Directories, and a series of extremely useful Advisory Circulars (ACs) issued in a numbered subject system corresponding to the subject areas of the Federal Aviation Regulations, thus the AC 150 series refers to airports. All the aeronautical publications carry on their covers a note to "Consult NOTAMs for latest information".

Urgent operational information

NOTAMs, as the name implies, are urgent notices for the attention of aircrews and operations personnel. They are given the fastest possible circulation. Most NOTAMs will refer to information already published in the AIP, and as such they are the usual means of transmitting the urgent AIP amendments noted earlier. If sent by telex, such NOTAMs are termed Class I NOTAMs and if distributed by mail, they are Class II. There is also provision for mail distribution of Aeronautical Information Regulation and Control (AIRAC) NOTAMs, which are used to give advance notification of changes in facilities, radio aids, and so on that have been planned in advance. In the

United States, Class I NOTAMs are distributed via the Service A network. For domestic purposes, the United States differs from the minimum international recommendations by further subdividing Class I NOTAMs into NOTAM-D (distant) and NOTAM-L (local) categories. The information contained in NO-TAM-D category is critical to flight and could affect a pilot's decision to make a flight.

This information is maintained in the Kansas City National Communications Center computer data base and is distributed automatically according to predetermined circulation lists. NOTAM-L information is considered less critical and retained on file only at those local air traffic facilities concerned with the operations at these airports. Additionally, if it becomes necessary to circulate urgent information of a regulatory nature, such as restrictions to flight or changes in Instrument Approach Procedures, these will be originated by the National Flight Data Center in Washington, DC, as a FDC NOTAM. These are distributed by the Kansas City Communications Center to all air traffic facilities with telecommunications access. The U.S. NOTAMs Class II consist of three parts:

1. Information that meets the criteria for NOTAM-D, which is likely to remain in force for an extended period (at least seven days). Occasionally NOTAM-L information and special notices are included in this part.
2. This part is in effect a check list of current FDC NOTAMs.
3. The third part contains information that would come under the category of information circular in the international context.

The U.S. NOTAM system also includes a biweekly Notices to Airmen Publication (NTAP) which is, in effect, a NOTAM Class 2. Three methods of notification of changes to the AIP are used in Great Britain, all involving mail transmission: amendment pages; Aeronautical Information Circulars (AIPs) and AIRAC Circulars. These latter notices are published some 2 to 3 months ahead of implementation date to provide sufficient lead time for the printing of new procedures and charts. The Aeronautical Information Circulars use a subject color coding—white, yellow, pink, mauve, green (e.g., white for administration matters, mauve relates to airspace restrictions, danger areas, etc). Figure 11.22 is an example of an AIC.

NOTAM code. With the very considerable amount of aeronautical information that is distributed as Class I traffic via teleprinter, it is necessary to condense it as far as possible, and this is achieved by use of a special code. An added advantage of using a code is that it overcomes language difficulties. One slight disadvantage is that small errors can render the whole message

UNITED KINGDOM	AIC 70/1995
AERONAUTICAL INFORMATION CIRCULAR	(White 220)
	27 July

Civil Aviation Authority
Aeronautical Information Service
Control Tower Building, LONDON/Heathrow Airport
Hounslow, Middlesex TW6 1JJ
Editorial: 0181-745 3456
Distribution: 01242-23515
Content: 0171-832 5700   INT 8   Cancels AIC 57/1995 (White 215)

THE TRAINING OF AIR TRAFFIC SERVICES PERSONNEL IN THE PREPARATION OF AVIATION WEATHER REPORTS (METAR)

1   At most aerodromes there is usually a requirement to provide weather reports to aircraft flying in the area. A weather report from a UK aerodrome will not be accepted by the CAA as a METAR for dissemination beyond the aerodrome unless the report has been compiled by a certificated meteorological observer. Furthermore, only reports from certificated observers should be used for the preparation of an aerodrome forecast (TAF).

2   The Meteorological Office (Aviation Services), on behalf of the CAA, arranges and provides training for ATS personnel in the preparation of aviation weather reports. Applications for the courses can only be accepted from sponsors (eg ATS Managers, SATCOs) prepared to accept responsibility for trainees while they are on MOD property. Those nominated must have experience in an Air Traffic Services environment. The duration of each course is two weeks. The first week is residential at the MET Office College, Shinfield Park, Reading. This is followed by a week of practical training (including night-time observations) at an appropriate operational meteorological office with an aviation observing commitment. Trainees are advised to familiarise themselves with the METAR code prior to commencement of the course. A copy of the course syllabus is shown at Annex A.

3   Examinations are set at the end of each week of the course. If successful, the trainee is awarded a certificate to confirm their completion of an approved training course for ATS personnel in the making and reporting of weather observations. Certificate holders become accredited observers whose observations can be accepted as 'official'.

4   The fee for the two week course is £650. This includes accommodation and food (Sunday evening until Friday lunch-time) during the residential week at the MET Office College. However, it is the responsibility of the sponsor/trainee to arrange and pay for accommodation and food during the second week of training. Payment will be requested in advance of each course and the invoice will normally be sent to the aerodrome operator or sponsor by MOD Accounts. In the event of a late cancellation when the vacant place cannot be re-allocated, a cancellation charge will be made. A full refund will be given if the Meteorological Office has to cancel a course or postpone it for an indefinite period.

5   After completion of the formal two weeks of MET Office training, the responsibility for ensuring that certificated observers fulfil their weather observing and reporting duties satisfactorily is that of the aerodrome management, usually the Air Traffic Services Manager/SATCO. Two weeks of professional training is the minimum required for the award of a certificate, **but no newly certificated observer can be considered fully competent without a period of further 'on-the-job' training at his/her home aerodrome under the supervision of more experienced colleagues/observers and in a variety of weather situations.** Parent meteorological offices monitor the quality of broadcast METAR observations and should bring any deficiencies to the attention of the duty observer/ATS personnel or aerodrome management. MET Office staff, if asked and/or when making liaison/inspection visits to aerodromes, on behalf of the CAA, will endeavour to answer any queries raised and be prepared to offer guidance and advice.

6   Dates for the residential first week of each course (held at the MET Office College, Shinfield Park, Reading) during the period September 1995 to June 1996 are as follows:

Course Reference Number	Date
AT1/95	11 to 15 Sep 1995
AT2/95	9 to 13 Oct 1995
AT3/95	13 to 17 Nov 1995
AT4/95	11 to 15 Dec 1995
AT5/95	8 to 12 Jan 1996
AT6/95	12 to 16 Feb 1996
AT7/95	4 to 8 Mar 1996
AT8/95	15 to 19 Apr 1996
AT9/95	13 to 17 May 1996
AT10/95	10 to 14 Jun 1996

Applications, in writing to the address in paragraph 7 below, should indicate the choice of location and commencement date (together with alternative(s) in case the first choice is unavailable) for the week of practical training. Ideally this should follow immediately after the week at Shinfield Park, but in some circumstances this may not be possible. A list of MET Office units prepared to accept trainees for practical training is shown at Annex B.

Figure 11.22   U.K. aeronautical information circular. *(U.K.A.I.S.)*

meaningless. The NOTAM code is used to enable the rapid dissemination of information regarding the establishment, condition, or change of:

- Radio aids
- Lighting facilities
- Airports
- Dangers to aircraft in flight or search and rescue facilities

All NOTAM code groups contain a total of five letters. The first letter of the group is always the letter Q. The second and third letters identify the facility, service, or danger to aircraft that is being reported on. The second letter is always either A,E,I,O, or U and has the following significance:

1. A and E refer to radio aids.
2. I refers to lighting facilities.
3. O and U refer to airports, search and rescue, danger to aircraft in flight.

The fourth and fifth letters indicate the STATUS OF OPERATION. Thus

- Q = NOTAM
- IY = taxiway centerline lights
- EO = obscured by snow

Airport operations staffs are required to initiate and internet NOTAM Class I material. To afford a complete understanding of the procedure, Figure 11.23 gives details of the NOTAM code.

Availability of Information

International recommendations regarding the use of aeronautical information are that it should be made available in a form suitable for the operational use of:

- Flight operations personnel including aircrew
- Services responsible for preflight information
- Air traffic service units responsible for flight information

The extent of the information made available at specific airports will depend on the stage lengths of flights operating out of that airport. Thus the information made available to these users at New York JFK would include international information whereas Omaha, Nebraska, would receive only domestic information within a certain range. This does not prevent special

NOTAM CODE

SECOND AND THIRD LETTERS

RADIO AIDS

AA ...(specify TWR, APP, ACC or FIC) air traffic services receiver ...KHz (or...MHz)

AB Inner marker, Instrument Landing System (IM) (for runway number ...)

AC ...(specify TWR, APP, ACC or FIC) air traffic services trans - mitter...KHz (or...MHz)

AD Middle marker, Instrument Landing System (MM) (for runway number ...)

AE Outer marker, Instrument Landing System (OM) (for runway number ...)

AF Fan - type marker

AG Glide path, Instrument Landing System (GP) (for runway number ...)

AH Non - directional radio beacon NDB

AI Instrument Landing System (ILS) (for runway number...)

AJ Radio range (other than VOR) and associated voice communications.

AK Radio receiving facilities

AL Localizer, Instrument Landing System (LLZ) (for runway number..)

AM Locator, outer, Instrument Landing System (LM) (for runway number..)

AN TACAN (UHF Tactical Air Navigation Aid)

AO Locator, outer, Instrument Landing System (LO) (for runway number...)

AP VOR, and associated voice communications

AQ VOR

AR Radio range (other than VOR) (RNG or VAR)

AS Radio range leg

AT Attention signal

AU Meteorological communications... KHz (or... MHz)

AV Voice communications... (KHz (or ...MHz)

AW VOLMET

AX Non - directional radio beacon (NDB) and voice facility

AY 200 MHz Distance - Measuring Equip- ment.

AZ Station location marker VHF (MKR)

EA En - route Surveillance Radar (RSR)

EB Broadcasting station (public) (BS)

EC CONSOL

ED DECCA or DECTRA

EE Ground Controlled Approach System (GCA)

EF Terminal Area Surveillance Radar (TAR)

EG Automatic Terminal Information Service (ATIS)

EH Elevation element of the Precision Approach Radar (PAR) (for runway number...)

EI Monitoring device associated with... (specify) radio aid

EJ All air ground facilities (except...)

EK Precision Approach Radar (PAR) (for runway number...)

EL LORAN

EM Azimuth element of the Precision Approach Radar (PAR) (for runway number...)

EN DME (1000 MHz Distance - Measuring Equipment)*

EO VOR/DME

EP Radar responder beacon (RSP)

EQ Surveillance Radar Element (SRE) of GCA

ER Radio transmitting facilities

ES All radio navigation facilities (except..)

ET Teletypewriter transmitting facility (ies)

EU Radio direction - finding station ... (frequency or type)

EV VORTAC (the combination of VOR and TACAN)

EW Ground interrogator, Secondary Surveillance Radar (SSR) System

EX SELCAL (Selective Calling System)

EY Surface movement radar (SMR)

* Where more than one DME is installed at the same location, the type of associated facility (e.g., ILS, VOR, TACAN, or VORTAC) should be indicated

Figure 11.23 Key to NOTAM code. *(ICAO 1991)*

N O T A M   C O D E

LIGHTING FACILITIES

IA Boundary lights
IB Aerodrome beacon
IC   -
ID Runway end identification lighting (for runway number...)
IE Marine light beacon
IF Floodlights
IG Visual approach slope indicator system (VASIS) (all types) (for runway number...)
IH Taxiway edge lights
II Hazard beacon
IJ Threshold lights (for runway number...)
IK Flares (for runway number...)
IL All landing area lighting facilities
IM Identification beacon
IN Range lights
IO Obstruction lights

IP Approach light system (type..(specify light intensity))(for runway number..)
IQ   -
IR Runway edge lights (type..(specify light intensity))(for runway number..)
IS   -
IT Taxiway guidance system
IU Runway center line lights (for runway number...)
IW Stopway lights (for runway number..)
IX Sequenced flashing lights (for runway number..)
IY Taxiway center line lights
IZ Runway touchdown zone lights (for runway number..)

AERODROMES - SEARCH AND RESCUE - DANGERS TO AIRCRAFT IN FLIGHT

OA Land aerodrome
OB Beaching facilities
OC Water aerodrome
OD Meteorological forecast service
OE Meteorological observation service
OF Meteorological watch service
OG Runway arresting gear
OH Heliport or helicopter alighting area
OI   -
OJ   -
OK   -
OL Taxiway - link
OM All runways (except number(s)..)
ON Stopway for runway number
OO Taxiway(s)
OP Rescue vessel
OQ Ocean station vessel
OR Refuelling (...type fuel(s) or... octane)
OS Search and rescue aircraft (specify VLR, LRG, MRG, SRG or HEL)
OT Rescue or fire fighting facility(ies).. specify type(s)
OU   -

OV ...(specify TWR, APP, ACC or FIC) air traffic services
OW Minimum flight altitude
OX   -
OY Pyrotechnics
OZ Warship

Figure 11.23 Continued.

SECOND AND THIRD LETTERS (Cont'd)

AERODROMES - SEARCH AND RESCUE
DANGERS TO AIRCRAFT IN FLIGHT

UA	Alighting area
UB	Mooring buoys
UC	Control area (airway)
UD	Prohibited, restricted or danger area designated as...(name or reference identification)
UE	Aircraft
UF	Fixed balloons
UG	Bombing or aerial depth charge dropping
UH	Air exercises (or flying displays)
U I	Gun or missile firing
UJ	Glider flying
UK	Demolition of explosives
UL	Landing direction indicator
UM	Mooring and docking facilities
UN	Parachute jumping exercises
UO	Wind direction indicator
UP	Free balloon
UQ	Apron
UR	Runway(s) number(s)
US	Strip..(number or magnetic direction)
UT	Grass landing area
UU	-
UV	Fog dispersal equipment
UW	Bird hazard
UX	Air refuelling
UY	Burning or blowing of gas
UZ	Runway threshold (number..)

Figure 11.23 Continued.

FOURTH AND FIFTH LETTERS

HAZARD OR STATUS OF OPERATION OR CONDITION OF FACILITIES

AA	–	AZ	Will take place
AB	Usable for length of... and width of.....		1) until further notice; or
AC	Covered by dry snow to a depth of... Note – This snow is not compacted		2) until...(date/time); or 3) from...(date/time) to...(date/time)
AD	Cleared of soft snow, full length and width		(on the days of...between the hours of...and...) at...(location)
AE	Totally free of snow and ice (cleared and dry)		(within the sector of...and a radius of...) at...height above
AF	Covered by (...type) ice to a depth of....		...(datum)
AG	Covered by dry snow (patches or partially cleared)		
AH	Covered by rime or frost (depth normally less than 1 mm)	EA	–
A I	Operating without tone modulation	EB	Location change to...effective...(date/time)
AJ	Operating without coding or without flashing	EC	Characteristics or identification or radio call sign changed to...
AK	Covered by compacted or rolled snow to a depth of....	ED	Operating frequency(ies) will be changed to...KHz (or ...MHz)
AL	Operating on reduced power		effective...(date/time)
AM	Snow clearance in progress (estimated time of completion is...(date/time)	EE	–
		EF	–
AN	Grass cutting in progress (estimated time of completion is...(date/time)	EG	–
		EH	Not heard
AO	Marked by	E I	Subject to interruption
AP	Work in progress(estimated time of completion is...(date/time)	EJ	–
		EK	–
AQ	Work completed	EL	Snow clearance will be completed...(date/time)
AR	Snow clearance completed		
AS	Grass cutting completed	EM	Military operations only
AT	Sanding is in progress (estimated time of completion is...(date/time)	EN	Not available due to...(specify reason) 1) until further notice; or
AU	Appears unreliable		2) until...(date/time); or
AV	Covered by ice(patches or partially cleared)		3) from...(date/time)to...date/time)
AW	Critical snow – banks exist along runways/taxiways(designation, height)	EO	Obscured by snow
AX	Braking action is...	EP	Available on prior permission (of...) only
	1 poor	EQ	Covered by wet snow or slush (patches or partially cleared)
	2 medium/poor		
	3 medium	ER	Covered by frozen ruts or ridges
	4 medium/good	ES	Out of service
	5 good)		1) until further notice; or
AY	Are to avoid area, radius of danger being...(about the point...)		2) until...(date/time); or 3) from...(date/time)to...(date/time) (due to the following condition(s)...)

Figure 11.23 Continued.

FOURTH AND FIFTH LETTERS (Cont'd)

HAZARD OR STATUS OF OPERATION OR CONDITION OF FACILITIES

ET   Test operation only. NOT for
      operational use
EU   -
EV   Changed to
EW Completely withdrawn
EX   -
EY   Is outside the limits of its assigned
      ocean station
EZ   Is within the limits of its assigned
      ocean station

IT   Aircraft restricted to runways and
      taxiways
IU   Unserviceable for aircraft heavier
      than...tons
IV   Unsafe
        1) until further notice; or
        2) until...(date/time);or
        3) from...(date/time) to...
           (date/time)
IW   -
IX   -
IY   -
IZ   -

IA   -
IB   -
IC   Report of apparent unreliability or
      track displacement hereby is cancelled
ID   Available on request to...
IE   -
IF   Flight checked and found reliable
IG   -
IH   -
I I   -
IJ   -
IK   Available on request(to...)immedi-
      ately (or at...(time period) notice)
IL   Hours of service now...
IM   -
IN   Operative (or re-operative), activ-
      ated (or re-activated)
        1) until further notice; or
        2) until...(date/time),or
        3) from...(date/time) to...
           (date/time)
IO   Operating normally
IP   Track(s) reported to be displaced
      (...degrees) (...direction) of
      published bearing(s), other tracks
      probably have shifted
IQ   To be used as radio beacon only
IR   Magnetic track(s) towards station
      is (are) now... (will be...at
      (date/time))
IS   Operative (or re-operative) subject
      to conditions/limitations already
      published

Figure 11.23 Continued.

FOURTH AND FIFTH LETTERS (Cont'd)

OC Caution advised or use caution

OG Operative but ground checked only, awaiting flight check

OH -

O I -

OJ -

OK Resumed normal operation

OL Track(s) ground checked, approved for instrument flying

OM Shut down for maintenance
    1) until further notice; or
    2) until...(date/time);or
    3) from...(date/time) to...
       (date/time)

ON -

OO On operational trials

OP Approved for operational use

OQ -

OR Previously promulgated shutdown has been cancelled

OS -

OT New facility in operation

OU Operating without interruption for voice transmissions
    1) until further notice; or
    2) until...(date/time);or
    3) from...(date/time) to...
       (date/time)

OV -

OW Wet (including standing water in pools)

OX Exercising at...(date/time, location and height above the specified datum)

OY Exercises completed

OZ -

OD Launch planned...(balloon flight identification or project code name) ...(launch site)..(planned period of launch(es) - date/time)...(expected climb direction)...(estimated time to pass 18,000 m (60,000 ft) or reaching cruising level if at or below 18,000 m (60,000 ft), together with estimated location)

OE Operation cancelled...(balloon flight identification or project code name)

OF Launch in progress...(balloon flight identification or project code name) ...(launch site)...(date/time of launch(es)..(estimated time passing 18,000 m (60,000 ft), or reaching cruising level if at or below 18,000 m (60,000 ft), together with estimated location)...(estimated date/time of termination of the flight)...(planned location of ground contact, when applicable)

Figure 11.23 Continued.

FOURTH AND FIFTH LETTERS (Cont'd)

HAZARD OR STATUS OF OPERATION OR CONDITION OF FACILITIES

UA Closed to all operations
    1) until further notice; or
    2) until...(date/time); or
    3) from...(date/time) to...
      (date/time)
UB Approach according to signal area
    only
UC Under calibration
UD Closed to all night operations
    1) until further notice; or
    2) until...(date/time); or
    3) from...(date/time) to...
      (date/time)
UE Closed to IFR operations
UF Closed for an unknown duration (or
    until...(date/time) due to flood
UG Closed for an unknown duration (or
    until...(date/time) due to ice or snow
UH Closed for an unknown duration (or
    until...(date/time) due to thaw
UI Closed
    1) until further notice; or
    2) until...(date/time); or
    3) from...(date/time) to...
      (date/time)
    for maintenance
UJ Daylight use only
UK Available for night operation
UL The facility status is now...
UM Operating in an unmonitored status
UN –
UO Operating with auxiliary equipment

UP Operating on auxiliary power supply
UQ Closed to VFR operations
UR Realigned
US Unserviceable
UT Operative but caution advised due to
    following conditions
UU Suitable for...(specify) equipped aircraft
    only
UV Covered by wet snow or slush to a depth
    of...
UW Covered by water to a depth of...
UX –
UY Migration in progress...(location of
    observation)...(date/time)..(species)...
    (direction of flight)...(height above
    specified datum)
UZ Concentration or local movement of...
    (species)...(nature of movement) at...
    (date/time)

————oOo————

Figure 11.23 Continued.

requests being made for aeronautical information for areas not normally covered at a certain airport.

Although international recommendations do not specifically call for the provision of information/briefing rooms at airports for the distribution of aeronautical information, many governmental agencies and airport authorities provide such facilities. These usually include suitable reference material such as the AIP and display charts as well as information bulletins, NOTAMs, and so on. An example of a typical information bulletin supplied by the German AIS at Frankfurt Airport is given in Figure 11.24. Major airlines will frequently use the information obtained from the Aeronautical Information Service together with their own operational information to publish company NOTAMs.

11.6 Summary

International agreements made through ICAO have laid down the basic standards for each of the technical services described in this chapter. Individual countries have shown a remarkable consistency in adopting these principles with only minor differences. When such differences do exist, they are well documented in both national and international publications including the various ICAO annexes to the convention. Implementation of these services on a regional basis is documented in the various air navigation regional plans published by ICAO.

As new technologies are developed, there will be inevitable changes but with a total membership of 183 nations[5], clearly these changes will not be accomplished overnight, MLS/GPS is just such an example. So far, the system of international standards negotiated through the aegis of ICAO has proved to satisfy the essential technical and safety requirements of the air transport industry.

The United States is in the forefront of those nations planning to take full advantage of technological developments, and the imaginative FAA plan for ATC technical services to the year 2000 provides a valuable insight into the kind of changes that are likely to take place (National Commission 1993).

References

Federal Aviation Administration (FAA). 1994. *Aeronautical Information Manual* (AIM). Washington, DC: Department of Transportation.
Aviation Weather Services. 1979. AC 00-45B (a supplement to *Aviation Weather* AC 00-6A). Federal Aviation Administration, Department of Transportation and National Oceanic and Atmospheric Administration, Department of Commerce, Washington, DC.
International Civil Aviation Organization (ICAO). 1985. *Air Navigation Plan—European Region, DOC 7754/23*. Montreal.

[5]As of January 1994.

| BUNDESREPUBLIK DEUTSCHLAND | *DFS* Deutsche Flugsicherung | PRE-FLIGHT INFORMATION |

```
EDDE    E R F U R T

C1558/95
A) EDDE
B) 9507010400   C) 9508302000
E) PAPI RWY 28 ON TEST.




EDDF    F R A N K F U R T

A1429/95
A) EDDF
B) 9504270000   C) 9510312400 EST
E) TRIGGER NOTAM
AIP ENROUTE/TERMINAL SUP 6/95 DATED 30 MAR 95.
DURING OUTAGE OF 'FFM' VOR WITH EFFECT 27 APR 95, THE MODIFIED
DEPARTURE ROUTES WILL BE IMPLEMENTED. VOR APPROACHES SUSPENDED.
RNAV DEPARTURE ROUTES ONLY FOR FMS/FMGC EQUIPPED AIRCRAFT.
DEPARTURE FREQUENCY 'FRANKFURT RADAR' 120.425. THE SHUT DOWN
OF VOR 'FFM' WILL BE ANNOUNCED BY FURTHER NOTAM.


A1567/95
A) EDDF
B) 9505080000   C) PERM
E) SID RWY18 KIR 3L, TURNING POINT CHANGED TO READ 7.5 DME RID.
AIP ENROUTE TERMINAL 7-13 AND 7-15 REFER.


A1617/95
A) EDFF
B) 9505081436   C) 9510312400 EST
E) FRANKFURT DME IN OPERATION, IDENTIFICATION FFM CH108Y (GHOST
FREQ 116.15MHZ) PSN 500318N 083817E, STATION ELEVATION 479FT.


A1620/95
A) EDDF
B) 9505081447   C) 9510312400
E) FRANKFURT DVORTAC FFM 114.20/CH89 OUT OF SERVICE FOR MAINT.
USE FFM NDB 320 AND FFM DME CH108Y/116.15MHZ INSTEAD. RADIALS
TO BE REPLACED BY BEARINGS.
AMEND SUP 6 DATED 30 MAR 1995 STANDARD DEPARTURE CHART
7-3, 7-7, 7-11 AND 7-15 ACCORDINGLY.


A1836/95
A) EDDF
B) 9506140515   C) 9507142400
E) RWY 07R: ILS LLZ OPR ON TEST. DISREGARD ALL SIGNALS.


A1837/95
```

102-5111-01.94

Figure 11.24 Frankfurt Airport AIS NOTAM, Class L. *(Deutsche Flugsicherung)*

ICAO. 1990a. Annex 2, *Rules of the Air*. Montreal.

ICAO. 1990b. *Air Navigation Plan—North Atlantic, North American and Pacific Regions, DOC 8755/13*. Montreal.

ICAO. 1991a. Annex 10, *Aeronautical Telecommunications*. Montreal.

ICAO. 1991b. Annex 15, *Aeronautical Information Services, 8th edition*. Montreal.

ICAO. 1992. Annex 3, *Meteorological Service for International Air Navigation*. Montreal.

ICAO. 1993. Annex 11, *Air Traffic Services*. Montreal.

National Commission to Ensure a Strong Competitive Airline Industry. 1993. *Change, Challenge and Competition, A Report to the President and Congress*. Washington, DC.

National Transportation Safety Board (NTSB) 1991. *Accident Report—Runway Collision of US Air Flight 1493, Boeing 737 and Skywest Flight 5569 Fairchild Metroliner, Los Angeles International Airport, February 1991, NTSB/AAR-91/08*. Washington, DC.

The Royal Institute of Navigation. 1995. *Navigation News*. London.

U.S. Department of Commerce. 1995. *US Terminal Procedures, Southwest (SW), Vol. 2 of 2*. Washington, DC.

Chapter

12

Airport Aircraft Emergencies

12.1. General

At any particular airport a number of different types of emergencies might occur, including an aircraft emergency, a building fire, or other major disruptions, such as a major spillage of flammable or poisonous liquids, or nonaccidental emergencies caused by bomb scares or other terrorist activities. This chapter will concentrate on the first category, the aircraft emergency, which is peculiar to airports and aviation and might involve a loss of life on a scale that is rightly termed disastrous. Air travel is not a particularly hazardous mode of transportation; indeed commercial air passenger transportation has a safety record that is bettered only by the railroads. Nevertheless every airport operator must recognize the possibility that an aircraft accident on the airport or in its vicinity can take place. For airports serving air carriers, this imposes a special responsibility to plan for the saving of a large number of lives through the provision of competent fire-fighting and rescue services, recognizing and even hoping that during the life of the airport they will never be employed to the limits of their capability.

12.2 Probability of an Aircraft Accident

Aircraft accidents are unlikely occurrences. A U.S. study established that the statistical probability of an accident in the United States was one accident in every 77,200 movements (FAA Advisory Circular 1970). A later study in the United Kingdom obtained rather similar results with an estimate of one accident per 67,000 movements in the United Kingdom (Aerodrome

Fire and Rescue Services 1972). In the latter study, it was determined that the probability of a fatal accident in which one or more persons is killed is one in 345,000 movements. French data indicate one accident in each 100,000 and a fire in each 200,000 movements (Ansart and Viel 1981). For air transport aircraft, the figures are very much lower; the probability of a catastrophic accident on a public transport flight is about one in a million per hour of flight. This low probability is due on a large part to the very demanding performance standards set by the aircraft certification bodies, which use a scale of probabilities, as shown in Table 12.1. The reliability of modern air transport aircraft has largely removed aircraft failure as a cause of accident; more than 90 percent of air transport accidents are attributable to some form of human error.

Even with low accident probabilities, aircraft disasters do occur at airports often with a large loss of life. On March 27, 1977, at Las Palmas

TABLE 12.1 Probabilities of Aircraft Failure by Type

Frequency	Probability	Examples
Frequent Likely to occur often during the life of the aircraft	↑ 10^{-3}	
Reasonably probable Unlikely to occur often but may occur several times in the life of the aircraft	↑ ↓ 10^{-5}	Engine failure
Remote Unlikely to occur to each aircraft in its life but may occur several times during the life of a number of aircraft of the type	↑ ↓ 10^{-7}	Low speed overrun Falling below the net takeoff flight path Minor damage Possible passenger injuries
Extremely remote Possible but unlikely to occur in the total life of a number of aircraft of the type	↑ ↓ 10^{-9}	High speed overrun Ditching Extensive damage Possible loss of life Hitting obstacle in net takeoff flight path Double engine failure on a twin
Will not happen	↑ ↓	Aircraft destroyed Multiple deaths

International Airport in Tenerife, a KLM B747 crashed while taking off, colliding with another B747 of Pan American back-taxiing on the runway. A total of 555 passengers and 25 crew members was killed, making this the worst air disaster in civil aviation history. A routine operating day had been converted into a day on which the airport administration had to cope with an aviation disaster of unprecedented scale. While casualty levels of this scale are unusual, with the size of modern airliners, accidents involving more than 100 persons are more common.

12.3 Types of Emergencies

The ICAO classifies aircraft emergencies, for which rescue and fire-fighting services might be required, into three categories (Airport Services Manual 1993).

Aircraft accident

When an aircraft accident has occurred either on or in the vicinity of the airport, air traffic control at the airport will alert the airport rescue and fire-fighting service giving details of the time and location of the accident and the type of aircraft involved. Other appropriate organizations such as the local fire department are notified in accordance with the airport emergency plan (see Section 12.9).

Full emergency

When an aircraft approaching the airport either is or is suspected to be in danger of an accident, the rescue and fire-fighting service is called to predetermined standby positions for the approach runway and is given details of the type of aircraft, number of occupants, type of trouble, runway to be used, estimated time of landing, and the location and quantity of any dangerous goods on board. In accordance with the procedure laid down in the emergency plan, the local fire department and other organizations are also alerted.

Local standby

When an aircraft has or is suspected of having some defect but the trouble is not sufficiently serious to cause any difficulty in landing, the rescue and fire-fighting service is alerted to its predetermined standby positions for the approach runway and is given all essential details by air traffic control. Table 12.2 indicates the relative frequencies of the various categories of emergencies for two large airports in 1979.

It can be seen that whereas aircraft accidents are a relatively rare occurrence, other forms of emergencies are much more frequent.

TABLE 12.2 Frequency of Aircraft Emergencies at London Heathrow, 1994

Emergency type	Frequency of Emergency in 1994 London Heathrow
Aircraft accident	0
Full emergency	63
Local standby	243
Total number of operations	411,172

12.4 Level of Protection Required

Not surprisingly, the level of rescue and fire-fighting protection depends on the size of the largest aircraft using the airport and the frequency of operation. Nine different levels of protection are designated by the ICAO 1993; the category into which the airport is assigned is determined from the following criteria based on airport movements in the busiest consecutive three months of the year:

1. When the number of movements of the longest aircraft in the same category in Table 12.3 totals 700 or more, that category should be adopted.

2. When the number of movements of the longest aircraft in the same category in Table 12.3 total less than 700, the airport category adopted should not be lower than one below that of the longest aircraft normally using the airport.

3. When there is a wide range in the lengths of the aircraft that are included in the 700 movements, the category adopted may be reduced to be no lower than two categories below that of the longest aircraft.

TABLE 12.3 Airport Categorization

Airport category	Airplane overall length			Maximum fuselage width
1	0	Up to but not including	9 m (29.53 ft)	2 m (6.56 ft)
2	9 m (29.53 ft)	Up to but not including	12 m (39.37 ft)	2 m (6.56 ft)
3	12 m (39.37 ft)	Up to but not including	18 m (59.06 ft)	3 m (9.84 ft)
4	18 m (59.06 ft)	Up to but not including	24 m (78.74 ft)	4 m (13.12 ft)
5	24 m (78.74 ft)	Up to but not including	28 m (91.86 ft)	4 m (13.12 ft)
6	28 m (91.86 ft)	Up to but not including	39 m (127.95 ft)	5 m (16.4 ft)
7	39 m (127.95 ft)	Up to but not including	49 m (160.76 ft)	5 m (16.4 ft)
8	49 m (160.76 ft)	Up to but not including	61 m (200.13 ft)	7 m (22.97 ft)
9	61 m (200.13 ft)	Up to but not including	76 m (249.34 ft)	7 m (22.97 ft)

SOURCE: ICAO 1983

TABLE 12.4 Minimum Usable Amounts of Extinguishing Agents

| Airport category | Foam meeting performance Level A | | Foam meeting performance Level B | | Complementary agents | | |
| | | | | | or | or | |
	water (L)	discharge rate foam solution/ minute (L/min)	water (L)	discharge rate foam solution/ minute	dry chemical powders (kg)	Halons (kg)	CO_2 (kg)
1	350	350	230	230	45	45	90
2	1000	800	670	550	90	90	180
3	1800	1300	1200	900	135	135	270
4	3600	2600	2400	1800	135	135	270
5	8100	4500	5400	3000	180	180	360
6	11800	6000	7900	4000	225	225	450
7	18200	7900	12100	5300	225	225	450
8	27300	10800	18200	7200	450	450	900
9	36400	13500	24300	9000	450	450	900

SOURCE: ICAO

Based on the airport category determined from 1, 2, and 3, the amount of extinguishing agents to be carried on rescue and fire-fighting vehicles is obtained from Table 12.4 and the minimum number of vehicles from Table 12.5. The "remissions" permitted by 2 and 3 above are being phased out by ICAO.

The quantities shown in Table 12.4 are computed from the amount of liquid required to control an area adjacent to the aircraft fuselage (the so-called critical area) in order to maintain tolerable conditions for the rescue of the occupants.

Extinguishing agents are two major types: *principal agents,* which are used for the permanent control of fire, and *complementary agents,* which have a high capability to "knock down" a fire, but which provide no expo-

TABLE 12.5 Minimum Number of Vehicles

Airport category	No of vehicles
1	1
2	1
3	1
4	1
5	1
6	2
7	2
8	3
9	3

sure or reflash protection. Modern principal extinguishing agents provide a fire-smothering blanket. ICAO recommends the use of:

1. *Protein foam (foam meeting performance level A)*. This is a mechanically produced foam capable of forming a long-lasting blanket.

2. *Aqueous film forming foam (AFFF)*. (foam meeting performance level B), which is effective on spill fires, providing faster extinguishing than protein foams. However, the liquid film over the fuel surface is destroyed by high temperatures. AFFF foams are not suitable for fires involving large masses of hot metal.

3. *Fluoroprotein foam*. which is a development from protein-base foams. The addition of a fluorocarbon to protein foam cuts down the amount of pickup of fuel on the surface of the foam bubbles. Although more expensive than protein foam, it is more suited as an extinguishing agent for fires where there is some depth of fuel.

Care must be taken to select a complementary extinguishing agent that is compatible with the principal agent. Although complementary agents do not have any significant cooling effect on liquids and other materials involved in the fire, they act rapidly in fire suppression and can ensure that a fire does not get out of control before permanent control can be achieved with the principal agent. Several complementary extinguishing agents are available:

- Carbon dioxide
- Dry chemicals
- Halocarbons

It is obviously important that there is a sufficient reserve supply of principal and complementary agents to ensure replenishment of the rescue and fire-fighting vehicles so that after an accident continued cover can be given to subsequent aircraft operations. It is recommended that a minimum reserve supply of 200 percent of the quantities of the agents shown in Table 12.4 should be maintained on the airport for the purpose of vehicle replenishment.

A rescue and fire-fighting service should have as its operational objective a response time of not more than three minutes, preferably not more than two minutes, to any part of the movement area in ideal visibility and surface conditions. This sort of performance level was achieved in the crash of the Eastern Airlines B727 at JFK airport on June 24, 1975. The first Port Authority appliance reached the aircraft two minutes after the alarm and three minutes after the crash. It took two minutes to control the main fire and three minutes to extinguish it. There were 11 survivors but 107 passengers and 6 crew members died. The New York City Fire Department was notified within four minutes of the crash, and the first units arrived within eight minutes.

Response time must be considered in conjunction with evacuation times. In a trial conducted by Douglas, 391 occupants were evacuated from a DC10 in 73 seconds in the dark using only emergency lighting (Aviation Safety Hearings 1977). This must be seen as an ideal time, achieved by subjects in a nonshocked condition, with no fear of danger of fire with all escape chutes in operation. In the Continental Airlines DC10 crash at Los Angeles in March 1978, the evacuation of 183 passengers and 14 crew took 5 minutes, not the 90 seconds demonstrated in the FAA's aircraft certification program.

12.5 Water Supply and Emergency Access Roads

Water for aircraft rescue and fire-fighting purposes can come either from the airport water supply or from natural water supplies within the airport area. It is desirable that the airport water supply is provided in apron and service areas and in the vicinity of administration areas. Fire-fighting vehicles are more easily replenished if the supply is extended to hydrants spread about the movement area where this is economically feasible. Natural surface water from rivers, lakes, streams, and ponds can be utilized only if the fire-fighting vehicles are adequately equipped to pick up and pump such supplies.

Although rescue and fire-fighting vehicles should have some all-terrain ability, reduction of the response time to an accident and the subsequent evacuation of casualties is made easier by the provision of emergency access roads to various areas in the airport and to areas beyond the airport boundary, especially in the final approach areas and the clearways designated for takeoff. The ICAO recommends particular attention to providing ready access to approach areas up to 3300 feet (1000 m) from the threshold. It is common for airport operators to assume that provided that obstruction clearances are observed in the approach and takeoff areas the airport's responsibilities have been met. The crash of a DC9 at Toronto on June 26, 1978, is an example of the very severe problems of rescue where terrain beyond the runway is virtually impassable. Rescue vehicles were able to reach the site in the clearway beyond the runway threshold only through the quick thinking of an airport bulldozer driver who immediately cleared a temporary road through an otherwise impassable wooded ravine.

12.6 Communication and Alarm Requirements

An aircraft emergency is an unplanned event, and accidents are very rare occurrences. Consequently, a reliable rescue and fire-fighting operation can be achieved only with a defined chain of command linked with effective communications. For each airport, the individual requirements for the communications network are likely to be different; however, in general there must be provision for the following:

Direct communication between the emergency activation authority (usually air traffic control) and the airport fire station. This should be in the form of two-way radio network and a direct telephone line not passing through an intermediate switchboard. The satisfactory operation of the equipment should be continuously monitored and arrangements made for 24-hour maintenance. The fire station itself should have alarm and public address system to alert the crew to emergencies and to permit general crew briefing of the details of the emergency. It is usual to have a device that silences alarm bells while broadcasts are being made over public address systems.

Communication between rescue and fire-fighting vehicles and both air traffic control and the airport fire station. Overall control of vehicles in the operational area must, for safety reasons, be under the direction of air traffic control and entry into the active areas can be made only with its permission. Logistically, this requires vehicles to be fitted with a two-way radio communication system. Desirably this is a multichannel discrete frequency system that permits vehicles to contact air traffic control, the airport fire station, and each other. Portable radio-telephone equipment should also be supplied to the officer in charge of the rescue and fire-fighting operation at the accident/incident site to permit the officer to move away from the vehicle without losing contact with the common frequency radio link. It is essential that the vehicles and air traffic control have unbroken communications en route to the accident and on site.

Other communication and alerting facilities. At a large airport, in the event of an emergency, a number of different parties are required to be informed and to take action. They include (ICAO 1993):

- Air traffic control
- Airport rescue and fire-fighting services
- Airport security
- Airport management
- Airline station managers
- Military units (at joint use airports)
- Local fire departments
- Medical services
- Local police

Simultaneous notification of interested parties can be achieved by the use of "series" or "conference" circuits in the emergency communications system. These require a trained response of strict communications discipline to ensure prompt and uninterrupted transmission of information and messages.

12.7 Rescue and Fire-fighting Vehicles

The overall level of protection recommended for airports has been covered in Section 12.4 from the viewpoint of the number of rescue and fire-fighting vehicles and the amount of extinguishants. Clearly, as traffic grows at an airport, its needs for protection change. Therefore, the carrying and discharge capabilities of the rescue and fire-fighting vehicle fleet should be based on the needs over the short- and medium-term future during the economic life of the vehicles. A forward-looking policy in the commissioning of vehicles permits a greater degree of standardization and overall long-term savings. Once commissioned, vehicles must be protected and maintained in a manner that permits immediate availability should an emergency arise. This implies not only the routine maintenance customary with all vehicles, but also maintaining the fitted equipment such as pumps, hoses, nozzles, turrets, two-way radios, searchlights and floodlights, which are all necessary to the adequate functioning of the vehicle. Providing the day-to-day protection for the vehicles must take account of emergency operational requirements, which necessitate that access to the movement area be unobstructed, that the vehicle running distance to the runways be as short as possible, and that the crew have the widest possible view of flight activity. In airports where water rescue vehicles might be required, it is normal to locate the craft on the airport and to provide launching sites so that the boats can be brought into action with a minimum response time.

In recommendations published in 1993, ICAO abandoned the concept of two types of RFF vehicles. The concept of the *small rapid intervention vehicle* (RIV) was based on the fact that major vehicles were unable to meet the response times specifications of a desirable two minutes and a maximum of three minutes to both ends of the runway and to any other part of the movement area in optimum conditions of visibility and surface condition. Advances in chassis design have produced major rescue and fire-fighting vehicles with adequate response times. RIVs are consequently being phased out.

The RFFS vehicle is designed to carry out the principal attack on an aircraft fire. Its design and construction should make it capable of carrying a full load at high speeds in all weather over difficult terrain. The recommended characteristics of the vehicle are shown in Table 12.6. The off-road performance of the major vehicle is a primary factor in equipment choice. It should have traction and flotation characteristics that accord with the terrain in which it is to be used. The minimum ground clearance and the angles of approach and departure must be adequate to permit the vehicle to cross depressions and slopes that could be obstacles to movement. Moreover, the design of the equipment should be such that operation is simple in order both to obviate delay and confusion should an accident occur. Moreover the design should minimize crew requirements for operation. A modern major

TABLE 12.6 Suggested Minimum Characteristics for Resource and Firefighting (RFF) Vehicles

	RFF vehicles up to 4500L	RFF vehicles up to 4500L
Monitor	Optimal for categories 1 and 2. Required for categories 3 to 9	Required
Design feature	High discharge capacity	High and low discharge capacity
Range	Appropriate to longest aeroplane	Appropriate to longest aeroplane
Hand lines	Required	Required
Under truck nozzles	Optimal	Required
Bumper turret	Optimal	Optimal
Acceleration	80km/h within 25s at the nominal operating temperature	80km/h within 40s at the nominal operating temperature
Top speed	At least 105km/h	At least 100km/h
All wheel drive capability	Yes	Required
Automatic or semi-automatic transmission	Yes	Required
Single rear wheel configuration	Preferable for categories 1 and 2 Required for categories 3 to 9	Required
Minimum angle of approach and departure	30°	30°
Minimum angle of tilt (static)	30°	28°

SOURCE: ICAO

vehicle is shown in Figure 12.1; Table 12.7 shows the manual and power tools to be carried to the accident site.

Many airports have approaches and departures over water. Water rescue vehicles must be chosen with similar care. They must have as high a speed as is practicable to reach possible accident sites, and the power unit must be reliable and capable of delivering maximum power in a minimum time under low temperature and high humidity conditions. The design of the vehicles must be related to the environment in which they are to operate (e.g., mud flats, tidal areas, and ice), and they must be capable of carrying the flotation equipment necessary to effect a rescue. It is not uncommon to find that airports with approaches over water provide a significantly lower level of rescue and fire-fighting capability for the accident in water than for that provided on land. This is unreasonable since the passengers and crews of aircraft using a facility expect a similar level of protection regardless of the airport into which they are flying or the particular approach they happen to be using.

Figure 12.1a A major all terrain vehicle.

Figure 12.1b A rapid-response vehicle.

12.8 Personnel Requirements

In determining the number and deployment of personnel required for the rescue and fire-fighting services, the ICAO recommends the use of the following criteria:

1. The rescue and fire-fighting vehicles can be staffed such that they can discharge both principal and complementary extinguishing agents at

TABLE 12.7 Manual and Power Tools Needed at Accident Site

Equipment for rescue operations	Airport category			
	1–2	3–5	6–7	8–9
Adjustable wrench	1	1	1	1
Axe, rescue, large nonwedge type	—	1	1	1
Axe, rescue, small nonwedge or aircraft type	1	2	4	4
Cutter bolt, 61 cm	1	1	1	1
Crowbar, 95 cm	1	1	1	1
Crowbar, 1.65 m	—	—	1	1
Chisel, cold 2.5 cm	—	1	1	1
Flashlight	2	3	4	8
Hammer, 1.8 kg	—	1	1	1
Hook, grab or salving	1	1	1	1
Saw metal cutting or hacksaw, heavy duty, complete with spare blades	1	1	1	1
Blanket, fire resisting	1	1	1	1
Ladder, extending (of overall length appropriate to the aircraft types in use)	—	1	1	2 or 3
Rope line, 15 m length	1	1	—	—
Rope line, 30 m length	—	—	1	1
Pliers, 17.8 cm, side cutting, prs	1	1	1	1
Pliers, slip joint 25 cm	1	1	1	1
Screwdriver, assorted (set)	1	1	1	1
Snippers, tin	1	1	1	1
Chocks, 15 cm high	—	—	1	1
Chocks, 10 cm high	1	1	—	—
Powered rescue saw complete with two blades; or pneumatic chisel, plus spare cylinder chisel, and retaining spring	1	1	1	2
Harness cutting tools	1	2	3	4
Gloves, flame resisting pairs	2	3	4	8
Breathing apparatus and cylinders	—	2	3	4
Spare air cylinders	—	2	3	4
Hydraulic or pneumatic forcing tools	—	1	1	1
Medical first aid kit	1	1	1	1

SOURCE: ICAO 1993

their maximum designed capability, and can be deployed immediately with sufficient personnel to bring them into operation.

2. Any control room or communications facility related to the rescue and fire-fighting service can continue operation until alternative arrangements are made under the airport emergency plan.

All personnel must be fully trained and familiar with their equipment and duties. Should an accident occur, they must be able to perform their duties without any limitation that could come from physical disability, and they must be mentally and intellectually capable of performing the responsibilities of a rescue and fire-fighting team. It is common from an economic standpoint especially at small airports to assign full-time rescue and fire-fighting personnel to other duties. These duties include fire prevention inspections, fire guard functions, bird scaring, grass cutting and frequently at small airports apron-related functions such as baggage loading and apron-equipment handling. In the event of a disastrous accident, the airport emergency plan will provide for alerting all personnel who can act in a support role to the full-time rescue and fire-fighting crew.

12.9 The Airport Emergency Plan

International aviation agreements, as set out in Annex 14 of the ICAO, require that each airport establish an emergency plan commensurate with the level of aircraft operations and other activities at the airport (ICAO 1995). Individual countries have their own regulations such as Federal Aviation Regulations Part 139 in the United States, which to some degree parallels the international requirement.[1] The purpose of the airport emergency plan is to minimize the effects of the emergency, particularly with respect to the preservation of life and maintaining aircraft operations by establishing a coordinated program between the airport and the surrounding community. In addition to setting up an agreed and recognized structure of command during the emergency, the plan should include a section of instructions to ensure immediate response of rescue and fire-fighting services, law enforcement, medical services, and other persons and agencies both on and off the airport. A comprehensive airport emergency plan considers:

1. *Preplanning before an emergency.* Defining organizational authority, testing, and implementing the plan.

2. *Operations during an emergency.* Defining what must be done at each stage and the structure of responsibilities as the emergency progresses.

3. *After the emergency.* Handling matters not usually having the urgency of preceding events, but necessarily defining the transition of command and operations back to normality.

The emergency plan should cover an on- or an off-airport incident. Normally the airport authority will be in command for an on-airport incident. A mutual and emergency agreement between the airport and the jurisdictions of the surrounding community will define command and responsibilities in the case of an off-airport incident. The purpose of an airport emergency plan is to ensure the following:

---

[1]For details of the requirements of the FAA Emergency Plan (12), refer to the end of this chapter.

a. orderly and efficient transition from normal to emergency operations

b. delegation of airport emergency authority

c. assignment of emergency responsibilities

d. authorization of key personnel for actions contained in plan

e. coordination of efforts to cope with the emergency

f. safe continuation of aircraft operations or return to operations as soon as possible (ICAO 1993)

Even an on-airport accident might be of such a magnitude that the airport services will not on their own be capable of coping with the situation. It is therefore essential that the airport authority arrange mutual aid emergency agreements with surrounding jurisdictions defining the responsibilities of each party, for example, in the provision of ambulances, additional fire-fighting staffs and equipment, and medical personnel.

The agreements should:

- Clarify the responsibility of each involved agency

- Establish an unambiguous chain of command

- Designate communications priorities at the accident site

- Designate an emergency transportation coordinator and indicate the organizational structure of emergency transportation facilities

- Predetermine the authority and liability of all cooperating emergency personnel

- Prearrange for the use of rescue equipment from available sources

A recommended outline of an airport emergency plan is given in Table 12.8. The airport and community agencies to be included in emergency plan are: air traffic services, rescue and fire-fighting services, fire departments, police, airport authority, medical services, hospitals, aircraft operators, government authorities, communication services, airport tenants, transportation authorities, civil defense, military harbor patrol or coast guard, clergy, public information office, and other mutual aid agencies (ICAO 1993). The flow of control will be different in the case of an accident depending on whether the site is on- or off-airport. Figures 12.2 and 12.3 are the ICAO-suggested flow control charts for these respective cases.

The ICAO further recommends that an emergency plan be tested by full-scale emergency exercises using all facilities and associated agencies at intervals not exceeding one year. The exercise should be followed by a full debriefing and critique in which all involved organizations participate. In

TABLE 12.8 **Example of Contents of Emergency Plan Document**

Section 1—Emergency Telephone Numbers

This section should be limited to essential telephone numbers according to site needs, including:

Air traffic services

Rescue and fire-fighting services (departments)

Police and security

Medical services
  Hospitals
  Ambulances
  Doctors—business/residence

Aircraft operators

Government authorities

Civil defense

Others

Section 2—Aircraft Accident on the Airport

Action by air traffic services (airport control tower or airport flight information service)

Action by rescue and fire-fighting services

Action by airport authority
  Vehicle escort
  Maintenance

Action by medical services
  Hospitals
  Ambulances
  Doctors

Action by aircraft operator involved

Action by emergency operations center and mobile command post

Action by government authorities

Communications network (emergency operations center and mobile command post)

Action by agencies involved in mutual aid emergency agreements

Action by transportation authorities (land, sea, air)

Action by public information officer(s)

Action by local fire departments when structures involved

Action by all other agencies

Section 3—Aircraft Accident off the Airport

Action by air traffic services (airport control tower or airport flight information service)

Action by rescue and fire-fighting services

Action by local fire departments

Action by police and security services

Action by airport authority

TABLE 12.8 Example of Contents of Emergency Plan Document (Continued)

Action by medical services
  Hospitals
  Ambulances
  Doctors

Action by agencies involved in mutual aid emergency agreements

Action by aircraft operator involved

Action by emergency operations center and mobile command post

Action by government authorities

Communication networks (emergency operations center and mobile command post)

Transportation authorities (land, sea, air)

Action by public information officer

Action by all other agencies

Section 4—Malfunction of Aircraft in Flight (Full Emergency or Local Standby)

Action by air traffic services (airport control tower or flight information service)

Action by airport rescue and fire-fighting services

Action by police and security services

Action by airport authority

Action by medical services
  Hospitals
  Ambulances
  Doctors

Action by aircraft operator involved

Action by emergency operations center and mobile command post

Action by all other agencies

Section 5—Structural Fires

Action by air traffic services (airport control tower or flight information service)

Action by rescue and fire-fighting services (local fire departments)

Action by police and security services

Action by airport authority

Evacuation of structure

Action by medical services
  Hospitals
  Ambulances
  Doctors

Action by emergency operations center and mobile command post

Action by public information officer

Action by all other agencies

Section 6—Sabotage Including Bomb Threat (Aircraft or Structure)

Action by air traffic services (airport control tower or airport flight information service)

TABLE 12.8 Example of Contents of Emergency Plan Document (Continued)

Action by emergency operations center and mobile command post

Action by police and security services

Action by airport authority

Action by rescue and fire-fighting services

Action by medical services
  Hospitals
  Ambulances
  Doctors

Action by aircraft operator involved

Action by government authorities

Isolated aircraft parking position

Evacuation

Searches

Handling and identification of luggage and cargo on board aircraft

Handling and disposal of suspected bomb

Action by public information officer

Action by all other agencies

Section 7—Unlawful Seizure of Aircraft

Action by air traffic services (airport control tower or airport flight information services)

Action by rescue and fire-fighting services

Action by police and security services

Action by airport authority

Action by medical services
  Hospitals
  Ambulances
  Doctors

Action by aircraft operator involved

Action by government authorities

Action by emergency operations center and mobile command post

Isolated aircraft parking position

Action by public information officer

Action by all other agencies

Section 8—Incident on the Airport

An incident on the airport may require any or all of the action detailed in Section 2, "Aircraft accident on the airport". Examples of incidents the airport authority should consider fuel spills at the ramp, passenger loading bridge, and fuel storage area; dangerous goods occurrences at freight handling areas; collapse of structures; and vehicle/aircraft collisions.

Section 9—Persons of Authority—Site Roles

To include, but not limited to, the following according to local requirements:

TABLE 12.8 Example of Contents of Emergency Plan Document (Continued)

Airport chief fire officer

Airport authority

Police and security—officer in charge

Medical coordinator

Off-airport

Local chief fire officer

Government authority

Police and security—officer in charge

The on-scene commander will be designated as required from within the prearranged mutual aid emergency agreement.

Experience indicates that confusion in identifying command personnel in accident situations is a serious problem. To alleviate this problem it is suggested that distinctive colored vests with reflective lettering and hard hats be issued to command personnel for easy identification. The following colors are recommended:

Red	Chief fire officer
Blue	Police chief
White (red lettering)	Medical coordinator
International orange	Airport administration
Lime green	Transportation officer
Dark brown	Forensic chief

An on-scene commander should be appointed as the person in command of the overall emergency operation. The on-scene commander should be easily identifiable and can be one of the persons indicated above or any other person from the responding agencies.

SOURCE: ICAO 1993

fact, full emergency drills of this scale are not as common as the ICAO recommends. For example, in many countries there are regulations that all airports must have a written airport disaster plan, but frequently the regulation does not require that the plan be tested. The first full-scale U.S. air carrier airport that used an air transport aircraft to increase the reality of its disaster drill was Oakland International Airport in May 1971. Since then many full-scale drills have been held that have emphasized the three Cs of disaster planning: command, communication and coordination (Stefanki 1977).

Command

Following through an on-airport accident, it can be seen that the captain is in command of the aircraft while in the air and immediately after the crash. Command changes as the airport fire trucks arrive and the flight crew evacuates passengers. In the U.S. the highest ranking fire officer assumes command on the disaster site (or the designated airport operations officer,

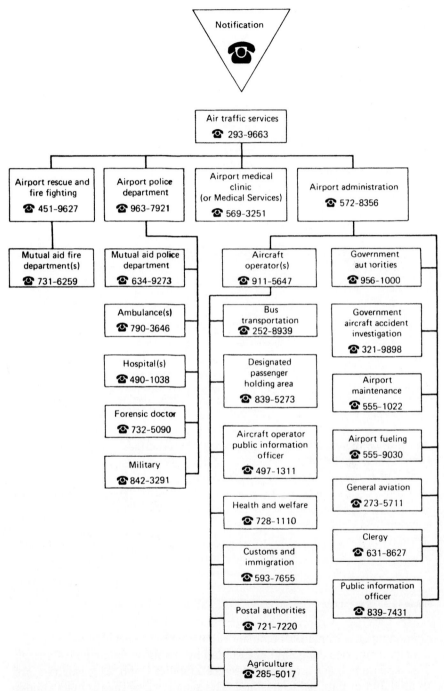

Figure 12.2 Flow control chart—aircraft accident on airport. *(ICAO 1993)*

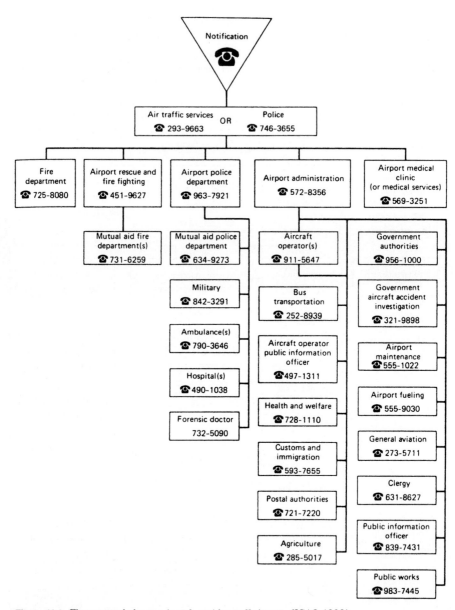

Figure 12.3 Flow control chart—aircraft accident off airport. *(ICAO 1993)*

depending on the chain of command set up in the airport emergency plan).
In some countries command is assumed by the most senior police officer.
This situation holds until all fires are stabilized and all casualties are
treated and dispatched to hospital. The accident site then remains under
the control of the airport operations officer until the arrival of the accident

investigation team; in the United States this is the National Transportation Safety Board. The "persons in charge" of the various agencies should wear distinctive colored vests as indicated in the emergency plan in Table 12.8.

Until the arrival of a medical officer, a paramedic should take control of the removal of casualties to the triage[2] area. Ideally the medical coordination designated under the emergency plan will be a trauma-trained doctor. Passengers are evacuated initially to the triage area, which is at least 300 feet (100 m) upwind of the crash. Here casualties are examined and tagged according to the severity of injury; the tags represent the priority for transportation and care:

Priority I	Immediate care	Red	Rabbit symbol
Priority II	Delayed care	Yellow	Turtle symbol
Priority III	Minor care	Green	Ambulance symbol
Priority 0	Deceased	Black	Cross symbol

The layout of an accident site is shown in Figure 12.4.

Communications

Communications from the accident site are achieved through the command post at the accident site. At well-equipped airports, this can be a field trailer, specially equipped with radio, radiotelephone, loudspeakers, elevated platform, and floodlighting. Experience has shown that there is usually little problem with communication between the fire and police departments and the airport authorities; the main difficulties have been found to lie among the medical coordination, hospitals, and ambulances (Stefanki 1977). Clear-channel radio networks, such as the Los Angeles County hospital administrative radio system, have been found particularly effective.

Coordination

Coordination of the many agencies and individuals to be involved in the case of an emergency or an accident requires planning, patience, and teamwork. It is essential that everyone involved, on- and off-airport, know what his or her responsibilities are when an emergency is declared. San Francisco International Airport, which has carried out many simulated disaster drills, has a separate building designated as a disaster building. In Jeddah it is part of the Haj terminal. Converted aircraft cargo trailers are used to transport disaster supplies and have been fitted out as a mobile command post and communications trailers.

---

[2] Triage is the sorting and classification of casualties to determine the order of priority for treatment and transportation.

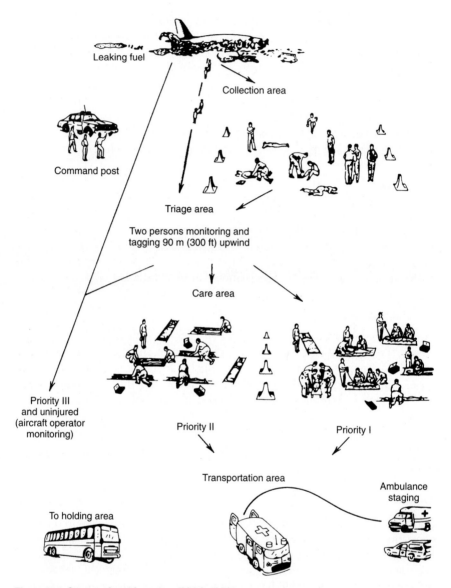

Leaking fuel

Collection area

Command post

Triage area

Two persons monitoring and
tagging 90 m (300 ft) upwind

Care area

Priority III
and uninjured
(aircraft operator
monitoring)

Priority II

Priority I

Transportation area

Ambulance
staging

To holding area

Figure 12.4 Layout of accident site. *(ICAO 1993)*

12.10 Aircraft Fire-fighting and Rescue Procedures

The principal objective of a rescue and fire-fighting service is to save lives in
the event of an aircraft accident (ICAO 1993). Because of the very large
amounts of inflammable material present at an aircraft accident, the likelihood of a fire is high. It is important to bear in mind that it is a characteris-

tic of aircraft fires that they reach lethal intensity in a very short time from outbreak. Coupled with the fact that the occupants of a damaged aircraft have suffered the shock of a high-speed impact this means that effective rescue from a survivable crash is possible only if:

1. The fire-fighting and rescue team is well trained and is familiar with equipment and procedures.
2. The equipment is effective for the purpose.
3. The accident site is reached in time

The details of fire-fighting are complex and beyond the scope of this text. However, a basic understanding of the techniques of rescue should be understood. First, rescue must be seen to include the protection of the escape routes from the aircraft. Although the rescue of the occupants is the primary objective, the overall requirement is to create conditions where survival is possible and rescue might be carried out. The prompt arrival of the RFFS vehicle at the site of a crash is most vital to effecting the rescue of victims. After four minutes of exposure to fire aircraft windows will melt and within one further minute, the cabin temperature will be sufficiently high to ensure no survivors. Therefore, it is often necessary to concentrate first on fire suppression prior to any rescue. Failure to suppress a fire or blanket a fuel spill might endanger many lives. Second, the rescue of those unable to escape without aid might be a time-consuming process. During this period, additional fire-security measures will be necessary and possibly the delivery of ventilation air into the fuselage.

Seen in this context, the RFFS vehicle will be used to suppress any fires and put down a precautionary blanket over the critical areas especially in fuel-wetted areas. Additionally, the first vehicle provides protection against ingress of fire into the fuselage when the aircraft doors and windows are opened. With the arrival of further vehicles, there is usually sufficient fire suppression capacity to permit the crew of the first vehicle to perform a number of functions:

- To form rescue teams to enter the aircraft to assist aircraft occupants who need help to evacuate
- To provide fire-fighting equipment within the aircraft for internal fires
- To provide lighting and ventilation within the fuselage

An accident of disaster proportions is something that few rescue and fire-fighting teams actually face. Therefore, adequate training and proper planning are essential so that when suddenly confronted with the reality of a accident, a crew is capable of coping with a very fast-moving situation that can rapidly develop from an accident into a disaster. Two examples of crashes with B727 in April 1976 serve to outline the problems of unpreparedness that result from insufficient training.

On April 5, 1976, an Alaska Airlines Boeing 727 crashed at Ketchikan International Airport. There was a full-time crash fire and rescue service under the control of the airport manager. However, it had no formal training in fire-fighting techniques or in operating the two crash trucks. The airport manager had neither the training to direct fire-fighting activities nor the experience and personnel to do it. Fire-fighting activities by the airport crew at the accident site were therefore described as minimal. The city fire department arrived at the site, but ran out of foam and hose in 20 minutes. The airport had not been inspected for almost 12 months by the FAA certification program.

On April 27, 1976, an American Airlines Boeing 727 crashed on takeoff from an aborted landing at St. Thomas, Virgin Islands. The crash fire and rescue crew responded immediately, but did not take all available emergency equipment; airpacks and some proximity suits were left behind. The CFR crew chief was the Port Authority duty officer. Consequently, no one was left to take overall charge of the operation. Flight operations were resumed shortly after the crash, but with no runway inspection and no crash fire and rescue cover for these operations. Subsequently, it was found that a direct line from the airport to outside emergency service had, unknown to the FAA, been removed, leaving only a commercial line.

In comparison with these two accidents, a potential disaster was averted by a well-trained and organized force when a DC10 Continental Airlines aircraft crashed after an aborted takeoff at Los Angeles International Airport on March 1, 1978, with 183 passengers and 14 crew members. The left main gear collapsed as the aircraft entered consolidated ground at the runway end at about 65 mph (105 kmh). A fire, fed by ruptured wing tanks, was attacked by firemen from a satellite station only 90 seconds after the crash. The vehicle from the main airport fire station 2.5 miles (4 m) away arrived after 4 minutes. Six exits were opened, and evacuation took 5 minutes. More than 100 persons were still on board when the first foam tender arrived. Forty-three persons were injured, but only two died, these being individuals who against the advice of the cabin attendant used the left overwing exit and jumped down into the fire. The rapid intervention of the airport rescue and fire-fighting services saved many lives as well as instruments and fittings from the aircraft interior worth more than (US) $10 million.

12.11 Foaming of Runways

Protein foam (Performance level A) has been applied to the runway in a number of emergency landings in the hope that the landing could be carried out in a safer manner. Fluoroprotein foam and aqueous film forming foam are not suitable for such an operation due to their short drainage time. Foaming has usually been carried out for wheels-up landings or aircraft

with defective nose gears. In some cases, foaming operations were successful; in others the purpose was not accomplished, mainly due to the aircraft
missing or overrunning the blanket.

Theoretically, there are four principal benefits from foaming that bear
some examination (ICAO 1993):

1. *Reduction in aircraft damage.* There is no firm evidence that the minimum damage attained in a number of well-executed landings on foam-
   coated runways would have been greater without foaming. Controlled
   emergency landings have also been carried out with minimum damage
   on dry runways. ICAO indicates that too many other variables are involved to draw valid conclusions from the small sample involved.

2. *Reduction in deceleration forces.* In the case of a wheels-up emergency landing, the reduced runway friction can be an advantage. Where
   the main gear is down, braking performance is only slightly worse than
   on a wet runway.

3. *Reduction of spark hazard.* Under the conditions of an emergency
   landing, aluminum alloys do not appear to be capable of generating
   sparks capable of igniting aircraft fuel. Magnesium alloys, stainless steel,
   and other aircraft steels can produce such sparks, and it is found that in
   the majority of tests a properly laid foam blanket did suppress sparks.
   Titanium friction sparks cannot be effectively suppressed.

4. *Reduction of fuel spill hazard.* Vapors above the foam blanket are un-
   affected by the application. However, fuel released over the blanket will
   fall through the foam and will spread under it, with a reduction of burn-
   ing area. A more intense fire might occur locally.

Other factors must be considered before deciding to foam a runway:

- Whether there is sufficient time to lay a foam blanket (this usually takes
  about an hour)
- The reliability on the landing techniques to be used: wind, visibility, pilot
  skill, visual and radio aids, aircraft conditions
- Adequacy of foam-making equipment
- Problem of cleanup to reopen operations
- Ambient conditions such as rain, wind, or snow
- Runway condition and slope

In cases where it has been decided to proceed with foaming the runway,
the following criteria must be met:

1. The pilot must be fully informed on how the operation is to be carried
   out and what protection is to be provided.

2. The vehicles providing the foam must not decrease the minimum level of airport protection required.

3. The foam used should be additional to the minimum level of protection plus replenishment needs.

4. The positioning of the blanket should recognize that with a gear-up landing the aircraft contacts the runway much farther from the threshold than usual.

5. In reduced visibility, the pilot must be given a reference point to indicate where the foam starts.

6. Unnecessary personnel must be cleared from the area.

7. Prior to use, the foam should be aged about 10 minutes to permit drainage and surface wetting.

8. The foam blanket should be continuous.

9. The blanket depth is about 2 inches (5 cm).

10. After laying the foam, all personnel and equipment are removed to emergency standby positions.

12.12 Removal of Disabled Aircraft

Annex 14 states that participating countries to the international convention are required to establish a plan for removing disabled aircraft on or adjacent to the movement area and recommends the designation of a coordinator to implement the plan (ICAO 1995). The reason for an aircraft's immobility can range from a relatively minor incident, such as a blown tire or frozen brake, to a major accident involving disintegration of the aircraft itself.[3] Whereas in the early days of aviation the task of removing a disabled aircraft was relatively simple, with large modern aircraft the job has become technically difficult and expensive, requiring special equipment and organization. Obviously if the aircraft is immobilized on a part of the airport where it interferes with operation, it is in the interest of the airport authority, other operators, and the traveling public that removal should be carried out rapidly. However, where a large aircraft is concerned, the task is complex and potentially dangerous, and it will not therefore always be possible to clear the airport as rapidly as might be desired. With large expensive aircraft, care must be taken to avoid further damage during the removal procedure itself. The responsibility for controlling the removal of an aircraft lies with the registered owner. In small aircraft is it usually possible, if nec-

---

[3]Except as specified in Annex 13, wreckage of aircraft after an accident must be left undisturbed until the arrival of the investigator in charge of the accident investigation (ICAO 1996).

essary, for the airport authority to undertake this with the agreement of the owner. With large aircraft, however, generally the airport does not have the expertise or experience necessary for its speedy and safe removal without secondary damage. This will require specialized teams, often involving not only the airline but the aircraft manufacturer and the insurer. For each type of aircraft, special knowledge of safe jacking techniques and other lifting procedures are necessary (Boeing Aircraft Company 1973).

The main purpose of the disabled aircraft removal plan is to ensure the prompt availability of appropriate recovery equipment and expertise at any incident site. The plan should be based on the characteristics of aircraft types that can normally be expected to use the airport. Such a plan should include:

- Itemization of equipment and personnel necessary together with location and time required to get to the airport

- Necessary access routes for heavy equipment

- Grid maps of the airport to locate accident site, access gate, and so on

- Security arrangements

- Arrangements for accepting specialized recovery equipment from airports in the technical pool

- Manufacturers' data on aircraft recovery for aircraft normally using airport

- Defueling arrangements with resident oil companies

- Logistics of supplying labor and special clothing

- Arrangements for expediting the arrival of the investigator in charge

In addition to heavy lifting equipment and general recovery equipment, which can be covered by advanced agreements with local companies and organizations, specialized lifting equipment such as pneumatic lifting bags and jacks will be necessary for some problems with heavy aircraft (Table 12.9). IATA has found it necessary to make such equipment available on a worldwide basis through the International Airlines Technical Pool (IATP). Around the world in 11 locations,[4] lifting kits are available in case of need. These kits, consisting of pneumatic lifting bags, large extension hydraulic jacks, and tethering equipment, are stored on pallets and are available for immediate shipment accompanied by skilled personnel to any accident location. They can be used by pool-member airlines and nonmembers on a fee basis. It is estimated that the kits can be at any accident site within 10 hours

---

[4]Sydney, Rio de Janeiro, Paris, Bombay, Tokyo, London, Johannesburg, Chicago, Honolulu, Los Angeles, and New York.

TABLE 12.9 Typical Methods of Recovery of Heavy Aircraft for Various Conditions of Damage

Conditions	Typical methods of aircraft recovery
Collapsed nose landing gear	Jacking and use of pneumatic lifting bags; hoisting with cranes and the use of specially designed slings; or by pulling down on tail tie-down fitting.
Collapsed or retracted main landing gear, but nose landing gear intact and extended	Jacks, pneumatic lifting bags, or cranes.
Collapsed main landing gear, one side only	Jacks, pneumatic lifting bags, or cranes.
Collapse of all landing gears	Jacks, pneumatic lifting bags, or cranes.
One or more main landing gear off pavement, no aircraft damage	Assuming that the aircraft has the landing gear bogged down in soft soil or mud, extra towing or winching equipment or use of pneumatic lifting bags will usually suffice for this type of recovery. It may be necessary to construct a temporary ramp from timbers, matting, etc.
Nose landing gear failure and one side of main landing gear failure	Jacks, pneumatic lifting bags, or cranes.
Tire failure and/or damaged wheels	Jacks and parts replacement

SOURCE: ICAO 1983

Figure 12.5 Lifting a damaged aircraft with airbags.

and at most locations within 5 hours. The complexity of the recovery operation is exemplified by Figure 12.5, which shows the lifting of a B707 aircraft by heavy crane equipment.

12.13 Summary

Disaster-scale accidents at airports seldom occur. Many airport operators will spend their entire careers at airports that, although having many emergencies, will have no accidents where there is a large loss of life. Safety in aviation, however, requires a continuous alertness to the possibility that a serious accident will occur unexpectedly at any time and will require a rapid and expert response from the airport rescue and fire-fighting services and aircraft recovery. The ability to respond promptly can be achieved only through preplanning and training.

Much of what has been covered in this is applicable as much to the United States as to other countries. However, U.S. operators must comply with the Airport Certification Program and should seek to follow Department of Transportation Order 1900.4, Emergency Planning Guidance for the Use of Transportation Industry. The FAA has therefore prepared literature that covers for U.S. airports (Airport Emergency Plan 1989):

- Scope and arrangement of plans
- Arrangements for mutual assistance and coordination
- Functions including management responsibilities, personnel assignments, and training
- An overall checklist for plan content
- Guidance summaries concerning aircraft accidents and incidents, bomb incidents, structural fires, natural disasters, crowd control and measures to prevent unlawful interference with operations, radiological incidents, medical services, emergency alarm systems, and control tower functions

The structure of U.S. requirements can best be exemplified by the FAA checklist shown in Table 12.10.

References

Aerodromes, Annex 14 to the International Convention on Civil Aviation, Vol. 1, 2nd edition. July 1995.

Ansart, F. and Andre Viel. 1981. *The S2000—A New Heavy Duty Crash Vehicle for Large Aerodromes.* ICAO Bulletin, Montreal: International Civil Aviation Organization (ICAO). October.

Aviation Safety Hearings before the U.S. House of Representatives, 95th Congress. July 1977.

Boeing Airplane Company. 1973. *Airplane Recovery 707, 727, 737.* As revised. Seattle.

Civil Aviation Authority. 1972. *Aerodrome Fire and Rescue Services.* Cheltenham, England. May.

Federal Aviation Administration (FAA). 1970. *Advisory Circular AC-150-5200-14*. Washington, DC: Department of Transportation. August 9.

FAA. 1970. *Removal of Disabled Aircraft, AC-150/5200-13*. Washington, DC: Department of Transportation. August.

FAA. 1989. *Airport Emergency Plan, AC-150/5200-31*. Washington, DC: Department of Transportation. January.

International Civil Aviation Organization (ICAO). 1980. *Airport Services Manual: Part 7 Airport Emergency Planning*. Montreal.

ICAO. 1986. *Removal of Disabled Aircraft, Aerodrome Services Manual, Part 5, 2nd edition*. Montreal.

ICAO. 1988. *Aircraft Accident Enquiry, Annex 13 to the International Convention on Civil Aviation, 7th edition*. Montreal.

ICAO. 1993. *Airport Services Manual: Part 1 Rescue and Fire Fighting, 2nd edition*. Montreal.

Stefanki, John X. 1977. Dealing with Disaster. *Airports International*. April/May.

TABLE 12.10 General Checklist for Preparing an Emergency Plan

Functional area	General requirements/ actions	Responsibility/ designations for performance and governmental agency/ element participation	Established programs/ reference material	Airport emergency assignments
Aircraft incidents and accidents	Establish notification procedures; plan to respond to and to cope with occurrences; establish alert and standby procedures for airport fire-fighting, ambulance, and rescue services; provide care for any injured passengers; provide services for uninjured passengers; and activate mutual assistance plan if needed.	FAA ATC/FSS personnel monitor communications, relay messages, and establish points of contact; airport management to establish means and procedures for overall notification and receipt of messages and for response to emergencies. Units to be notified include airport management, airport operations, fire department, emergency medical service, the airport security office, the airline, and the NTSB.	FAA handbooks 7110.8B *(Terminal Air Traffic Control)*, 7110.9B *(En Route Air Traffic Control)* 7110.10A *(Flight Services)*, 7210.3 *(Facility Management)*, and 8020.4A *(Aircraft Accident Notification Procedures and Responsibilities)*. AC 150/5200-15 *(Availability of the International Fires Service Training Association's Aircraft Fire Protection and Rescue Procedures Manual)* also procedures established for emergency alarm system	Entries in this column to be made by person in charge of emergency plan.

TABLE 12.10 General Checklist for Preparing an Emergency Plan (Continued)

Functional area	General requirements/ actions	Responsibility/ designations for performance and governmental agency/ element participation	Established programs/ reference material	Airport emergency assignments
Bomb incident procedures including a designated parking area for aircraft suspected of having a bomb aboard	Carry out procedures to ensure that airport authorities, airport security offices, airline, etc., are notified; and park aircraft in isolated designated area. *Note: The plans for parking such aircraft are expected to be treated as privileged information*	ATC; airport security office; airport management; the airline; and public affairs.	The program administered by the Office of Air Transportation Security, Headquarters, FAA, SE-1.	
Structural fires	Establish procedures for notification of the fire department; evacuate structures in the vicinity of the fire; have employee designated to report to structure, facility, etc., to cut off power supply source as necessary for safety during fire-fighting operations. In addition, make sure that emergency crews can gain access to locked areas such as electrical vaults.	The fire department; airport security office; the emergency electrician; medical services; and airport management.		

| Natural disasters | Set up plans for protecting the public during storms, such as hurricanes, floods or tidal waves; curbing operations as necessary during storms; and the utilization of fallout shelters, storm shelters, etc. | Combined efforts of airport disaster control organization | The Natural Disaster Warning System established by the National Oceanic and Atmospheric Administration. |
| Crowd control and measures to prevent unlawful interference with operations. | Make arrangements with law enforcement or other authorities to get intelligence reports; carry out procedures for crowd control, and take security measures to preclude sabotage; control of motor traffic, gates, access areas where pedestrians may enter the airport and block entry through underground service ducts, sewers, or tunnels. | Airport management; airport security office; local and state police forces. | |

TABLE 12.10 General Checklist for Preparing an Emergency Plan (Continued)

Functional area	General requirements/ actions	Responsibility/ designations for performance and governmental agency/ element participation	Established programs/ reference material	Airport emergency assignments
Radiological incidents or nuclear attack.	These requirements are in two categories of radiological incidents connected with the air transportation of radioactive material and nuclear attack; Under the first category, carry out notification procedures and establish security measures around the area by the use of guards, ropes, barricades, etc.	ATC; airport security office; various state organizations capable of coping with nondefense types of nuclear incidents; and the tripartite agreement between AEC-DOD-HEW to cover assistance following radiological emergencies.	FAA Handbooks 7110.8B *(Terminal Air Traffic Control)*, 7110.9B *(En Route Air Traffic Control)*, 7110.10A *(Flight Services)*, and 7210.3 *(Facility Management)*. AEC *Radiological Assistance Program Handbook; Information summary on Interagency Radiological Assistance Plan; Monograph on AEC Radiological Assistance Program; Monograph on US Interagency Radiological Assistance Plan; Radiological Emergency Procedures* for the nonspecialist (available from AEC, Division of Operational Safety, Washington, DC 20545).	

Under the second category, follow the procedures for operating under the Disaster Control Organization and establishing contact with DOD organization for military support of civil defense functions	The airport emergency organization and DOD military forces.	The training programs of the Office of Civil Defense (OCD) for radiological monitors and for fallout shelter management.
Medical services	Provide for using facilities located on the airport and/or arrange to get ambulances, services, and other mutual assistance from hospitals, clinics, etc. that are located off the airport.	The designated airport medical officer.	The local and state health department's emergency medical service programs.
Removal of disabled aircraft	Establish agreements between airport management and the airlines indicating for the quick removal of aircraft with tire, wheel, or gear failures. This involves removing the aircraft from surface maneuvering areas of the airport and performing the maintenance elsewhere.	The NTSB/FAA representatives for release of the aircraft to owner/operator for	AC 150/5200-13, *Removal of Disabled Aircraft*; ATA Aircraft Recovery Committee; and

SOURCE: Federal Aviation Administration

Airport Access

13.1 Access as Part of the Airport System

Up to a few years ago it was customary for airport operators to consider that the problem of getting to the airport was chiefly the domain of the urban or regional transportation planner and of the surface transport operators. But congestion and difficulties in accessing airports have, as will be seen, very strong implications on their operations. Therefore, the airport administrator has an unavoidable vital interest in the whole area of access and accessibility, perhaps one of the most difficult problem areas to face airport management. An administration might have to watch severe deterioration in its own operations due to problems outside the limits of the airport itself, conditions over which the airport operator appears to have less and less direct control.

Figure 13.1 is a conceptualized diagram indicating how potential outbound passengers and freight traffic through an airport will be modified by capacity constraints at the various points in the system; a similar chain operates in reverse for inbound traffic. It is important to realize that should any of the potential constraint areas become choke points, throughput is reduced. Lack of access capacity is far from being a hypothetical occurrence. Several of the world's major airports already face severe capacity constraints in the access phase of throughput. Using direct traffic estimation methods, urban transport planners can show that the most severe access problems can occur at airports set in the environment of large metropolitan areas, if these airports depend largely on road access. In fact, three of the world's largest airports, Los Angeles, Chicago O'Hare, and London Heathrow, have for some time displayed severe symptoms of access congestion. In Los Angeles, in the 1970s

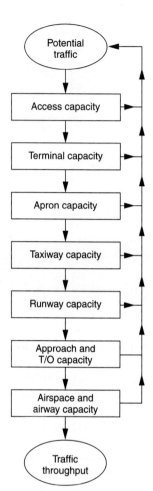

Figure 13.1 Sequential capacity constraints on outbound airport throughput.

before the double decking of the landside circulation road, the environmental capacity of the airport was declared from a determination of the landside access capacity. In 1980 the Department of Airports proposed that the total number of aircraft operations should be determined as follows:

$$\text{MTAO} = \frac{365 \times 0.90 \times \text{RCAP} \times \text{PPV}}{\text{ASOP} \times \text{CHTF} \times \text{ANPO}} \tag{13.1}$$

which is a reformulation of

$$\text{AEDT} = \frac{\text{MTAO}}{365} \times \text{ASOP} \times \frac{\text{ANPO}}{\text{PPV}} \tag{13.2}$$

and

$$CHTF \times AEDT = RCAP \times 0.90$$

where

AEDT = average number of vehicles entering the central terminal area in previous 6 months

ANPO = average number of annual passengers per actual air operation in the previous 6 months

ASOP = actual number of air operations divided by the proposed number for the prior 6 months

CHTF = critical hour traffic factor: the three-hundredth highest hour of vehicular traffic during previous 12 months divided by the average number of vehicles entering central terminal area daily

PPV = average number of air passengers per inbound vehicle

RCAP = entering central terminal area roadway capacity in terms of vehicles per hour

0.90 = constant.

This procedure was an attempt to ensure that the scheduled airside activities would not impose unacceptable loads on a landside access system. It is likely that similar procedures will have to be adopted by other airports in the future.

Chicago O'Hare, which generates in the mid 1990s just under half-a-million person-trips per day, is principally served by the Kennedy Expressway. The Kennedy Expressway is one of the world's busiest freeways and is frequently severely congested. As a consequence, as long ago as 1981, O'Hare made plans to connect to downtown Chicago with an urban transit line planned to accommodate more than 24,000 passengers daily. This has since been built but attracts fewer passengers than this figure. Heathrow Airport, has a severe access problem even though slightly less than two-thirds of its passengers arrive by public transport or taxi. Difficulties stem largely from the configuration of three of its terminals, which are crowded into a small "island" site between the two main runways and which can be accessed by road only by a tunnel beneath one of the runways. To limit the level of traffic into the terminal area, passengers are encouraged by a pricing mechanism to use remote long-term parking. There are also severe restrictions on taxis in the central terminal area.

At nearly all airports, much of the access system in terms of the highways, the urban bus and rail systems, and taxis, is outside the control of the airport administrator, both financially and operationally. However, the interface of the access system is very much within the administrator's control. A poor operational interface can discourage travelers from using what would otherwise be an excellent and (from the airport's viewpoint) desir-

able system. On the other hand, by imposing selective operational constraints on a mode that is becoming less desirable (e.g., automobiles in a
congested road access situation), significant if not large changes in the traveler's modal choice can be brought about.

13.2 Access Users and Modal Choice

Airport passengers often, but not always, constitute the majority of persons
entering or leaving the airport. Excluding individuals making trips as suppliers to the airport, the airport population can be divided into three categories:

- Passengers—originating, destined, transit, and transfer
- Employees—airline, airport, government, concessionaires, and such
- Visitors—greeters, senders, sightseers, and such

All but the transit and transfer passengers make use of the access system. There is no single figure for the division of the airport population between these categories. The split varies considerably among airports and
depends on such factors as the size of the airport; the time of day, week, and
year; the airport's geographical location; and the type of air service supplied. Large airports with large based-airline fleets have extensive maintenance and engineering facilities. London Heathrow and Chicago O'Hare
both had more than 50,000 employees on site in 1995. Most of these were
airline workers and staff. Airports serving international rather than domestic operations tend to attract large numbers of senders and greeters. Similar
numbers of visitors are not found, for example, at many U.S. airports that
serve mostly domestic and business flights although there are exceptions
such as Kennedy, which provides service to the Caribbean ethnic flights.
Table 13.1 lists the spread of breakdowns of airport "population" that have
been found by a number of surveys. It can be seen that the range of values
is very large.

Any reasoned consideration of airport access must rest on a clear understanding of the distribution of the true origins and destinations of the air
traveler. Over the last 25 years, many superficial solutions have been proposed for the access problem, many of which have involved the use of some
dedicated high-speed tracked technology to link the airport with the city
center in an effort to reduce the demonstrated dominance of the automobile. These proposals fail to recognize that the reason that the automobile is
widely used to access the airport is that, except for a handful of very large
metropolitan areas with dominant central business districts, air travelers do
not for the most part begin or end their journeys in city centers. Table 13.2
lists the percentage of passengers originating from or destined to arrive in

TABLE 13.1 Proportion of Passengers, Workers, Visitors, and Senders/Greeters at Selected Airports

Airport	Passengers	Senders and Greeters	Workers	Visitors
Frankfurt	0.60	0.06	0.29	0.05
Vienna	0.51	0.22	0.19	0.08
Paris	0.62	0.07	0.23	0.08
Amsterdam	0.41	0.23	0.28	0.08
Toronto	0.38	0.54	0.08	Not included
Atlanta	0.39	0.26	0.09	0.26
Los Angeles	0.42	0.46	0.12	Not included
New York JFK	0.37	0.48	0.15	Not included
Bogota	0.21	0.42	0.36	Negligible
Mexico City	0.35	0.52	0.13	Negligible
Curacao	0.25	0.64	0.08	0.03
Tokyo	0.66	0.11	0.17	0.06
Singapore	0.23	0.61	0.16	Negligible
Melbourne	0.46	0.32	0.14	0.08
US Airports	0.33–0.56	—	0.11–0.16	0.31–0.42 (includes senders (and greeters)

SOURCE: Institute of Air Transport Survey, 1979.

TABLE 13.2 Percentage of Airline Passengers with Origin or Destination in the Central Business District

	Distance from Airport to CBD (miles)	% Passengers Oriented to CBC
United States		
Los Angeles (LAX)	11.0	15
New York (JFK)	11.5	47
Atlanta	7.5	24
San Francisco	12.0	25
Miami	10.0	35
Washington (DCA)	2.0	25
Boston	2.5	14
Philadelphia	6.3	14
Denver	7.5	30
United Kingdom		
Liverpool	6	37
Manchester	8	11
Glasgow	6	28
Birmingham	7	25
Newcastle	6	17
London (Heathrow)	15	29
London (Gatwick)	24	21

SOURCE: FAA and CAA

the central business district (CBD) of the city for a selection of airports in the United States and the United Kingdom. The decision therefore to concentrate effort on providing for CBD-oriented access could force the traveler into the city center, resulting in a circuitous, expensive, and time-consuming access journey. Moreover, as will be seen later, the traveler is very likely to end up accessing on the already congested radial urban highways at peak times, thus further complicating the urban transport problem and lengthening travel time. The convenience of the auto mode is the principal reason for its popularity in those countries where there is a high level of car ownership or, more important, car availability. Table 13.3 indicates for a few selected U.S. and European airports the general popularity of the car or taxi even in the presence of public transport (Coogan 1995).

In countries where the intercity public-transport system (i.e., train and bus) is weak, highway-oriented private modes become essential for those accessing the airport. With increased emphasis on public transport in urban areas since the 1970s, access patterns are however changing in the U.S. At Boston Logan Airport, for example, the combined use of taxi and public modes has grown from 16 percent in 1970 to 46 percent by 1990. Some European airports, such as Zurich, Frankfurt, Brussels, and London Gatwick, are connected to the intercity rail network. In theory all such rail links pro-

TABLE 13.3   Access by Car or Taxi for Selected Airports

	% Car or Taxi	Public Transport Available
Zurich	59	Yes (rail and bus)
Munich	59	Yes (rail and bus)
London Gatwick	62	Yes (rail and bus)
London Heathrow	64	Yes (rail and bus)
Frankfurt	65	Yes (rail and bus)
New York JFK	66	Yes (bus)
New York La Guardia	67	Yes (bus)
London Stansted	70	Yes (rail and bus)
Amsterdam	70	Yes (rail and bus)
Copenhagen	72	Yes (bus)
Helsinki	73	Yes (bus)
Chicago O'Hare	73	Yes (rail and bus)
Paris CDG	74	Yes (rail and bus)
Brussels	74	Yes (rail and bus)
San Francisco	74	Yes (bus)
Boston	75	Yes (rail and bus)
Washington National	75	Yes (rail and bus)
Paris Orly	78	Yes (rail and bus)
Philadelphia	79	Yes (rail and bus)
Miami	79	Yes (bus)
Los Angeles	80	Yes (bus)
Dusseldorf	84	Yes (rail and bus)

SOURCE: Coogan 1995

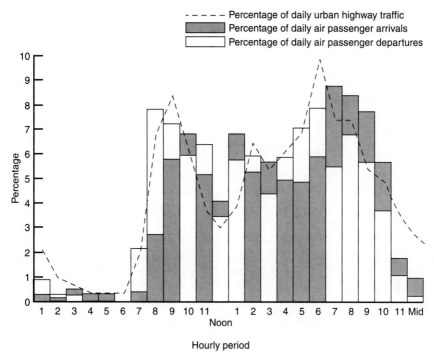

Figure 13.2 Daily highway and air passenger traffic patterns in the San Francisco Bay area.

vide direct connections to all parts of the country, and surface rail should be an attractive alternative to the car as the access mode. In reality, the form of the network can materially affect the efficacy of rail as a viable alternative. At Paris Orly, connections are not good, and rail usage is consequently low. The Gatwick-Victoria link has proved successful, not necessarily for reasons of conventional network connectivity but more because Victoria Station in central London serves as a link to the city-wide Underground system and connections between main-line rail stations and to the West End are easily made by taxi. Other than Gatwick, Zurich Kloten is perhaps one of the first airports to be truly connected to an intercity rail network; the Swiss authorities originally hoped that eventually between 50 and 60 percent of access trips would be made by rail. In practice, the figure has fallen far short of this goal attaining only 34 percent by 1990.

A major problem that must be considered in all questions of access is the degree of coincidence between traffic peaks for urban transport, both road and rail, and airport passenger traffic. This is exemplified in Figure 13.2, which shows the combined arrival and departure passengers at San Francisco Airport and the vehicular traffic variations in the Bay area. As explained in Chapter 2, the variations in airport peaks are related to the diurnal rhythms of urban life in that airline schedules are often set to coincide with peak travel generated by the eight-hour working day (see Chapter 2). Consequently, the airport traveler competes with the urban dweller for road space and transit capacity during peak hour periods. For the passenger using the automobile, taxi, and bus this means delay through congestion; for those using urban and intercity rail systems, it means possible difficulties in finding seats and handling baggage in crowded facilities.

13.3 Access Interaction with Passenger Terminal Operation

The method of operation of the passenger terminal and some of the associated problems of terminal operation depend partly on access in as much as this can affect the amount of time that the departing passenger spends in the terminal. Short dwell times in terminals require few facilities. For example, provincial domestic air terminals in Scandinavia often take the place of intercity rail and bus stations. Consequently, they are rightly designed and operated as very functional buildings and are relatively spartan facilities using half the space norms of many other European airports, since passengers are not expected to spend much time in the terminal. Facilities where longer dwell times are expected must provide a high level of comfort and convenience (e.g., restaurants, bars, cafes, relaxation areas, shopping, post offices, and even barbers). Naturally more terminal revenue to pay for such facilities can be generated in the longer dwell time. It is the departing passenger who places most demands on the airport terminal system. Departing dwell times depend chiefly on the length of access time, reliability of access time, check-in and security search requirements, airline procedures, and consequences of missing a flight.

Length of access time

It is likely that the amount of time for a particular access journey is a random variable that is normally distributed about its mean value. It is reasonable to assume that the variance of the individual journey time about the mean is in some way proportional to the mean. This is shown conceptually in Figures 13.3a and 13.3b, where two access journeys of mean length t_1 and t_2 with respective variances of σ_1 and σ_2. If the access times t_1 are truly normally distributed about the mean, all but a negligible proportion (0.5 percent) are contained between $t_1 \pm 3\sigma_1$ As a result, if all but 0.5 percent of trips are to arrive a standard time K before scheduled time of departure (STD), then the cumulative curve (Figure 13.3c) shows that the average time spent in the terminal is $3\sigma_1 + K$ for the longer access time and $3\sigma_2 + K$ for the shorter. Since there is strong evidence that journey times are random, cumulative arrival patterns of the form of Figure 13.3c are frequently observed.

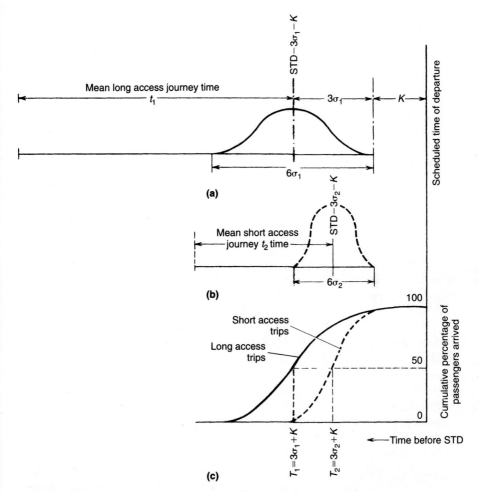

Figure 13.3 Comparison of passenger terminal dwell times for long- and short-access journey.

Reliability of access trip

The effect of reliability on departing terminal dwell times is shown in Figure 13.4. If there are two access trips each with the same mean trip time of t but with variances of σ_A and σ_B it can be that the mean terminal dwell time, under assumptions of normality and 99.5 percent arrivals by K minutes before STD, are $3\sigma_A + K$ and $3\sigma_B + K$, respectively. The effect on the cumulative curve is demonstrated to be a more gradual slope for low reliability access. Access routes with very low reliability can result in very long average passenger dwell times in the terminal.

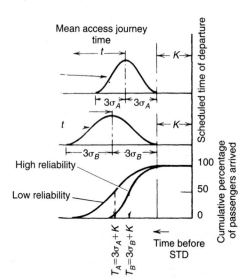

Figure 13.4 Effect of reliability of access times on passenger terminal dwell times.

Check-in procedures

Check-in requirements are not the same for all flights. For many long distance international flights, check-in times are a minimum of 1 hour before scheduled time of departure,[1] while for domestic and short-haul international this is usually cut to 30 minutes. The effect is a leftward shift of the cumulative arrival curve; an example of this can be seen on the passenger data from Manchester International Airport (Figure 13.5), with long-haul passengers spending an average of 22 more minutes in the terminal than short-haul passengers. Similar differences are often observed between check-in procedures for chartered and scheduled passengers. It is not unusual for a passenger on a charter flight to receive instructions to check in at least 90 minutes before scheduled time of departure. Figure 13.6 shows observed differences in check-in behavior between charter and scheduled passengers at a European airport.

Consequences of missing a flight

Depending on the type of flight and the type of ticket, the passenger will have a very different attitude toward arriving after the flight has closed out and consequently missing the aircraft. This can be exemplified by considering a hypothetical tripmaker making three different flights from Tampa International Airport. The first flight is on a normal scheduled ticket at full fare to Miami; the second is on a normal scheduled full fare ticket to Buenos

---

[1]For international flights to the U.S., FAA procedures in many parts of the world require a reporting time of at least 2 hours before the scheduled departure time.

Aires; the third is a special chartered holiday flight to London. The implication of missing the three flights is not at all the same. Should the passenger miss the first flight, there will soon be another flight, and there is no financial loss. In the case of the second flight, the ticket remains valid, but because the connections will now be lost and there might not be an alternative flight rapidly available, there is serious inconvenience and maybe some financial loss. Missing the third flight, however, could cause much inconvenience through a spoiled holiday and serious financial loss because the ticket is no longer valid. The passenger therefore will arrange his arrival at the airport in such a way that the risk of missing each flight is different. Subconsciously, the risk levels might well be set at one in one hundred for

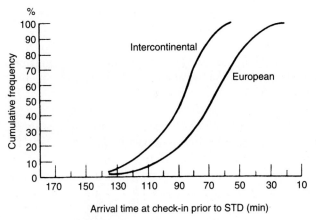

Figure 13.5 Effect of flight length on passenger dwell times. *(Ashford et al 1976)*

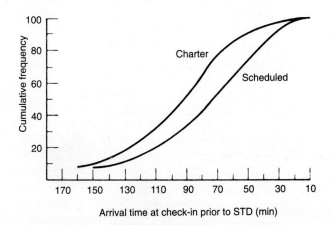

Figure 13.6 Effect of flight type on passenger terminal dwell time. *(Ashford et al 1976)*

missing the first flight, one in a thousand for missing the second, and one in ten thousand for the third. Figure 13.7 shows the variation in arrival time that could be expected for an access journey of 60 minutes mean duration with a standard deviation of 25 minutes at these risk levels, assuming for each a close-out time of 20 minutes before STD. It can be seen that the average time the passenger spends in the terminal is 59, 69, and 76 minutes, respectively.

In practice, the arrival patterns at individual airports are a mixture of all these factors. The variation between arrival times can be seen in Figure 13.8, which shows the cumulative arrival curves for four European airports. At the time these data were collected, prior to a German reunification, Berlin served a relatively small access cachement. Its access times were reasonably predictable, and most flights were short-haul. Paris and Schiphol both served a mixture of short- and long-haul flights, and access times were less predictable. Heathrow also served short- and long-haul flights, but road access time varied a great deal and are subject to considerable congestion. The cumulative curves are a measure of the impact of these variables on terminal dwell times.

Recent work examining the effect of the length and reliability of access times has confirmed that unreliable access times can cause congestion in the check-in areas and long dwell times in the departure lounges (Ashford

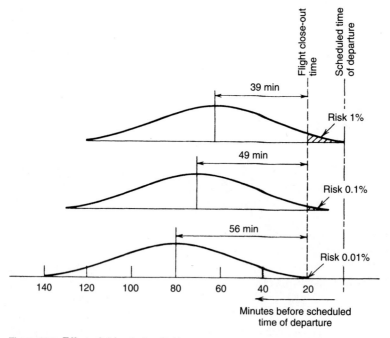

Figure 13.7 Effect of risk missing flight on average passenger terminal dwell times.

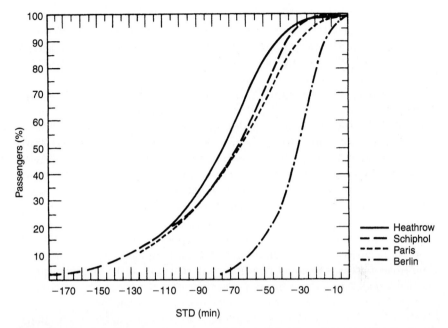

Figure 13.8 Check-in times for passengers prior to scheduled time of departure for European flights at four selected airports.

and Taylor 1995). While such dwell times might be favorable to developing commercial income for the airport, they also reduce the terminal's capacity to cope with flow.

13.4 Access Modes

Automobile

In most developed countries, the private car is the principal method of accessing airports. This has been the case since the inception of commercial air transport, and the situation seems most unlikely to change in the foreseeable future. As a consequence, airports must integrate a large parking capability into their design and operation. Large U.S. airports, such as Kennedy and Chicago O'Hare, have extensive parking areas in both close and remote positions.

As airports grow in size, it becomes difficult to provide adequate parking space within reasonable walking distance of the terminal. This is particularly true for largely centralized terminals such as O'Hare and Gatwick, less so for decentralized designs, such as Dallas, Paris Charles de Gaulle, and Kansas City. In the case of centralized operations, it is common to divide the parking areas into short-term facilities close to the terminal long-term parking areas often served by shuttle services. Normally the pricing mechanism is a sufficient incentive to ensure that long-term users in fact utilize the remote parking areas. Serious internal circulation congestion can limit the airport's capacity if too many cars attempt to enter the facilities close to the terminal, a condition that has caused problems with the operation of the first terminals at Toronto and Paris Charles de Gaulle airports where parking is integral to the terminal and access is via a tunnel under the apron. At London Heathrow, the constraints on space within the "central area" are so severe that short-term parking rates are set at approximately four times long-term rates to discourage cars from entering this central site. Figure 13.9 and Table 13.4 show recommended criteria for providing long- and short-term parking that have been found useful in the United States and Canada (Whitlock and Clearly 1976; Ashford and Wright 1992). Although parking requires considerable land area and space, there are substantial profits to be made from its provision. As airports increase in size the relative importance of the contribution of the parking facilities to overall revenue also increases to about one-fifth of all revenues at the largest airports. At many major airports in the U.S., car parking is almost as large a contributor to total revenue as the landing fees from the aircraft.

Major airports relying overwhelmingly on the automobile as the major access mode find that it is not solely in the matter of supplying and operating car parking that this decision materially affects the operation of the passenger terminal. Use of the space-extensive car mode requires the provision of substantial lengths of curbside space in front of the terminal for dropping off and picking up passengers; de Neufville indicates that for U.S.

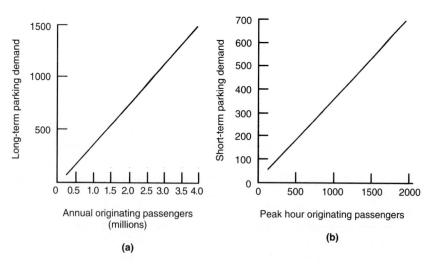

Figure 13.9 (a) Long-term parking demand related to annual originating passengers; (b) short-term parking demand related to peak hour originating passengers. (*La Magna et al 1980*)

TABLE 13.4 Parking Requirements Recommendations

Roads and Transport Association of Canada (smaller airports)	1.5 spaces per peak hour passenger (short term) 900 – 1200 spaces/million enplaned passengers (long term)
FAA (non-hub airports)	1 space per 500 – 700 annual enplaned passengers

airports this averages 1 foot per 3000 annual passengers (Ashford and Wright 1992), while for European airports the figure is 1 foot per 4000 annual passengers (Ashford 1982). Table 13.5 gives some examples of facility provision. Large volume terminals must be designed in one of two ways to accommodate this requirement; either they must become long and linear, a form now familiar at Dallas-Fort Worth, Kansas City, and Terminal 2 of Charles de Gaulle Airport, or the space must be provided by separating departures and arrival traffic onto separate floors, such as at London Gatwick, Singapore, Johannesburg, and Tampa. The first solution leads to a highly decentralized passenger terminal complex with possible difficulties in interlining, especially for baggage-laden international passengers. The second solution will almost certainly lead to the segregation of departing and arriving passenger flows throughout the terminal building and to considerable duplication of circulation space. In the early 1980s, Los Angeles International, which previously had operated a one-way one-level access system, had to undertake a very expensive modernization scheme that involved double-decking the access road and substantial modification of the terminals to permit two-level operation. Only in this way could access and terminal capacity be brought up to airside capacity.

Strict policing of the way passengers are picked up or dropped outside the terminal can have a substantial effect on curb requirements. Table 13.6

TABLE 13.5 Curbside Access Length Per Million Passengers/Annum at Declared Terminal Capacity

	Curbside Access Length (m)	Annual Declared Capacity (mppa)	Curside Access Length per mppa at Capacity
US Airports	—	—	100 (typical)
London Heathrow (LHR)	3100	30	103
LHR (terminal 1 only)	1500	14.7	102
LHR (terminal 2 only)	1000	8.6	116
LHR (terminal 3 only)	600	10.1	59
Amsterdam Schiphol	1200	18	66
Frankfurt	900	30	30

SOURCE: Reference 6

TABLE 13.6   Curbside Access Activity Times for Four U.S. Airports

Departures	Miami Unloading/ Loading Time (min)	Miami Dwell Time	Denver Unloading/ Loading Time (min)	Denver Dwell Time	New York La Guardia Unloading/ Loading Time (min)	New York La Guardia Dwell Time	New York JFK Unloading/ Loading Time (min)	New York JFK Dwell Time
Auto	1.3	3.0	1.0	2.3	0.6	1.2	1.2	2.5
Taxi	1.0	1.8	0.7	1.2	0.5	1.1	0.8	1.3
Arrivals								
Auto	2.8	4.3	2.9	4.2	1.2	2.4	1.6	3.3
Taxi	0.9	—	1.0	—	0.3	NA	—	—

SOURCE: Reference 8

shows substantial differences between the loading/unloading times and dwell times of four U.S. airports (La Magna, Mandle and Whitlock 1980). Because the time need only be in the order of 2 to 3 minutes (the time to get out of the car, reenter, and move off), it would appear that vehicles are in fact actually parked for a short time rather than merely waiting. This time variation is emphasized when comparison is made with taxis, which use the curb front in an efficient manner. The efficient utilization of the high volume curbside access space requires an active presence of some form of traffic policing to keep vehicles moving (Mandle, La Magna, and Whitlock 1980).

Taxi

For the air traveler, the taxi is perhaps the ideal method of accessing the airport, from all aspects except one—cost. In general this mode involves the least difficulty with baggage, is highly reliable, operates from real origin or destination, and provides access directly to airport curbside. Unfortunately, it can be comparatively expensive, although not necessarily so if hired by a party of travelers or by an individual who would otherwise consider using a personally owned car and is likely to incur high parking charges for an extended parking stay. The airport operator normally has two principal interests with respect to taxi operations at the airport: first, the balance of supply and demand and, second, the financial arrangements with the taxi operator. It is likely that these two matters will warrant simultaneous consideration.

The airport has an interest in maintaining a reasonable balance of supply and demand of taxis on the airport. Taxis must be available at unsociable hours, such as at night when perhaps most other public transport is not operating, and during peak hours of operation, passengers should not have to wait an unreasonable amount of time for a taxi. Equally important there should not be so many taxis within the airport terminal area that they cause a congestion problem. To achieve these ends, the airport needs control. Many U.S. and other airports do not permit taxis to pick up a fare on airport property without a special license, for which the taxi operator must pay annually. The annual license fee gives this operator the privileged but controlled right to operate at such airports as Washington National. The license fee adds to the airport's income, and the airport operator can ensure that supply and demand are in reasonable balance. Some airports, such as Schiphol, do not charge a license fee but award and renew licenses based on the performance of the operator. The license can be withdrawn if the operator fails to supply sufficient vehicles. At Heathrow London, the central terminal site at one time suffered from too many cruising taxis, which caused congestion on the terminal access roads. This was controlled by a scheme that holds taxis in a taxi park away from the central area until dispatched by radio communication to the required points in the terminal area.

Limousine

Limousine services, which are reasonably common in the United States and a number of other countries, are either minibuses or large automobiles that provide connection between the airport and a number of designated centers (usually hotels) in the city. The limousine company pays the airport operator in exchange for an exclusive contract to operate a service to provide access according to an agreed-on schedule. The actual form of service varies. In small cities, the limousine usually operates to only one central location, and in larger cities, to designated multiple locations. In some very small cities, on paying a supplement to the standard fare, the limousine operates a multiple origin-destination service, often to the home, and in that respect becomes very similar to taxi service.

Operationally, the limousine is similar to the bus, and where bus services are feasible it is unusual to have limousines also. Services that have multiple pick-up and drop-off points in the urban area have gradually disappeared in the United States, their place being taken by a combination of bus and taxi services. From the airport operator's viewpoint, limousines require very few facilities. Because they use small vehicles, loading and unloading is simple and rapid. It can be carried out at the normal curbside. The only facilities necessary are signs to direct passengers to where they should congregate and wait. For the passenger, limousine service is usually relatively inexpensive, yet it gives a level of service very similar to that of the taxi, which can be up to five times as expensive. The contracts are lucrative to the limousine operator because passenger load factors are high, and therefore the concessionary fees that go to the airport operator can be high in comparison with the cost of providing facilities. Because limousines are in fact a form of public transport, they relieve road congestion and the need for parking.

Rail

The connection of an airport to an existing urban rail rapid transit system potentially overcomes a major problem of dedicated airport-city center links (i.e., that most travelers are not destined to the central city). The network gives the opportunity of traveling to many destinations in the urban area. Obviously if the rail network is very limited in size, the attractiveness of the mode is likely to be low.

In both design and operation, the transfer onto the mode at the airport end must be easy. An example of poor design was the Washington National rapid transit station, which was a considerable and inconvenient distance from the terminal. This fault has been corrected in the reconstruction of the airport landside during the early 1990s. Even a bus shuttle connection to the train appears to create a perceived barrier at those airports where it is provided. Experience of Paris Orly, and Boston Logan with 8 percent by rail, indicates that passengers do not like to use rail systems that require further change of mode. The replacement of a bus shuttle service to the rail station with a VAL people mover at Orly has failed to raise rail modal split above 4 percent. Over time this type of design is likely to effect the successful operation of the rail link to Paris Charles de Gaulle, which will have a centralized station away from the various terminals served by a bus shuttle service to the terminals. Currently the rail share of traffic is only 13 percent.

Where specialized links between the airport and the city are provided, it is essential that the town end of the line be well sited. Designers of facilities in Zurich, Munich, and Amsterdam have taken note of these problems. The Heathrow Underground rail connection was successful for two reasons: the ease of connection at the airport terminal and the fact that there is direct connection to 250 stations on the Underground network and easy connections to suburban and intercity rail lines. However, as passenger traffic increased by more than 60 percent between 1980 and 1995, the mixing of air passengers encumbered by luggage and urban commuters became unacceptably crowded except for those air passengers with very little luggage. Consequently, British Rail and BAA jointly funded Heathrow Express which provides direct non-stop conventional rail service to a downtown London main line terminal. Figure 13.10 shows the layout of the airport rapid transit station in relation to the three terminals in the central area. The more remote Terminal 4 has its own rapid transit stop.

Very few airports are served by intercity trains. Frankfurt, Amsterdam, and London Gatwick are among the best known; more recently Zurich Kloten, and Birmingham, England, have initiated services. The success of a conventional rail line will depend on the level of connection it provides to the rest of the surface transport system. If connections are to a very limited network or a network on which very few trains are operated or to a poor

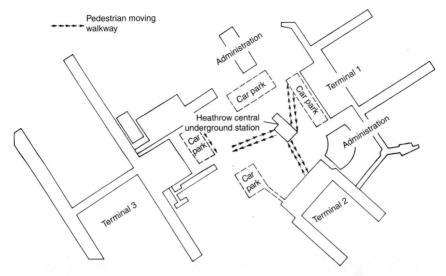

Figure 13.10 Location of Heathrow Underground station (rapid transit) in relation to Terminals 1, 2, and 3 at London Heathrow Airport.

surface bus and taxi system (as formerly with Orly Rail), ridership will be low. The Gatwick rail link, which carries more than one-quarter of the airport's passengers, is successful because it also links well to the London Underground, the London Transport bus system, and taxis at its town end.

Three other general factors are noteworthy. First, baggage handling can be a problem at both ends of the rail journey link and might present difficulties of storage en route if the rail carriages are not specially designed or modified. Second, rail links seldom attract large percentages of airport employees. Because of the size of airports, employees' destinations on the airport can be a long way from the passenger terminal; also employees will not necessarily select a residential location that gives a good public transport link to the airport. This problem will have to be solved in the massive New Seoul International Airport, which is designed to handle ultimately 100 million passengers per year and seven million tons of freight. Rail modal split estimates for employee journeys have been made at the very high level of 75 percent, yet the perimeter of the airport itself will be 23 miles (37 km). Third, access journey time does not appear to be critical to air travelers except in the very shortest hauls with competitive surface modes.

The selection of an access mode is much more affected by ability to cope with inconvenient and heavy baggage. A relatively minor flaw in design, such as an inconvenient bus transfer, a long walk, or a bad flight of stairs that makes it difficult to use a mode with reasonable comfort will result in low modal utilization.

Frequently, the construction of a rail link does not materially increase the use of public transport. The Heathrow Underground link can be termed a success, but the traffic that accrued to it came mainly from other public transport modes. Car and taxi usage was not dramatically affected (Figure 13.11).

Bus

Virtually all airports carrying reasonable volumes of passengers by scheduled operators are connected by bus to the city center. Normally this is arranged by a contract between the bus operator and the airport authority whereby the bus company usually pays the airport a concessionary fee or percentage for the exclusive right to provide an agreed scheduled service. Service is supplied to a number of points in large cities, but perhaps to only one point for a small urban area.

At the larger European airports, the bus is an extremely important mode of access. Many airports therefore emphasize bus access and supply sophisticated curbside bus bays and bus unloading arrangements, such as those at London Heathrow (Figure 13.12).

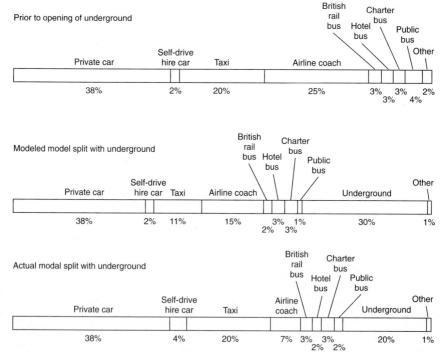

Figure 13.11 Effect of providing Underground (rapid transit) link on access modal split to London Heathrow Airport. *(London Transport, Civil Aviation Authority, and British Airports Authority)*

Figure 13.12 Bus bays at London Heathrow.

The bus becomes extremely important at airports serving many vacation destinations. For example, in Almeria, Spain, the overwhelming majority of travelers are vacationers arriving from northern Europe who travel in organized groups and are brought to and from the airport by chartered buses. Because very few passengers are not leisure travelers, and because the use of the personal auto is very low even among this group, at "vacation" airports the landside access is designed almost entirely around the accommodation of bus passengers. Bus loading and unloading areas are designated and must be kept clear of taxis and automobiles. Bus parks are as important as car parks, and the airport operator has an interest in ensuring that the bus parks are kept operational and clear. Figure 13.13 shows a bus park that caters to chartered buses for vacation passengers.

Dedicated rail systems

In the area of airport access, nothing has caught the public imagination more than the concept of some form of high-speed[2] tracked vehicle that will convey the passenger from the airport to the town center unimpeded by surface road or rail traffic (Ashford and McGinity 1975). Many such schemes have been suggested and investigated at sites all over the world.

---

[2]High speed in this context means technology capable of speeds of 180 mph (300 kmh) which is attained by modern European conventional trains (TGV).

Figure 13.13   Remote bus park at Vienna Airport.

The realities of the economics of dedicated systems, especially high speed systems, have been less sanguine. Monorail systems, such as that linking Tokyo Haneda airport to central Tokyo, have had a mixed reception, taking many more years than anticipated to achieve financial viability. Studies indicate that dedicated systems are unlikely to reach economic viability at riderships of less than 10 million passengers per year with reasonable right-of-way costs (McKelvey 1984). However, with city center attractions of the order shown in Table 13.2 and estimated modal splits of the order of those shown in Table 13.3, riderships of the level of 10 million passengers per year are feasible only at the very largest airports. These are normally sited some considerable distance from the central business districts of the large cities, requiring long stretches of very expensive right-of-way. Construction costs of surface or elevated dedicated lines in urban areas are very high, but these costs might be small in comparison with the cost of the purchase of central urban right-of-way. Normally such schemes are possible only if some abandoned right-of-way becomes available, such as in Chicago or Atlanta. High-speed tracked airport-access vehicles on dedicated rights-of-way are not likely to be built anywhere in the world where the economics of access costs are correctly considered. They are unnecessary, are likely to cost half as much as the remote airport they purport to serve, and can move passengers only to and from the central city, where the travelers probably do not wish to go. Moreover, if they require public subsidy, they raise an ethical question as to whether the air traveler has any right to expect to travel in the urban area at a higher speed than any other traveler. Even so, it is likely that they will continue to receive a disproportionate amount of public and media interest.

13.5 In-town and Other Off-airport Terminals

Experience with in-town terminals with check-in facilities has been very varied. When the West London Terminal was closed, only 10 percent of passengers were using the facility. As well as being uneconomic, it was difficult to have reliable connections between the check-in terminal and the airport due to increasing road congestion on the airport access routes. Similarly, the Port of New York Authority closed its West Side Terminal because of poor usage. In contrast, at Victoria Station in central London, a very successful town check-in operation for the rail served Gatwick Airport has been operated for more than 20 years. As a rule, airlines are not in favor of in-town check-in facilities. They find that the inherent duplication of staff leads to high operational costs that are untenable in the financial climate of the modern airline. There are also considerable problems with the baggage from the viewpoint of aviation security.

Examples of successful, in-town airline bus terminals with no check-in facilities are more numerous. For example, Air France has had very successful bus terminals at the west and south sides of central Paris for many years.

For some time, San Francisco and Atlanta airports have been experimenting with remote parking areas scattered throughout their regions connected by shuttle minibus service to the airport. This park-ride-fly operation has met with reasonable success and is likely in the long term to be extended to other airports.

Since the use of the in-town bus terminal is part of the modal choice process, it depends on the factors that affect mode choice enumerated in Section 13.6. It is important to realize, when examining how off-airport terminals operate, that cost and time do not seem to be very sensitive factors in choice of access mode. An otherwise attractive service can perform poorly if the convenience level of the traveler is debased by long walking distances with baggage, frequent changes of level by stairs, crowded vehicles, and inadequate stowage space.

13.6 Factors Affecting Access Mode Choice

The level of traffic attracted to any access mode is a function of the traveler's perception of three main classes of variables:

- Cost
- Comfort
- Convenience

Decisions in terms of these variables are made not only on the level of service provided by a particular mode, but also on the comparative level of service offered by competing access modes. In addition to making an out-of-pocket price comparison, the traveler makes a decision based on the level of comfort and convenience provided by the various modes. The principal considerations are shown in Table 13.7.

In work carried out in the United Kingdom, which examined passengers' perception of the level of service provided by access to a major airport, it was found that passengers placed the highest importance on such factors as ease of luggage handling, convenience with which the access mode connected to the check-in area and journey time (Ashford, Ndoh, and Bolland 1993). Delay and congestion, cost of travel, and parking costs were not ranked highly assessing level of service. Table 13.8 shows the rankings as expressed by travelers at London Heathrow.

Transportation planners have numerous models ranging from the simple to complex to explain the modal selection procedure. For details of the modeling approach, the reader should consult planning references (Ashford 1976).

13.7 General Conclusions

From the foregoing discussions, it can be seen that the access condition observed at any individual airport is location specific. Within a large country

TABLE 13.7 Factors Affecting Model Choice

Car	Bus	Rail
Ease of loading and unloading	Location of bus terminal within airport	Location of terminal airport
Distance baggage must be carried and difficulties such as stairs	Speed and reliability of service	Necessity for shuttle bus to reach terminal at airport end
Ease of finding long-term or short-term parking	Whether specialized express service of part of urban bus network	Difficulty in handling baggage
Access route congestion and travel time reliability	Difficulty in handling baggage	Siting of station or stations in town relative to ultimate destination or to taxis, buses and other train terminals
Shuttle arrangements for long-term car parks	Siting of in-town terminal relative to ultimate destination or to taxis, buses, and trains	
Vulnerability of car to vandalism and theft		

TABLE 13.8   Ranked Importance of Selected Attributes in Passengers' Choice of Access Mode

Attribute	Rank
Ease of luggage handling	1
Convenience of transfer to check-in area	2
Expected journey time	3
Comfort of mode	4
Parking space availability	5
Convenience of interchanges where more than one vehicle or mode used	6
Actual journey time	7
Delay and congestion	8
Cost of mode	9
Overall opinion of access	10
Access information	11
Parking cost	12

with widespread access to air transport via many large and small airports, the conditions in many small and medium airports are very similar, and their access provisions are consequently very much alike. In general, however, the access condition depends on the nature and volume of airport traffic; the location and geographical setting of the airport and the urban areas it serves; and the economic, social, and political structure of the country in which it is situated. The diversity of road and rail-based access provisions at a number of the world's larger airports (Table 13.9) indicates that with respect to the public transport modes, no generalization is possible. It is possible, however, to discern that road-based access modes are currently vital to the operation of most airports and are likely to remain so in the foreseeable future. Except at a few very large airports in major metropolitan areas where urban traffic congestion is severe, the car and taxi will continue to dominate all other modes of airport access.

References

Ashford, N. and R. Taylor. 1995. *Effect of variability of access conditions on airport passenger terminal requirements.* Unpublished report. AAETS. Loughborough University of Technology.

Ashford, N. et al. 1976. *Passenger Behaviour and the Design of Airport Terminals.* Transportation Research Record, No. 588, Washington, DC: Transportation Research Board.

Ashford, N., N. Ndoh, and S. Bolland. 1993. *An Evaluation of Airport Access Level of Service,* Transportation Research Record 1423, Transportation Research Board, National Research Council, Washington, DC.

Ashford, N.J. 1982. *Heathrow Terminal 5 Inquiry,* BA81. London: British Airways. October.

Ashford, N.J. and P. McGinity. 1975. Access to Airports Using High Speed Ground Modes. *High Speed Ground Transportation,* 9, No. 1. Spring.

Ashford, N.J. and P.H. Wright. 1992. *Airport Engineering, 3rd edition,* New York: Wiley Interscience.

Coogan, M. 1995. *Airport Ground Access by Rail at Various International Airports. Report to the FAA.*

Gorstein, M. 1980. *Airport Ground Access Planning Guide, Report FAA-EM-80-9,* Federal Aviation Administration, U.S. Department of Transportation.

TABLE 13.9   Characteristics of Public Transport Modes at Selected Airports

Airport	Type of Coach/Bus or Rail Type	Distance (km)	Number of Stops	Frequency	Departure Points	Arrival Point	Trip Length (min)
				Road-Based Modes			
Vienna	Coach	16	0	every 20 min	Downtown terminal and railway stations	Air terminal	25–35
Copenhagen	Bus & coach	10	0	15–20 min	Town center & Malmo, Sweden	Air terminal	40
Madrid	Bus	12	0–7	12–20 min	Town center, various stops	Air terminal	30–50
Paris CDG	Bus & coach	25	0	12–15 min	Coach terminal/ rail terminals	Air terminal	30–60
Luxembourg	Bus & coach	6	0–3	15–20 min	Central rail station, city center, coach station	Air terminal	15–20
Atlanta	Minibus, coach, limousine	13	0 to several	20 min	Town center, suburbs, hotels, & motels	Air terminal	35–40
Dallas-Forth Worth	Coach	37 (average)	over 30	15–30 min on demand	Town centers	Airport	45–60
Los Angeles	Bus	24 or more	0–5	On demand	Town center and suburbs	Air terminal/ Airport	75
Montreal Mirabel	Coach & bus	55	0	30–60 min	Central station, hotel, Dorval Airport, Montreal Metro	Air terminal	25
Hong Kong	Bus & limousine	5–10	0–6	15 min	Various hotels, Central business district	Air terminal	30

Airport	Type of Coach/Bus or Rail Type	Distance (km)	Number of Stops	Frequency	Departure Points	Arrival Point	Trip Length (min)
				Road-Based Modes			
Melbourne	Minibus, bus & coach	24	0	30–60	Downtown terminal	Air terminal	20–40
				Rail-Based Modes			
Frankfurt	Train	12		up to 60 per day town centers	Central city or surrounding	Air Terminal	11
Vienna	Train	16		up to 20 terminal	Downtown	Air terminal	30
Paris CDG	Train/coach	25		15 min stations	Central city	Airport	35
Amsterdam	Train	15		15 min	Central station	Air terminal	17–20
London Gatwick	Train	43		15 min station	Central city	Air terminal	30
London Heathrow[a]	Rapid urban rail transit	25		3–7 min in rapid transit network	Multiple stations (special terminal	Air terminal	40–60
New York JFK	Rapid rail transit/coach	26	8	20 min	57th Street Station	Surburban station/link by coach	48
Washington National	Rapid rail transit	7	6	5–10 min	Central city rapid rail stations	Airport rapid rail station (special terminal)	20 (average)
Tokyo Haneda	Monorail	29	5	7 min	Airport	Off center station	15

[a] London Heathrow is also linked by airport coaches to main line regional conventional rail stations.

La Magna, F., P.B. Mandle, and E.M. Whitlock. 1980. *Guidelines for Evaluating Characteristics of Airport Landside Vehicular and Pedestrian Traffic.* Transportation Research Record No. 732, Washington, DC: Transportation Research Board.

Mandle P., F. La Magna and E. Whitlock. 1980. *Collection and Validation of Data for an Airport Landside Dynamic Simulation Model, Report TSC-FAA-80-3,* Federal Aviation Administration, U.S. Department of Transportation. April.

McKelvey, F.X. 1984. *Access to Commercial Service Airports, Final Report,* Federal Aviation Administration, U.S. Department of Transportation. June.

Transportation Research Board. 1975. *Airport Landside Capacity.* Special Report 159. National Research Council, Washington, DC.

Transportation Research Board. 1987. *Measuring Airport Landside Capacity.* Special Report 215. National Research Council, Washington, DC.

Whitlock, E.M. and E.F. Clearly. 1976. *Planning Ground Transportation Facilities.* Transportation Research Record, No. 732, Washington, DC: Transportation Research Board.

Operational Administration and Performance

14.1 The Airport as a Service

Previous chapters have described various elements of the airport operational system that come together to provide a service to the airlines and to the traveling public. It is important to recognize the difference, in management terms, between providing a service and producing a "good." Goods are tangible objects; they can be physically examined by potential customers, transferred from the point of creation to some other point of sale and, very importantly, stored or held in inventory against peak demands. Services, however, are created and issued simultaneously, and cannot be stored.

Airports differ from many other enterprises in a number of aspects that must affect the manner of evaluating management performance:

1. The end product is a service rather than a manufactured good.
2. They operate in a highly regulated and technologically sophisticated environment.
3. They operate in a highly political framework.
4. They operate in an international environment.
5. Operation is frequently on a continuous, 24-hour basis.
6. Emergencies can be routinely anticipated at any time.
7. Although they provide services for the air traveler or the cargo shipper, this is done indirectly. The contractual relationship for transport lies with the airlines, and many terminal services are provided by concessionaires

8. Investment decisions are relatively infrequent. The costs involved (e.g., in runways, taxiways, aprons, terminals, and access) are very high, and the results of investment decisions are long lasting.

In some ways, of course, airports are very similar to some businesses in that they require:

- Substantial amounts of capital investment

- Continuous and expensive maintenance

- A careful control of finances, although at many airports profitability in the commercial sense is not required

- A long-term planning capability to ensure that the operation can respond to changes such as those that occur in the areas of demand, technology, and working practices

14.2 Operations Within the International Context

Many of the airports that operate flights for the transport of passengers function within an international context. The earliest recognition of the international implications of civil aviation came from Europe with so many nation states, where a flight of only 200 miles (320 K) could be an international journey. As World War II drew to a close it was apparent that the potential for civil aviation was now intercontinental as well as international. In early 1944 the United States approached its allies and number of neutral nations with a view to discussing postwar civil aviation. As a result, in November 1944 at the Chicago Convention on Civil Aviation attended by 52 nations, a basic framework for civil aviation was agreed upon. The Convention set out this framework in the form of 96 articles that provided for the establishment of international recommended practices. Twenty-six national states ratified the convention, and on April 4, 1947, the International Civil Aviation Organization, ICAO, came into being, with headquarters in Montreal. ICAO functions with a sovereign body, the Assembly, and a governing body, the Council. In the Assembly, which normally meets every three years, each member has one vote; decisions of the Assembly are taken by a simple majority of votes, except as provided otherwise in the Convention. The 33-member Council is the essential governing body, which, by its makeup, reflects those states most important in civil aviation and those that ensure a geographical balance. One of the main duties of the Council is to adopt international standards and recommended practices for safety, environment and infrastructure (ATC and airports). Once adopted, these are incorporated into the Annexes to the Convention on International Civil Aviation. There are 18 Annexes (Table 14.1). Airport administrators will find that Annex 14 is of prime impor-

TABLE 14.1   Annexes to the ICAO Convention on International Aviation

Annex 1–Personnel licensing	Annex 11–Air traffic services
Annex 2–Rules of the air	Annex 12–Search and rescue
Annex 3–Meteorological service for International Air Navigation	Annex 13–Aircraft accident investigation
Annex 4 –Aeronautical charts	Annex 14–Aerodromes
	Volume I Aerodrome design and operations
	Volume II Heliports
Annex 5–Units of measurement to be used in air and ground operations	Annex 15–Aeronautical Information Services
Annex 6–Operation of Aircraft	Annex 16–Environmental protection
Part I Internatioal commercial air transport - aeroplanes	Volume I Aircraft noise
Part II International general aviation - aeroplanes	Volume II Aircraft engine emissions
Part III International operations - helicopters	
Annex 7–Aircraft nationality and registration marks	Annex 17–Security - safeguarding International Civil Aviation against Acts of Unlawful Interference.
Annex 8–Airworthiness of aircraft	Annex 18–The safe transport of dangerous goods by air
Annex 9–Facilitation	
Annex 10–Aeronautical Telecommunications	
Volume I (Part I - Equipment and Systems and Part 2 - Radio frequencies)	
Volume II (Communication procedures including those with PANS status)	

Note: Annex 9 is the responsibility of the Air Transport Committee, Annex 17 is the responsibility of the Committee on Unlawful Interference and the remaining Annexes are the responsibility of the Air Navigation Commission.

tance in the operation of their facilities, but parts of other documents, such as Annexes 9 to 13 and 16 to 18 also deal with operations.

By 1995 the original 26 member states of ICAO had increased to 183. The intervening years have also seen a quantum leap in civil air transport, highlighted in 1970 by the introduction of the first wide-body, high-capacity, long-range jet, the B747, followed closely by the L1011 and DC10, with consequent major developments at airports. Such growth does not make the work of ICAO easier, rather the reverse. With so many members trying to reach consensus, progress can be slow while industry development surges ahead. States with highly developed civil aviation industries, such as the United States, can and do exert pressure for speedier action, not entirely devoid of commercial considerations, but generally producing a beneficial effect.

14.3 Domestic Regulation

Since ICAO lacks inspection capability, and does not carry out any enforcement activities, the international standards and recommended practices of ICAO are not binding on the member states. It is, however, normal for member states to incorporate these standards into their own aviation legislation in as far as adoption is desired. Many states have their own standards, which might vary slightly from those of ICAO; on occasions they might be even higher. For example, aviation in the United States must conform to Federal Aviation Regulations (FARs), which are administered by the FAA. The Federal Aviation Act of 1958, as amended in 1979, states that the FAA Administrator is empowered to issue airport operating certificates to airports serving air carriers certificated by the Civil Aeronautics Board and to establish minimum safety standards for the operation of such airports. (The Civil Aeronautics Board has since been abolished and the authority transferred to the Department of Transportation.) Any person desiring to operate an airport serving air carriers certificated by the certification authority may file with the Administrator an application for an airport operating certificate. If the Administrator finds, after investigation, that such person is properly and adequately equipped and able to conduct a safe operation . . . he shall issue an airport operating certificate.

Part 139 of the Federal Aviation Regulations adopted in 1987 applies to airports that serve any scheduled or unscheduled passenger operation of an air carrier that is conducted with an aircraft having seating capacity of more than 30 passengers. No one may operate such an airport without an airport operating certificate. Operation must be in accordance with an up-to-date Airport Certification Manual[1] that includes procedures for ensuring safe operation under normal and emergency conditions.

In a similar way, in the United Kingdom control of airport standards is achieved by the Air Navigation Order, which requires that all flights for the public transport of passengers and for flying instruction take place at airports that conform to government standards. Airports meeting these published standards are licensed by the CAA. The standards published in CAP168 incorporate the Standards and Recommended Practices of Annex 14 insofar as these have been adopted by the United Kingdom. By conforming to British legislated requirements, airport operators are essentially in conformity with the international standards of ICAO.

14.4 Organizational Intent

Unless an organization, and those who work in it, have a clear understanding of its main purpose and its goals and objectives and are committed to a

---

[1]See Chapter 15, The Airport Operations Manual.

planned program drawn up to achieve them, it will be an inefficient organization. In this respect at least, airports are no different from any other organization. Examples of mere lip service to such a management philosophy abound, and the signs of the superficial approach are clear:

- Multiplicity and diversity of objectives
- Vagueness and intangibility of objectives
- Tendency towards conflicting goals
- Lack of innovation

This can be avoided if airport administrations adopt a systematic approach to setting down management intent. This statement of intent, which would normally be provided by a policy board or committee, can be refined down from a broad statement of purpose through a limited number of goals to detailed objectives. It is these objectives that then form the basis for an operating plan drawn up by airport executive management against which judgments of performance can be made. While it is often stated that managers should not be held responsible for events over which they have no control, the implication being that this refers only to poor performance, the corollary is that this applies also to good performance if it results from events outside management control (e.g., governmental directives). The statement of objectives and the existence of a formal plan of action to implement them assists in identifying those results that can be credited directly to management. The advantages of such a systematic approach are:

1. Subjective assessments are more easily avoided
2. Personality becomes a lesser consideration
3. It assists in the planned progress of management efficiency

14.5 The Framework of Performance Measurement

The process of establishing performance criteria commences with a statement of purpose from the airport's governing body. The statement of purpose, sometimes referred to as a *mission statement,* establishes the general framework within which the airport operates. Within this general framework the controlling body/board sets out broad guidance in the form of policy to assist executive management in making operational decisions without the necessity to refer back to the Board or Committee of Control. Goals are established on the basis of the airport's stated purpose and, arising out of these, objectives can be set to specify what steps management will take to reach the goal(s). The level of achievement of each objective can be prescribed by a standard, using references such as quality, quantity, cost, and time frame. The standard, therefore, represents an acceptable performance for a particular area and represents a measurable value. It has been said that "what gets measured gets done."

A program based on performance standards can then set out the action steps necessary to achieve the objective(s) at a particular level of performance. Although purpose and goals are usually interlinked, the former will tend to remain unchanged as the basic reason for the airport's existence. Goals, on the other hand, might change with circumstances. For this reason, goals have to be timely and realistic. Some might be seen as being particularly critical to the success of the airport; these will be classified as "key result" areas. One such area at an international airport might be the efficiency of immigration and customs control facilities (U.S. Federal Inspection Services). A more detailed example of goal setting and performance measurement with respect to a key result area is given in Table 14.2.

Governing bodies vary widely in the ways in which they set out *purpose*, yet it is an essential starting point. The statement need not be elaborate or lengthy; in fact in most instances, the shorter, the better. Thus the stated purpose of Toronto International Airport is that it: "will continue to be an important component of the Canadian Civil Air transportation system. It will be a major national hub for domestic, transborder and international air travel and the primary terminus for major air carriers serving the Toronto area." In the Articles of Association of Amsterdam's Schiphol Airport Authority, the stated purpose is: "the performance of airport operations, including in particular the construction, maintenance, development, and operation of Schiphol in the widest sense of the words and furthermore the performance of all other acts of a commercial or financial nature, either directly or indirectly related to aviation, both for its own account and for account of, or in union or conjunction with third parties." Somewhat narrower goals were stated for the former British Airports Authority when it was created in 1965 to operate a number of British airports. These were: "to plan, develop and operate these airports to meet the present and future needs of the international and domestic air transport system in order that travelers and cargo may pass through them as safely, swiftly and conveniently as possible." The purpose of the Port of Portland, Oregon, which operates both the seaport and the airport, is stated in the following way: "Under the applicable laws of the State of Oregon, the object, purposes and occupation of the Port of Portland are to promote the aviation, shipping, maritime, commercial and industrial interests of the port for the convenience of the public."

Aeroport de Paris, founded at the end of 1945, is a multiairport authority responsible for those airports within 31 miles (50 km) of Paris, including Charles de Gaulle, Orly, and le Bourget; a number of light-aircraft fields, and a heliport. Its defined objectives are: "to organize, operate and develop all civil aviation transport installations centered on the region of Paris, to facilitate the arrival and departure of aircraft, to provide navigational guidance

TABLE 14.2   Example of Goal Setting and Performance Measurement—Key Result Area

Key Result Area: Physical plant and equipment	Goal: To provide the highest degree of maintenance possible for the physical facilities and to provide for timely, cost effective, capital improvements for future needs
Objective (including standard)	**Performance Review**
To design and manage construction of all maintenance projects in financial year on schedule and within budget.	Two projects in the total amount of $67,107 were completed in the first quarter. Other maintenance projects under design and construction.
To process all tenant projects in financial year while reducing the controllable processing time by 25 percent.	A procedure has been initiated for projects not affecting public or operational areas that eliminates two review steps. Processing time has been reduced at least 50 percent. Working on streamlining approval procedures for major tenant project s to cut down review time.
To design and manage (over $1,000,000) nonterminal capital construction projects so that the *average* cost of change orders does not exceed 5 percent of awarded contracts.	Average cost of change orders on contracts completed during first quarter was 3.13 percent. Percentage calculated on six nonterminal capital projects completed during quarter.
To achieve a cleanliness satisfactory rating for 80 percent of the inspections conducted by custodial services.	99.2 percent of all inspections conducted by custodial services unit received a satisfactory rating. Rating procedures are being reviewed to determine validity of monitoring methods and inspection procedures.
To provide 70 percent of the available maintenance manhours for planned and scheduled work in financial year.	Only 7.1 percent of the available maintenance manhours was reported as devoted to planned and scheduled work. Only two shops (i.e., plumbing and painting) are currently operating under limited centralized planning and control. As the system is completely implemented and becomes fully operational during the last six months of this fiscal year, a significant increase in percentage of craft time assigned to preventive maintenance should occur. (The 7.1 percent report is based on *total* maintenance manhours available.)

to ensure loading, unloading and surface routing of passengers, merchandise and material transported by air as well as all auxiliary installations."

The Port Authority of New York and New Jersey operates four airports and two heliports in the New York–New Jersey metropolitan region and also marine terminals, bridges, tunnels, and bus and truck terminals. The

Authority's published objectives are: "to plan, develop and operate terminals and other facilities of transportation, economic development and world
trade that contribute to promoting and protecting the commerce and economy of the Port district." Clearly, the goals of an authority as set out in the
original statement of purpose are often too vague or too narrow for detailed
management purposes. Management objectives become defined as an organization matures, and evaluation of management must be undertaken in the
light of a realistic set of goals. The true working sets of goals differ between
airports as each airport sits in its own institutional and physical environment handling its own mix and volume of traffic. It is apparent that goals
can be set in a number of areas, and it is likely that any single airport will
have more than a single goal. Table 14.3 defines the principal goal areas and
indicates a number of objectives within each goal area that might be used as
a management target. It is emphasized that the list of objectives is not
meant to be exhaustive, merely exemplary. These illustrate that an airport
administration may pursue a wide range of objectives leading to a number
of goals and that some of these objectives might be contradictory.

14.6 Management Structures

Figure 14.1 shows in simple terms the structure of the management system
for a single airport. Policy is determined by the policy board, which through
the political process is responsible in some way to the public. The board
might be directly elected, as is frequently the case in the United States, or it
could be appointed by the elected officials, which is a more common situation in European countries. The major airport policy decisions of the board
give overall guidance of the chief executive (airport director) in the management of the airport. The chief executive directs the running of the airport within these constraints reporting back to the policy board, which
monitors that the policies are being carried out. In the ideal situation, a policy board sets policy only and does not interfere with the actual running of
the airport. Conversely, the chief executive concentrates on running the airport, but does not determine major policy decisions, which are rightly considered a political area. Inevitably, however, there is a blurring of this rather
simplistic demarcation of responsibilities, and often the Chief Executive is a
member of the policy board itself, for example, at Lambert, Saint Louis
International, United States. Table 14.4 outlines the principal functions of a
policy board. These include deciding the goals and objectives that management should pursue and defining the level of resources that it is able to com-
mit. Concurrently, it must explicitly define the level of service that it aims to
provide and the nature of the traffic it expects to serve. Its final function is
to monitor management performance to ensure that the interests of the
public and the airport are being properly served. Here it might well en-
counter some difficulty due to the wide diversity of opinions coming from

TABLE 14.3 **Examples of Goal Areas and Objectives for Airport Administrations**

Goal	Objective
Financial	To maximize profit or minimize loss To maximize turnover To minimize financial risk To minimize investment
Economic	To generate industry and commerce for the region To provide high levels of employment at the airport To promote high income levels for those working at airport To promote a particular type of industry for the region To minimize user costs
Operational	To promote high levels of aviation safety To provide for training flights To segregate charter from schedule traffic To provide 24-hour operations To provide all-weather operations To provide general aviation facilities To provide a wide range of commercial activities in the passenger terminal To provide luxurious facilities in waiting areas To provide very stringent security controls
Social	To maximize the accessibility of the airport to the region it serves To maximize access to the overall air transport system To provide easy connection (e.g., shuttles) to particular destinations To alleviate unemployment To spread the benefits and disadvantages of the airport evenly across the community
Political	To provice an airport within a particular political jurisdiction To provide a base for the national carrier To provide a showpiece "gateway" to a country To demonstrate national technological capability To provide a military airbase in times of war
Environmental	To minimize land take To minimize noise impact over populated areas To spread noise impact evenly over populated areas To minimize effect of airport traffic on surface routes To minimize air and water pollution To eliminate air and water pollution To eliminate noise impact at night To provide a pleasant visual environment on the landside of the terminal

airlines, government agencies, and the general public (including the local community). It might be that the expression of these various opinions on performance will change the stated goals.

In turn, the airport executive management is only too well aware that comment arises not so much in relation to goals as it does in relation to the implementation of these goals on a day-to-day basis. The actual day-to-day management of the airport is carried out by the component departments, who report through their department heads to the airport director. A typical

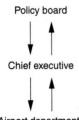

Policy board

Chief executive

Airport departments

Figure 14.1 The airport management system.

TABLE 14.4   Functions of Policy Board

Policy establishes a pattern for management to follow in making decisions and reduces the need to refer back to the governing body for decisions.

Goal setting	Operational
	Financial
	Economic
	Social
	Political
	Environmental
Resources	Fiscal
	Physical
	Manpower
	Scale of operation
Level of service	Quantity and quality
	Client structure
Management monitoring	Scale of operations
	Competence
	Fiscal
	Operational
	Service levels
	Planning
	Medium term
	Long term
	Goal achievement

departmental structure with principal responsibilities is shown in Figure 14.2, which shows four main departments: operations, administration, engineering, and planning and finance. Many airports combine the administration and finance operations into one department. There is no compelling reason for separating the function although the four-department structure shown does separate audit from the finance department; this is quite desirable. The structure shown is reasonably widely used, and the reader is invited to compare this conceptual structure with the actual structures shown in Chapter 1.

Figure 14.3 shows how individual reports within the four departments are conveyed, preferably on a continual basis, to the airport director. Material from the subject area reports enable the airport director to inform the policy board of the current status of the airport. This is the only way that the board can monitor the performance of the airport administration. It is clear that the policy board itself must have a clear idea of its goals and objectives, otherwise performance monitoring is devalued to a relatively meaningless examination of profit or loss and a comparison of current traffic figures with previous years. In fact, too often this is the only performance monitoring carried out at many airports where increases in traffic flow are automatically regarded as good, belying the service nature of the airport function. The annual report of an airport authority is the management's published statement of achievement for a working year. Too often these reports concentrate far too heavily on the financial aspects of performance. An examination of Figure 14.3 makes it

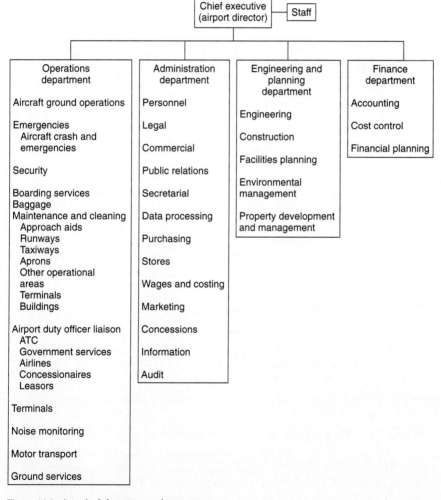

Figure 14.2 A typical departmental structure.

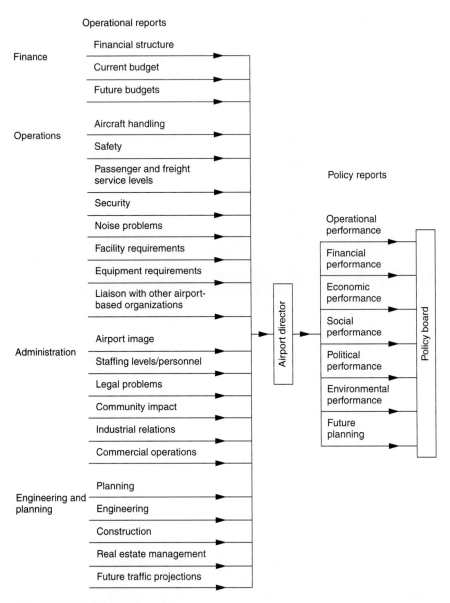

Figure 14.3 Performance monitoring.

clear that it is quite possible for an airport authority to have very favorable financial performance yet provide a poor overall level of service to the community. However, there is a growing awareness of the need to balance the emphasis on financial performance with an equal attention to service for both passenger and airline customers, and the need to meet the ever growing challenges of competition within the air transport industry. Both of these aspects were addressed in the recent annual report of the Chicago Airport System where early mention is made of "providing passengers with the highest level of customer service," followed by the statement that:

> Our focus on customer service was not only good for the passengers who used Chicago's airports, but it also recognized the competitive nature of the airport business. Airlines can move their operations from one city to another, and with them go connecting passengers, jobs and air service.

In much the same vein, San Francisco airport, in its annual report for 1991, devoted considerable space at the beginning of its report to cataloging the "Passenger Services" available and introduced during the year. It was noticeable that both airports followed the widespread trend among the world's major airports of providing the means for passengers to provide their assessments of the airport's performance.

But there are many hub airports in a monopoly position, sustained by public investment. A policy of overcharging, under investment, and excessive commercialization can produce the appearance of a well-managed airport if the evaluation is made solely in terms of financial performance. However, the public is entitled to ask whether these policies best serve the original public intentions as stated in terms of promoting either the general economic development of a region or civil aviation. Airport management is often placed in the "no win" situation of suffering criticism for being too commercial or not sufficiently commercial in the operation of the facilities. In such situations one very useful procedure is the management audit by means of which management performance is assessed by an impartial external body.

14.7 External Assessment of Management Performance

A management audit involves a thorough examination and evaluation of an organization's purpose, goals, and objectives and degree of success in achieving those ends. The aims of such an audit are to provide the appropriate information to enable management to improve its methods of operation. Although the process has its roots unmistakably in the financial area, the scope has now extended to all aspects of management. The idea is a good one because it enables a fresh view to be brought to bear on the operation of an organization.

In 1972, the General Accounting Office (GAO) of the U.S. government suggested three broad areas for the management audit:

- The use of the organization's financial resources
- The economy and efficiency of the organization
- The level of achievement of desired results—objectives

In some states of the United States, for example California and Pennsylvania, state legislation requires that certain public bodies be subjected to periodic management audits, usually every five years. Among these public bodies are the local government, the police, and, in some cases, airports.

Because every airport has a unique physical, social, and institutional environment, there is no set form for a performance review. For example, in many developed countries, the problems of environmental impact and community response are seen to be extremely important elements of the management task. In other parts of the world, these concerns are considered less critical than the generation of air traffic and the economic development that it is hoped will follow. The structure of a performance review is specific to the airport involved. Therefore, the reader should not be led to conclude that the following example from a U.S. airport represents a generalizable format that is universally applicable.

Example of the structure of a performance review

Areas examined

- Overview
- Organization and management
- Planning and development
- External relations
- Financial operations
- Financial planning and management
- Physical operations
- Goals and objectives

Overview

- General perception of operation, including level of service to travelers, motivation of employees, industrial relations, and financial performance
- Degree of budgetary control
- Utilization of staff in relation to traffic carried
- Trends in operating income and expense

Organization and management

- Reflection of priorities and requirements in organizational structure
- Role of chief executive and duplication of activities of deputy directors
- Scope of responsibilities of deputy directors
- Location of functions with individual divisions, overlap of responsibilities or areas inadequately covered
- Internal capability for formal financial and management analysis

- Ability of existing management structure to cope with future conditions
- General status of management training program
- Particular training problems associated with vacancies caused by staff close to retirement
- Management of general airport maintenance
- Design and usage of management information systems
- Currency status of airport operations manual

Planning and development

- Professional capability and training of staff
- Status of planning philosophy—initiative or reactive
- Integration of long-term master plan and short-term project planning
- Suitability of capital project program in terms of time frame used, statement of priorities, realism of cost and time estimates, and usefulness in disseminating information

External relations

- Perception of primary community concerns and administrative reaction to these concerns
- Effectiveness of community relations bureau in publicity and community participation areas
- Appreciation of community interest and concerns at all levels of the airport management
- Community opinion of airport
- Sufficiency of staff for task

Financial operations

- Modernity and acceptability of accounting practices—automation
- Billing procedures—manual versus automated
- Periodic reporting, analysis, and control of budget variances
- Availability and use by management of routine periodic accounting reports

Financial planning and management

- Time horizon used for forecasting expenditures and revenues
- Coordination of capital budgeting
- Provision for future investment sources
- Formal structure of the financial planning and management function

Physical operations

- Service levels in processing and accommodating inbound and outbound passengers

- Availability of space for medium-term traffic growth
- Adequacy of governmental inspection facilities and procedures
- Utilization of facilities in terms of number of aircraft per stand per day, apron traffic congestion
- Baggage handling performance (time, congestion, damage, and loss)

Management by objectives (MBO)

- Necessity of establishing a MBO program
- Key elements of the MBO program
- Community relations
- Affirmative action
- Passenger services
- Safety and security
- Resource management
- Staff development
- Cost control
- Physical plant and equipment
- Long-range planning and development
- Revenue generation
- Goal development
- Objectives development
- Performance measures

The scope and complexity of this type of review point clearly to the need for more than one person to conduct it and also to a need for the review audit team to have a sound knowledge of business practice, be familiar with airport management, and have the necessary communications skills. It should then be possible to:

- Identify the objectives of the sponsoring group and the airport management and ensure that they are fulfilled.
- Identify the needs of those most likely to carry out the recommendations of the audit and see that these are met.
- Stay within the time scale and budget for the audit.
- Maintain an objective and businesslike approach to all aspects of the audit.
- Create an awareness of the importance of the audit.

As an illustration of the workings of an audit, it is useful to review the results of an actual airport audit from examples of some of the areas covered.

Organizational structure

The following areas were reviewed:

- External and internal environment in which the airport operates
- Current departmental activities
- Present organizational structure
- Management competence
- Operating philosophy

The audit team's findings included:

- Increasing pressures on the airport from external sources such as community and special interest groups
- A large-scale redevelopment plan about to be inaugurated
- Changes of user airlines as a result of deregulation
- Lack of clarity in defining functions of operations department

Recommendations included:

1. Organizational structure should be changed by dividing management into two broad areas: internal and external. The chief executive should take over direct responsibility for all external matters in addition to performing an overall coordinating role. "Internal" management would be assumed by an airport director who would directly oversee three primary functional divisions: administration, operations (including maintenance), and facilities planning. Each of the three functional areas would have a deputy director as its head.

2. Consideration should be given to separating community relations from the public relations department and having it report to the chief executive.

3. The airport board of control should assist the chief executive in dealing with external affairs.

4. The airport operations manual should be revised and republished.

14.8 Implementation

Although consultants' reports, including management audits, are widely accepted in the business world, it is also an accepted fact that the recommendations arising from them are not always implemented. This might be as a result of a conscious decision by the policy-making board that they are unacceptable, or it could be as the result of internal resistance of special-interest groups within the organization itself. The latter are usually easily overcome by a resolute management.

Nevertheless this whole area of performance review, whether on an individual or organizational basis, is a sensitive one and the implementation of change is inevitably disturbing. The disturbance can be minimized and implementation of recommendations thereby assured provided that:

- A framework of performance is published
- The performance criteria are clear and understood by all
- Departments contribute to the establishment of criteria
- The audit/review team have well established qualifications and experience of airports
- The audit process covers a reasonably wide spectrum of airport personnel

It is as well to recognize that such performance reviews are limited in scope and by the time taken to conduct them. To this extent it might be that certain recommendations or details cannot easily be implemented and it is usual for senior management to make counter proposals to their policy-making body when they judge this to be the case.

14.9 Internal Assessments

The full scale external management audits are by no means the only way of assessing an airport's performance, and in any case they are not universally applicable. It is conventional management practice for self-auditing to be carried out in order to compare objectives and their associated standards with actual achievements. It can be argued that because there are internal reviews they are subjective with all the shortcomings that this implies. This need not be so, provided that the participants are committed to the process as a means of improving the airport's operational and service competence. Again the stated purpose, goals and objectives are a necessary starting point. The objectives themselves with the associated action programs will provide the basis for a checklist to facilitate a systematic approach to the assessment. The compilation of detailed performance records will be necessary for review and analysis so that future action can be determined (e.g., a revision of objectives/standards). In this way airport management is both user and observer.

It could well be that records of performance can include a consumer audit; many airports have adopted the practice of seeking feedback from passengers. Chicago has its customer service hotline to the airport customer service division; San Francisco has an annual survey in which passengers are asked by consultants to rate the airport's services. Customer comment cards are often provided in airport terminals. London's Heathrow Airport, which was one of the earliest to introduce this system, enjoys a well-earned reputation for the detailed responses it provides to passengers' comments. Airport operational management has also to assess its performance in relation to its other customers—the airlines. This can usually be provided through the airports' operators committee at monthly meetings. Some knowledge of the airlines' own assessments of their passengers' concerns are useful for airport management. One major U.S. carrier, in listing areas in which passengers are particularly sensitive includes:

- On-time departures and arrivals
- Baggage handling
- Cleanliness of airport facilities

In the climate of deregulation and privatization, which has intensified in recent years, competition among airlines is a well-established fact of life and has already accounted for the disappearance of some well-known names. It is only with the introduction of the newer highly efficient extended-range aircraft that competition has emerged as a factor that airports, too, must consider. As one airport commissioner remarks: "Chicago's airports are not monopolies. We're in direct competition for air travel business with other mid-continent hubs—not to mention competition as an international gateway for the U.S." Like airlines, airports are also facing a new challenge—meeting the rising demand for quality in service. The incentives are there to look anew at the ways in which airport operational performance can be assessed and measures taken to meet the continuing challenge which air transport presents.

The Airport Operations Manual

15.1 The Function of the Airport Operations Manual

The complexities of today's airports cover a great variety of disparate elements from special purpose buildings to complex equipment needed to meet the needs of aircraft, and passenger and cargo handling. Procedures to enable airports to function are equally complex and involve many different organizations, all of which need to have the fullest possible information if the airport is to operate efficiently. It is for this reason that an *Operations Manual* can provide such a valuable tool for management. Since one of the primary considerations for the air transport industry is safety, in the air and on the ground, a substantial part of any Operations Manual will be devoted to this subject. In most countries the primacy of safety is recognized by a requirement that airports must be licensed before they can be used for air transport. As part of the licensing process, airports might be required to produce and maintain an *airport manual*. An airport manual required by national regulations is usually limited in its contents to those elements of the airport related to the safe conduct of aviation. A particular example is described in Section 15.4.

In Section 15.2 a document of much broader scope is discussed. The Airport Operations Manual here is conceived as a document that describes the operation of the airport across its entirety. In scope it will contain all those aspects required as part of the licensing process, but in addition, all those additional aspects that contribute to the working of the airport *system* (e.g., the facilities and procedures necessary for the safe and efficient movement of passengers, baggage and cargo). The Operations Manual can therefore serve as a basic reference document in which the facilities and procedures of the various operational areas are recorded. It can also serve as a useful introduction training aid for newly hired or transferred personnel.

15.2 A Format for the Airport Operations Manual

Ideally the operations manual will embrace the airport in its entirety, outlining all procedures to be adopted under the various conditions of operation. It will also contain details of the principal infrastructure and equipment for reference purposes. The manual should be seen as the updated standard reference that describes the condition of the airport and how it is operated in all its aspects. A suggested format for a comprehensive operations manual for an airport follows:

A. Introduction

Purpose and scope of manual
Format and table of contents
Licensing bodies
Type of license, terms, and conditions of operation

B. Technical administration

1. Status

Legal identification, addresses of airport, and licensee.
Operating hours.
Latitude and longitude of airport, grid reference, location of airport reference point, elevations of aprons and runway thresholds, airport reference temperature.
Runway lengths, orientation.
Description of aprons and taxiways.
Level of rescue and fire-fighting service protection.
Level of air traffic control.
Level of airport radio and navaids.
System of aeronautical information service available.
Organizational structure of airport administration: names and addresses of key personnel.
Airport movement statistics—passengers, operations, airlines served, based aircraft, aircraft types served.
Airport statistics—staff numbers, visitors, physical areas of facilities by type.

2. Administrative procedures

Procedures for complying with requirements relating to accidents and incidents.

Procedures for promulgation of operational status of facilities.

Procedures for recording aircraft movements.

Procedures for control of construction affecting safety of aircraft.

Procedures for control of access to airport and for control of vehicles while on airport operational areas.

Procedures for control of apron activities.

Procedures for reception, storage, quality, control, and delivery of aviation fuel.

Procedures for the removal of disabled aircraft.

C. Airport characteristics

Plans showing the general layout of the airport, preferably at some scale in the region of 1:2500. Elements to be shown include the airport reference point; layout of runways, taxiways and apron; airport lighting including VASIs and obstruction lighting; location of navigational aids within the runway strips; location of terminal facilities and fire-fighting rescue facilities.

Data necessary for INS equipped aircraft, i.e., INS coordinates for each gate position.

Description, height, and location of obstacles that infringe on the protected surfaces.

Survey data necessary for the production of the ICAO Type A chart, which provides an aircraft operator with the necessary data to permit compliance with performance limitations within the runway, runway strip, clearway, stopway, and takeoff areas.

Data for and method of calculation of each of the airport runway cleared distances (TORA, EDA, TODA, and LDA) with elevations at the beginning and end of each distance. Where reduced declared distances are accepted due to temporary objects infringing on the runway strip, transitional approach or takeoff surfaces, the method of calculating the distances should be indicated.

Operational limitations due to obstructions and so on.

Details of the bearing strengths and surface conditions of runways, taxiways, and aprons.

Temporary markings of displaced thresholds.

D. Operational procedures

Ground movement control procedures—rules of access, right-of-way, and vehicle obstruction marking. Structure of control and duties

Responsibility and procedures for routine operations on apron, including marshaling, docking stand allocation, aircraft servicing

Routine inspection procedures of movement areas, including runway strips, runways, aprons, taxiways, grassed areas, and drainage

Procedures for measurement and notification of runway surface condition—responsibility for reporting, measurement of runway braking action, depth and density of snow and slush, frequency of reporting.

Details of the airport snow plan, including categories of snow warnings: preliminary and final with duration and intensity forecasts; organization and chain of responsibility; standing instructions for procedures in snow conditions; equipment including maintenance and care; training of personnel; procedure or clearing runways, taxiways, aprons, airport roads, and domestic areas; hiring additional equipment.

Procedures for the general cleaning and sweeping of runways, taxiways, and aprons.

Details of bird hazard control plan including methods of control: technical measures, killing, and environment modification. Treatment of runways, taxiways and aprons, grassland, built-up areas and special areas, such as garbage dumps. Control of hazard outside airport boundaries

Procedures for determining the availability of grass runways after heavy rains or flooding.

E. Rescue and fire-fighting services (RFFS) (or emergency orders)

Purpose of emergency orders.

Categories of emergency: aircraft accident, aircraft ground incident, full emergency, local standby, weather standby, aircraft bomb warning, building bomb warning, hijack, act of aggression on ground, building fire, approach alert within 10 miles (15 km) of airport.

ICAO category of the airport.

Range of on-airport and off-airport rescue and fire-fighting services.

Details of appliances and extinguishing media necessarily available on-airport to meet the ICAO category requirements.

Structure of airport emergency services: organization, qualified staffing and chain of command.

The airport emergency procedures, including in-shore rescue where necessary. Rendezvous points, grid map, designation, and maps of remote search areas.

Degree of response of service.

Training program for RFFS.

Procedures and limitation of aircraft operations during temporary depletion of RFFS.

Other responsibilities: fire prevention, storage of dangerous goods.

Procedures for removal of disabled aircraft: general procedures alternative runways, emergency removal, designation of recovery team and responsibilities; formulation of recovery plan; phasing of recovery.

Medical facilities: quantities of equipment available, list of qualified personnel, structure of medical organization and control, arrangements for obtaining external medical assistance, medical procedures in case of accident, allocation of casualties to hospital.

Responsibilities in case of emergency: air traffic control, airport fire service, telephone switchboard, airport security, ground services, motor transport pool, airport engineering, airlines and handling agents, airport authority staff, off-airport services (e.g., fire services, police, medical services)
Location of ministers of religion.
Emergency communications systems; emergency direct lines, radio channels, direct ex-directory lines, radio-equipped vehicles, intercom channels, posted instructions.

F. Airport security plan

General description and rationale of security plan.
Responsibilities for airport security: airport administration and airport director, airlines, and other authorities.
Structure of airport security organization, police, and definition of diversion responsibilities.
Degrees of security (1) nominal, (2) period of heightened tension, (3) alert.
Persons authorized to prevent a flight or order an inspection or aircraft search.
Responsibilities for action in case of security alert at airport: general responsibilities, airport switchboard, airport fire service, police, air traffic control, ground services, airlines, and handling services.
Search procedure for aircraft: positioning, conduct of search, baggage identification, and cancellation of alert.

G. Airport lighting

The lighting scale of each runway, including approach and threshold lights.
The airport lighting system: method of operation, general layout diagram and individual circuit loops, ancillary lighting control, control console, constant current regulators, ac distribution switchboard.
Maintenance and routine inspection procedures.
Procedures for operation, including various brilliancy settings.
Failure maintenance and fault-finding procedures.
Standby power arrangements.
Technical diagrams of all lighting units.
Location of obstacle lighting on and off airport; responsibility for maintenance.

H. Meteorological Services

Organization and structure of meteorological services; staff structure and responsibilities.
Class of service and information provided—form of messages.
Equipment—uses and maintenance.
Supply and use of meteorological services.

I. Air Traffic Control Services

Description of the system for managing air traffic in airspace in the vicinity of the airport including: organization and responsibility of air traffic services; aeronautical information service.

Rules of the air and air traffic control; general flight rules.

Rules governing the selection of the runway in use and the circuit direction.

Standard procedures.

Noise abatement procedures.

Search and rescue alerting.

Method of obtaining and disseminating meteorological information, including RVR, visibility, and local area forecasts.

J. Communications and navaids

Description and procedures for the use of general communications channels; AFTN, SITA, ARINC.

Description and procedures for use of air/ground and operational ground radio where these are not covered by air traffic control procedures.

Description of and operating procedures for radio and radar navigation aids; installation procedures; inspection and maintenance.

K. Signals and marking

Location of the signals area; procedures for display of temporary and permanent signals.

Location of wind socks.

L. Access provision

Airport location plan showing regional road network and public transport routes.

Access routes in immediate vicinity of airport.

Provision of parking: number and location of spaces; contract and obligations of parking operators.

Taxis and car hire: names and contractual obligations of operators.

Passenger Access Information System.

M. Passenger terminal

Organizational structure of terminal administrative staff: description and responsibilities.

Passenger Information System: scale, equipment, maintenance procedure, standard information formats.

General information: airlines operating from airport, passenger and aircraft handling companies, airline interline handling agreements.

Layout of passenger terminal.

Arrivals, method of processing.

Departures: method of processing, definition of permissible cabin baggage, security arrangements, and definition of dangerous goods.

Special handling of passengers: procedures and equipment for handling disabled persons; unaccompanied minors, and VIPs.

Airport services and concessions: procedures for operating airline ticket desks, mishandled baggage procedures; listing of, location of, and contractual obligations of concessionaires such as: duty-free shops, catering, banks, car hire, insurance desks, valet service, shops, hotel booking facilities, postal and telephone services.

Description and operational arrangements of facilities for meeters, senders, and spectators.

Description and operational arrangement of support facilities for airport airline and other staffs working in the passenger terminal area.

N. Cargo handling facilities

Organizational structure of airport authority, airlines, and other cargo handlers in cargo area.

Location of facilities and layout.

Cargo handling procedures: export and import, including customs clearance requirements.

Description of airport and customs computer facilitation system.

Procedures for handling mail.

Procedures for handling airlines' company freight.

O. General Aviation

Description and layout of general aviation facilities.

General regulations relating to general aviation operations.

Inbound procedures.

Outbound procedures.

Scale of fees, method of collection.

Aeronautical Information Services, meteorological services, and lounge arrangements for general aviation.

P. Typical appendices

Amendment procedures.

Distribution list of operations manual.

Necessary telephone numbers, airlines, concessionaires, agents airport staffs, other airports.

Bylaws and regulations of airport.

Two-letter airline codes.

Three-letter airport codes.

Standard information symbols used at airport.

15.3 Distribution of the Manual

The airport operations manual should be regarded by those administering the airport as the definitive document containing a statement of standard operating procedures and a description of airport equipment and facilities. An updated version of the manual should be held by the chief administrator of the principal operations areas of the airport. At a very small airport, the manual would be a rather simple document, and probably only one copy would exist, which would be kept by the director. In a large complex operation, a more likely distribution list for the manual and its subsequent amendments would be:

- Airport director
- Deputy director
- Assistant director operations
- Assistant director engineering
- Assistant director administration
- Chief fire officer
- Chief air traffic control officer
- Airline station managers of based airlines

Whether simple or complex, the updating and amendment of such a manual requires close attention. Sources of information should be carefully checked. It is possible, as in the case of one major U.S. airport, for an accident investigation to discover uncertainty regarding the origin of information on the precise length of a runway.

Updated sections of the manual that are pertinent to their own sections should be available to the line managers of the operations of the passenger terminal, the cargo terminal, the passenger and cargo aprons, air traffic control, engineering meteorology, fire fighting and rescue, and maintenance. It is also useful to use appropriate sections of the manual as instructional aids in the training of the various grades of manual workers.

15.4 U.S. Example: FAA Recommendations on the Airport Certification Manual

In a number of countries, the Airport Operations Manual or the Aerodrome Manual, in some form or another, is linked to certification of the facility as an airport. In the United States, Federal Regulations Part 139, which deals with the certification of land airports serving scheduled carriers (Code of Federal Regulations 1994) requires that each applicant for an airport operating certificate must submit an Airport Certification Manual to the FAA[1]. For the operation of an airport serving unscheduled carriers a *Limited Airport Operating Certificate* can be issued based on the submission of *Airport Certification Specifications.* The administration sets out a number of guidelines on how a manual is to be prepared, recognizing that the suggested format is not all inclusive but is intended to provide broad guidance so that each operator will have flexibility in developing a manual to fit the particular airport. The recommendations are largely related to a description of the data that must be supplied for eligibility for certification and other data to indicate compliance with operational rules. The suggested format for the FAA manual is substantially less broad than that indicated in Section 15.2, and includes the following:

Suggested airport certification manual—FAA format

1. Introduction

   (*a*) Name and location of airport
   (*b*) Mailing address of airport manager or operator

2. Personnel

   (*a*) List of key personnel and job titles
   (*b*) Brief description of functions
   (*c*) Organizational chart showing operational lines of succession

3. Airport familiarization

   (*a*) Brief description of airport and primary activities (e.g., named carriers/equipment)
   (*b*) Maps and charts showing such things as general layout of the airport terrain features and runway and taxiway system

4. Pavement areas

   (*a*) Movement areas available for air carriers with their safety areas; use of a map or diagram is recommended
   (*b*) Describe the procedures for maintaining the paved and safety areas

5. Marking and lighting

   (*a*) Describe runway and taxiway systems of identification
   (*b*) Indicate which types of operable lighting systems are installed on surfaces used by air carrier aircraft

---

[1]For detailed guidance reference should be made to the code of Federal Regulations, Title 14. Parts 77, 105, 139, together with Advisory Circulars in the 139 and 150 series.

(*c*) Describe the markings systems in use at the airport (e.g., runway markings, taxiway centerline and edge markings)

(*d*) Indicate the location of each obstruction required to be lighted or marked within the airport's area of authority

(*e*) Describe and list the procedures for maintaining the marking and lighting systems

6. Snow and ice control

(*a*) Describe equipment available for snow and ice removal and indicate whether an airport or external municipality/contractor

(*b*) Specify the selection and application of approved materials for snow and ice control to avoid possible engine ingestion

(*c*) Provide details of the snow and ice control plan

7. Aircraft rescue and fire fighting

(*a*) Specify the airport's fire-fighting index and name the largest applicable aircraft

(*b*) Provide a list with descriptions of the fire-fighting and rescue equipment and agents to be used to meet the certification requirements, including the results of the response time as determined by test runs

(*c*) Describe methods of alerting RFF crews to an emergency

(*d*) Show on a grid map the location of the fire station(s) on the airport and primary traffic routes for the fastest response to all air operations areas; also show the designated emergency access roads

(*e*) Specify the operational requirements relating to on-airport and off-airport responses to emergencies

(*f*) Name personnel authorized to dispatch (off airport), reduce and recall ARFF resources

(*g*) Procedures for notifying air carriers of any changes to normal complement of ARFF unit

(*h*) List the channels of communication available to the ARFF unit

(*i*) List and describe RFF training programs

8. Handling and storing of hazardous substances and materials

(*a*) Identify the location of personnel designated to receive and handle hazardous articles and materials (in those cases where the airport operator acts as a cargo handling agent)

(*b*) Describe the controls and procedures in use to assure that the cargo shipped can be handled and stored safely, including any special handling procedures required for safety

(*c*) Indicate the location of the special areas on the airport for the storage of flammable liquids and solids, corrosive liquids, compressed gases, and magnetized or radioactive materials.

(*d*) List the procedures and devices used for safely storing, dispensing, and otherwise handling fuel, lubricants, and oxygen on the airport (other than articles and materials that are, or are intended to be, aircraft cargo) including:

(*i*)   Grounding and bonding

(*ii*)   Public protection

(*iii*) Control of access to storage areas

(*iv*)   Fire safety in fuel farm and storage areas

(*v*)   Fire safety in mobile fuelers, fueling pits, and fueling cabinets

(*vi*)   Level of training required for fueling personnel

(*vii*) The fire code of the public body having jurisdiction over the airport

9. Traffic and wind direction indicators

(*a*) Indicate the location(s) of wind direction indicators on the airport, and

(*b*) Arrangements for displaying and maintaining signals and markings, both permanent and temporary

10. Airport emergency plan

(*a*) The Airport Certification Manual must include a comprehensive emergency plan

Note: In view of the extensive technical information to be included, detailed guidance is provided by a separate advisory circular (AC 150/5200-3 1), which was issued January 1989.

11. Self inspection program

(*a*) Details of self inspection program, including frequency, personnel training, information dissemination, and corrective action procedures for unsafe conditions

(*b*) Appropriate checklists for: continuous surveillance, periodic condition evaluation, special inspection.

12. Ground vehicles

(*a*) Indicate on a map of the airport the limits of access to movement areas and safety areas

(*b*) Designate those ground vehicles approved to have access to the movement area and safety areas, and the procedures necessary for their safe and orderly operation

(*c*) Detail the arrangements made for communications with and control of ground vehicles operating in the movement areas and safety areas

13. Obstructions

   (*a*) Identify each object within the area of authority that is identified in
       Federal Regulations, Part 77
   (*b*) Describe the marking and lighting of each of the obstructions
   (*c*) Provide an airport layout plan locating all lighted obstructions that
       fall within the airport authority, keyed to a narrative description
       that also identifies any parties, in addition to the airport authority,
       responsible for their maintenance

14. Protection of navaids

   (*a*) Designate the airport department responsible for surveillance of
       all proposed construction of facilities on the airport to avoid any
       possible derogation of a navaid
   (*b*) Describe procedures to be used to provide protection or assistance
       to the owner (if another person) for the protection of navaids lo-
       cated on the airport
   (*c*) Procedures and assignments for security patrols, fence mainte-
       nance, etc. (where this is judged to be necessary due to the place-
       ment of navaids)

15. Public protection

   (*a*) Detail the procedures, devices or obstacles used to prevent inad-
       vertent entry of persons or vehicles into any area containing haz-
       ards for the unwary trespasser
   (*b*) Provide an airport layout plan indicating the type and location of
       fencing and fence gates also showing those areas restricted from
       use by the general public

       Note: The prevention of intentional infiltration of airport security
       areas is within the preview of the regulation on airport security,
       Federal Regulations, Part 107.

16. Wildlife hazard management

   (*a*) Indicate the nature of the existing conditions on the airport and
       the control techniques to be employed if a wildlife hazard exists, or
       show why there is no wildlife hazard problem
   (*b*) If a problem exists, include a wildlife hazard management plan

17. Airport condition reporting

   (*a*) Describe the procedures used for identifying, assessing, and dis-
       seminating information to air carrier users of the airport
   (*b*) Describe the various internal means of communication that may be
       available for urgent dissemination of information
   (*c*) Document any system of information flow agreed with airline ten-
       ant(s)
   (*d*) Describe the conditions and procedures for issuing NOTAMS

18. Identifying, marking, and reporting construction and other unserviceable areas

    (*a*) Describe how construction areas and unserviceable pavement and
safety areas are marked and lighted

    (*b*) Describe the provisions made for identifying and marking any ar-
eas on the airport adjacent to navaids that, if transversed, could
cause emission of false signals or failure of the navaid

    (*c*) Describe how construction equipment and construction roadways
on the airport are to be marked and lighted when on or adjacent to
aircraft maneuvering areas

    (*d*) Describe procedures for the routing and control of equipment, per-
sonnel, and vehicular traffic during periods of construction on the
aircraft maneuvering areas of the airport.

19. Non-complying conditions

    (*a*) Conditions under which air carrier operations will be halted

    (*b*) Designate the airport department to be informed if someone dis-
covers an uncorrected, unsafe condition on the airport

15.5 United Kingdom Example: CAA Recommendations on Aerodrome Manual[2]

A less comprehensive format than recommended in Section 15.2 and by
the FAA (as shown in Section 15.4) is the *Aerodrome Manual* required by
the CAA in the certification of airports in the United Kingdom (Licensing
of Aerodrome 1990). An abridged form of the CAA's suggested contents
follows:

Suggested aerodrome manual—CAA format

Technical administration

(*a*) Name and address of aerodrome

(*b*) Name and address of licensee

(*c*) Details of the following:

    (i)   Latitude and longitude of aerodrome

    (ii)   Grid reference

    (iii) Location of aerodrome reference point

    (iv) Elevations of aerodrome and apron

    (v)   Availability of aviation fuel

(*d*) Name and status of official in charge of the day-to-day operation of
the aerodrome

(*e*) List of key personnel, their functions, and channels of responsibility

---

[2]Crown Copyright where applicable.

(f)   Procedures for complying with regulatory requirements relating to accidents and mandatory occurrence reporting

(g)   The system of aeronautical information service available

(h)   Procedures for promulgating information on the aerodrome operational state, temporary withdrawals of facilities, runway closures and so on

(i)   Procedures for recording aircraft movements

(j)   Procedures for the control of works, including trenching and agricultural activity, which may affect the safety of aircraft

(k)   Procedures for the control of access to the aerodrome and its operational areas, including the location of notice boards, and the control of vehicles on the operational areas

(l)   Procedures for maintaining apron control, including marshalers' instructions

(m)   Procedures for the reception, storage, quality control, and delivery of aviation fuel

(n)   Procedures for the removal of disabled aircraft

Aerodrome characteristics

(a)   Plans, preferably to a scale of 1:2500 showing the position of the aerodrome reference point, layout of the runways, taxiways and aprons; the aerodrome markings and lighting (including VASIS or Low Intensity Two-color Approach System [LITAS] and the obstruction lighting); the siting of navigational aids within the runway strips, and their degree of frangibility

(b)   Description, height, and location of obstacles that infringe the standard protection surfaces, and whether they are lit

(c)   Location, reference number, and date of the survey plans from which the data at (a) and (b) were derived and details of the procedures for ensuring they are maintained and updated

(d)   An up-to-date copy of the survey data related to the surfaces and areas on aerodrome and in airspace

(e)   Survey data necessary for the production (where appropriate) of ICAO Type A charts

(f)   Data for, and the method of calculation of, declared distances and elevations at the beginning and end of each declared distance

(g)   Method of calculating reduced declared distances when there are temporary objects infringing on the runway strip, transitional, approach, or take-off surfaces

(h)   Details of the surfaces and bearing strengths of runways, taxiways, and aprons

Operational procedures

(a)   Procedures for routine aerodrome inspections and reporting including the nature and frequency of inspections

(b) Procedures for inspecting the apron, the runways and taxiways following a report of debris on the movement area, an abandoned takeoff due to engine, tire, or wheel failure, or any accident likely to result in debris being left in a hazardous position

(c) Procedures for the measurement and promulgation of water and slush depths on runways and taxiways

(d) Procedures for the measurement and/or assessment and promulgation of runway surface friction conditions

(e) Procedures for the measurement and reporting of RVR

(f) Procedures for protecting the runway during low visibility takeoffs

(g) Procedures for the safe integration of other aviation activities such as; gliding, parachuting and banner towing

(h) Details of the aerodrome snow plan

(i) The bird hazard control plan

(j) Procedures for sweeping runways, taxiways, and aprons

(k) Procedures for determining the availability of grass runways following heavy rains or flooding

Rescue and fire-fighting services (RFFS)

(a) The RFFS category of the aerodrome, details of the appliances and extinguishing media to be available to meet the requirements for that category, and the quantities of operational reserves to be maintained

(b) The number of trained and qualified duty personnel

(c) The aerodrome emergency procedures, including inshore rescue, where appropriate

(d) The training program for RFFS services

(e) Procedures and limitations relating to temporary depletion of the RFFS

Medical

(a) Quantities of equipment available

(b) List of personnel qualified in first aid

(c) Details of arrangements for summoning external medical assistance

Lighting

(a) Lighting scale of each runway

(b) The aerodrome lighting system, its method of operation and type of fittings, arrangements for recording inspection and maintenance

(c) Procedures for the operational use of the lighting, including brilliancy settings

(d) The location of and responsibility for obstacle lighting on and off the aerodrome

(e) Standby power arrangements

Signals and markings

(a) The location of the signals area if any, and the arrangements for displaying signals and markings, both permanent and temporary
(b) Location of windsleeves

Air traffic services

Details of the following:

(a) The system for the management of air traffic in the airspace associated with the aerodrome, including procedures for the coordination of traffic with adjacent aerodromes except information of procedures already published
(b) Rules governing the selection of the runway in use and the circuit direction
(c) Noise abatement procedures
(d) Search and rescue alerting
(e) Method of obtaining and disseminating meteorological information, including RVR and meteorological visibility, and local area forecasts

Communications and navaids

(a) Description of and instructions for use of air/ground and operational ground radio communications where these are not covered in other manuals
(b) Description of and operating procedures for navigational aids

References

Code of Federal Regulations, Part 139. Certification and Operations: Land airports serving certain air carriers. January 1994. Washington, DC.
Airport Certification Manual (ACM) & Airport Certification Specifications (ACS), AC 139.201-1. July 1988.
Licensing of Aerodromes, CAP 168. 1990. Civil Aviation Administration, London.

Index

Illustrations are in **boldface.**

About the authors

Norman Ashford is Professor of Transport Planning at Loughborough University in England, where he is a Chartered Engineer. He is a registered engineer in the States of Georgia and Florida as well as the Province of Ontario, Canada. He formerly was Director of the Transportation Institute for the State of Florida. He is Managing Director of The Loughborough Airport Consultancy.

H.P. Martin Stanton (deceased) was an airport operations expert of international renown who worked for International Civil Airports Association, among others, The Frankfurt Airport Authority, The Ministry of Civil Aviation Britain; and the International Civil Airports Association in Paris. He was a qualified pilot and an air traffic controller.

Clifton A. Moore is president of Lianoconsult Inc., an airport consulting service. His background features 34 years of wideranging experience, including CEO for the Southern California Regional Airport Authority. He played a central role in the development of the Los Angeles International Airport Terminal and the modernization of LAX in the early 1980s. He was world president of the International Civil Airports Association for eight years.